THE PROBLEMS OF PHILOSOPHY

Each volume in this series is devoted to the exploration of a single philosophical problem or group of problems. The books are large enough to allow adequate space to all major viewpoints. Selections are from contemporary as well as from classical philosophers, and whenever the issues under discussion involve ideas of other disciplines, extracts from scholars in these fields have also been included. Thus, several of the volumes will contain selections from physicists, mathematicians, psychologists, theologians, historians, and others. Each volume is edited by a specialist who has written a detailed introduction and supplied an annotated bibliography. If there is a sufficient public response, it is our aim to revise the volumes periodically and bring the bibliographies up to date.

We hope that these books will prove useful to readers of very different backgrounds. Teachers of philosophy who wish to discuss a given topic in depth have been handicapped by the absence of anthologies of this kind and by the inaccessibility of much of the material now made easily available. Scholars in related fields who wish to acquaint themselves with what philosophers have said on a given topic should also find these volumes very helpful. Above all, it is hoped that this series will be of value to the constantly growing "lay public" interested in serious subjects. The reader who wants to understand the rival philosophical positions can learn far more from studying the philosophers themselves than from the colorless and frequently inaccurate summaries contained in general histories of philosophy. The aim throughout has been to present only material distinguished for its clarity and intelligibility. If there is any presupposition shared by all the editors, it is the conviction that in order to be profound it is not necessary to be obscure.

PAUL EDWARDS, General Editor

Problems of Space and Time

READINGS SELECTED, EDITED
AND FURNISHED WITH
AN INTRODUCTION BY

J. J. C. Smart

PROBLEMS OF PHILOSOPHY SERIES
Paul Edwards, GENERAL EDITOR

MACMILLAN PUBLISHING CO., INC.
New York
COLLIER MACMILLAN PUBLISHERS
London

Fifth printing 1979

Library of Congress catalog card number: 64-15842

Macmillan Publishing Co., Inc.
866 Third Avenue, New York, N.Y. 10022

Collier Macmillan Canada, Ltd.

Printed in the United States of America

Acknowledgments

Grateful acknowledgment is hereby made to the following
for permission to use material reprinted in this volume:

Harvard University Press, Cambridge, Mass., for material
reprinted by permission of the publishers from Max Jammer,
Concepts of Space, Cambridge, Mass.: Harvard University Press,
copyright, 1954, by The President and Fellows of Harvard College;
and for material reprinted by permission of the publishers from
Nelson H. Goodman, *The Structure of Appearance,* Cambridge,
Mass.: Harvard University Press, copyright, 1951, by The President
and Fellows of Harvard College.

The author, and the editor of *Mind,* Vol. 65 (1956), for
permission to quote, with additions and omissions, from G. E. M.
Anscombe's article, "Aristotle and the Sea Battle."

The Westminster Press, Philadelphia, and Student Christian
Movement Press, Limited, London, for material from *St. Augustine:
Confessions and Enchiridion,* Vol. VII. The Library of Christian
Classics. Published in 1955, The Westminster Press. SCM Press.
Used by permission.

Thomas Nelson and Sons Ltd., Edinburgh, for material from
Descartes' Philosophical Writings: Principles of Philosophy, Part II,

tr. and ed. by Elizabeth Anscombe and Peter Thomas Geach (1959).

University of California Press, Berkeley, Calif., for material from Sir Isaac Newton's *Mathematical Principles of Natural Philosophy*, English translation revised and edited by Florian Cajori (1934).

Manchester University Press, Manchester, England, for material from *The Leibniz-Clarke Correspondence*, ed. by H. G. Alexander (1956); and for *Prolegomena to Any Future Metaphysics*, by Immanuel Kant, tr. with an introduction and notes by P. G. Lucas (1953).

Macmillan & Co. Ltd., London, St. Martin's Press, Inc., New York, and The Macmillan Company of Canada Ltd., Toronto, for material from *Critique of Pure Reason*, by Immanuel Kant, tr. by Norman Kemp Smith (1929).

Macmillan & Co. Ltd., London, and The Author's Estate, for material from *An Introduction to Metaphysics*, by Henri Bergson, tr. by T. E. Hulme (1913).

The Open Court Publishing Company, La Salle, Ill., for material from *The Science of Mechanics: A Critical and Historical Account of Its Development*, by Ernst Mach, tr. by Thomas J. McCormack, sixth edition (1960), and with a new introduction by Karl Menger; from "The Philosophical Significance of Relativity," by Hans Reichenbach; "Geometry as a Branch of Physics," by H. P. Robertson; "Autobiographical Notes," by Albert Einstein, all from *Albert Einstein: Philosopher-Scientist*, ed. by P. A. Schilpp, Library of Living Philosophers, second edition (1951); and also selected passages from Adolf Grünbaum's essay, "Carnap's Views on the Foundations of Geometry," in *The Philosophy of Rudolf Carnap*, ed. by P. A. Schilpp, Library of Living Philosophers (1962).

The Clarendon Press, Oxford, for material from *Appearance and Reality*, by F. H. Bradley, second edition (1930).

Cambridge University Press, Cambridge, England, for material from *Space, Time and Gravitation*, by A. S. Eddington (1920); from *Examination of McTaggart's Philosophy*, Vol. II, Part I, by C. D. Broad (1938).

George Allen & Unwin Ltd., London, for material from *Our Knowledge of the External World*, by Bertrand Russell (1922).

Charles Scribner's Sons, New York, for the use of Chapter 6 (with some omissions) from *The Laws of Nature* by R. E. Peierls, reprinted with the permission of Charles Scribner's Sons. Also by permission of George Allen & Unwin Ltd., London.

Harcourt, Brace & World, Inc., New York, for material from *The Structure of Science: Problems in the Logic of Scientific Explanation* by Ernest Nagel, © 1961 by Harcourt, Brace & World, Inc. and reprinted with their permission. Also by permission of Routledge & Kegan Paul Ltd., London.

Dover Publications, Inc., New York, for material from *The Philosophy of Space and Time*, by Hans Reichenbach, tr. by Maria Reichenbach and John Freund, with introductory remarks by Rudolf Carnap (1958).

Philosophical Library, New York, for material from *The Philosophy of Nature*, by Moritz Schlick, tr. by Amethe von Zeppelin (1949).

Dover Publications, Inc., New York, and Methuen & Co. Ltd., London, for permission to reprint "Space and Time," by H. Minkowski, from *The Principle of Relativity*, by Albert Einstein and others, tr. by W. Perrett and G. B. Jeffery, with notes by A. Sommerfeld (1923).

The author, and the editors of *The Philosophical Review*, Vol. 66 (1957), for permission to reprint a revised version of Adolf Grünbaum's essay, "The Philosophical Retention of Absolute Space in Einstein's General Theory of Relativity."

The author, and the editor of the *Australasian Journal of Philosophy*, Vol. 19, No. 3 (1941), for permission to reprint J. N. Findlay's article, "Time: A Treatment of Some Puzzles."

John Wiley & Sons, Inc., New York, and the Massachusetts Institute of Technology Press, Cambridge, for material from *Word and Object*, by W. V. Quine (1960).

The author, and the Directors of *The Journal of Philosophy*, Vol. 52 (1955), for permission to reprint Richard Taylor's essay, "Spatial and Temporal Analogies and the Concept of Identity."

CONTENTS

P A R T   I I I

Space-Time and Relativity

P A R T   I V

Recent Philosophical Analyses

Problems of Space and Time

want to insist that it is not the *same* two things which are at different places separated by different temporal intervals, simply on the ground that the things involved *are* at different places, and that they therefore in no sense *move* closer to-

INTRODUCTION

PLAN OF THIS VOLUME

Though I have tried to include in this volume as many selections as possible from earlier philosophers, it is inevitable that a collection of writings on the philosophy of space and time should involve a high proportion of contributions from writers who have flourished in the last hundred or even the last fifty years. The publication of systems of non-Euclidean geometry by Bolyai and Lobachevsky in the first half of the nineteenth century, and a little later—and in a more general form—by Riemann, has led to profound changes in our views about the nature of geometry and of the philosophy of space. Similar great developments have come to pass in the present century, with Einstein's special theory of relativity and Minkowski's elegant theory of Space-Time, which has forever made it impossible to write on the philosophy of space and that of time in entire abstraction from one another; and with Einstein's general theory of relativity, in which the geometrical ideas of Riemann have found physical application. On account of these scientific developments, philosophical questions have come to be asked that could not have been envisaged in earlier periods. A book of readings in the philosophy of space and time must therefore lay very great emphasis on the last hundred, indeed on the last fifty, years. I have, however, tried to keep the selections as untechnical as possible, and even where I have strayed somewhat over the borderline into technicality, as with Minkowski's indispensable paper, the mathematics required is very easy (on about freshman level), and I have sometimes also included more popular accounts of the same subjects, as in the selections from Schlick and Peierls, which should be quite intelligible to the entirely non-mathematical reader.

Part I of this book takes us through the history of the subject. It begins with a beautiful account of "The Concept of Space in Antiquity" by Max Jammer. This seemed to me preferable to presenting an inevitably scrappy collection of excerpts from the Greek writers themselves, some of whom

1

would in any case be hard to interpret from mere translations and without scholarly assistance. A passage from Aristotle is, however, included in the article by Miss Anscombe. This is on Aristotle's puzzle about the sea fight tomorrow. Aristotle held that it was necessarily true that either there will or will not be a sea fight tomorrow, but not necessarily true that there will be and not necessarily true that there will not be a sea fight. On the other hand, he held that propositions about present or past events *are* necessarily true, if they are true, and necessarily false, if they are false. Miss Anscombe discusses the notion of "necessarily true" which occurs here, and besides giving a new translation of the passage from Aristotle gives an elucidation of what Aristotle meant which seems to me to be more correct than some which have been given. There has indeed been a vast and stimulating literature on this question, and the reader is referred to the Bibliography for some of the more important items.

In the passage from St. Augustine's *Confessions* we find an able and subtle expression of philosophical puzzlement: the sort of puzzlement that can be felt only by a highly intelligent thinker. In recent times Wittgenstein has sometimes alluded to this passage from St. Augustine, and it is particularly appropriate that in the article by J. N. Findlay, which will be found in Part IV of this volume, there is a discussion of some of St. Augustine's puzzles. At the time when he wrote this article Findlay was very much influenced by Wittgenstein's conception of philosophical method.

Next follow extracts from Descartes, Leibniz, Newton, and Locke. The tension between Descartes' absolute theory of space (and his identification of matter with space) on the one hand and Leibniz's relational theory of space on the other hand, recurs throughout the history of the subject. See, for example, Grünbaum's article in Part III and the selection from Quine in Part IV. We shall find that the issue between Descartes and Leibniz is still unresolved and will depend on the future development of physics. Newton combined an absolute theory of *space*, which is metaphysical and not really essential to his theory, with what is in effect a relational theory of *place*, and this last is also beautifully expressed in the passage from Locke. The selection from Mach, which contains an analysis of Newton's experiment with the rotating bucket, shows how Newton's ideas can be fitted into a relational theory of space and time. The rotating bucket experiment is also discussed by Nagel in our selection. The selections from Kant's *Critique* and Bradley's *Appearance and Reality* contain arguments of a purely metaphysical kind which purport to show the unreality of space and time. One of the selections from Kant is his "First Antinomy," an alleged contradiction in the notions of space and time, and it is not sufficiently realized that, as Kant himself has said, this antinomy was as much a motive for his metaphysical system as was his discovery of Hume's analysis of causality. The selection from Bertrand Russell is intended to indicate how Kant's first antinomy may be refuted. Russell also discusses the rather similar but more subtle puzzles which were propounded by Zeno, and shows how they should be resolved. The selection from Kant's *Prolegomena* consists of his famous passage about the left hand and the right hand, whereby he tried to show that space (even though not a real thing in itself) is absolute rather than relative.

Part II deals with the relation between space and geometry: how much our choice of a geometry is determined by the nature of physical space and how much it is a matter of convention that we use a certain geometry rather than another. It is of interest that the second selection from Reichenbach together with the essay by Robertson were both highly praised by Einstein, when he said: "I can hardly think of anything more stimulating as the basis for discussion in an epistemological seminar than this brief essay by Reichenbach (best taken together with Robertson's essay)."[1] The reader will also be helped by the selection from Nagel, which forms part of a lucid contemporary treatise. The dialogue by Eddington is extremely stimulating and entertaining.

Part III is particularly concerned with the notion of space-time and the theory of relativity. The selection from Peierls is an excellent account of the special theory of relativity for the general reader. It may enable the entirely nonmathematical reader to understand more of some of the more difficult papers in this part of the book. It should be remembered that the *special* theory of relativity, unlike the general theory, is a thoroughly well-tested and indispensable part of modern physics. Some accounts of special relativity make it look as though it rests entirely on a few optical experiments, such as the Michelson-Morley experiment. A great merit of Peierls' chapter is that he shows the way in which special relativity is thoroughly embedded in modern physics and is confirmed by

all sorts of experiments. The selection from Einstein forms
part of an intellectual autobiography, and in a fascinating way
he reconstructs the genesis of his special and general theories
of relativity.

It is in my opinion of the highest importance that those
interested in the philosophy of space and time should be
acquainted with Minkowski's beautiful paper on "Space and
Time," which was given as a semipopular lecture in Cologne
in 1908. For those readers who may find it a little difficult
(though the mathematical apparatus contained in it is not
really very difficult) the selection from Schlick on "The
Four-Dimensional World" should be of help. In his paper
Minkowski says, with justification, "Henceforth space by
itself, and time by itself, are doomed to fade away into mere
shadows, and only a kind of union of the two will preserve an
independent reality." The first paper by Grünbaum differs
from the rest in that it is especially concerned with the
general theory of relativity, and he argues persuasively that
Einstein's general theory of relativity does not enable us to
dispense with an absolute space-time. In fact the issue between
Descartes and Leibniz, over whether matter has to be explained in terms of space or whether space consists simply in
relations between the things "in space," has not by any means
been settled in favor of Leibniz. I shall have more to say
on this subject in the second part of this introduction.

Part IV contains a number of recent analyses by philosophers of concepts of space and time. Quine and Goodman
both show how the notions of past, present, and future,
together with grammatical tenses, can all be eliminated from
language. In effect they show how illusory is our common
notion of the passage of time, of the future moving into the
past. Broad comes near to this position too, but it will be seen
that he shies away from it and returns to the notion of what
he calls "the transitory aspect of time." Findlay's acute article
contains occasional expressions suggesting the notion of time
flow, but he skillfully unravels some of the confusions we can
get into when we think about time. Taylor's article brings out
convincingly the analogies between space and time; these are
often not appreciated on account of confusions which he
exposes. Grünbaum's second paper in our volume is largely
concerned with the apparent temporal directionality of the
universe. Why is the universe asymmetrical in the time direction? For example, there are traces of the past, but nothing

like traces of the future. Grünbaum explains the reasons for this puzzling fact. He also convincingly argues against the notion that the passage of events from future through present to the past is an objective feature of the physical world. I would in fact myself wish to go even further and say that it is not a feature of the mental world either, especially since on other grounds I would argue for a physicalist theory of mind.

THE PHILOSOPHY OF SPACE AND TIME

Space

When we naïvely begin to think about space we most naturally think of it as though it were either some all-pervading stuff or some sort of receptacle. For example, we may wish to say that this table is now in the place previously occupied by the bookcase, and so there seems to be something—namely, place or volume of space—which is distinguishable from the material bodies which may or may not fill it. It is then tempting to think of places as though they were distinguishable parts of an ethereal jelly, something like a material body and yet not a material body, in which material bodies are to be located. As the first selection in this volume shows, some of the Pythagoreans identified so-called empty space with air, and Parmenides and Melissus also denied the existence of truly empty space. The atomists, on the other hand, made a clear distinction between atoms and the void that separated them. Even so, it was still tempting to think of space as though it were something, and Lucretius thought of space as though it were a receptacle. Since he also held that space is infinite, it would seem that he was thinking of space as a receptacle not in the way in which a box is a receptacle for toffees but in the way in which the sea is a receptacle for fish.

The notion of space as a stuff is expressed in its most extreme form by Descartes, according to whom the essence of matter is to be extended. On this view matter and space are identified and there is no such thing as empty space. An obvious objection, which Descartes considered but never satisfactorily answered, is as follows: If matter is simply extension, how can we distinguish one part of space from another? How are we to analyze such a statement as that one bit of matter has moved relatively to another, or that one part of space is more densely occupied with matter than is another part of space? Though Descartes did not possess the mathematical tools to deal with this objection, we shall see that a reply can be made to it by means of the Riemannian conception of a space of variable curvature.

According to Leibniz space is no sort of stuff. He held that it was merely a system of relations in which indivisible "monads" stand to one another.[2] This relational theory of space, divested of the elements idiosyncratic to Leibniz's general metaphysics, has ever since vied with the view of space as a sort of ethereal stuff, the "absolute" theory of space. We shall see that the issue between absolute and relational theories of space has by no means been decisively settled, though it has been considerably clarified. At this stage it is worth considering an argument which occurs in our selection from Kant's *Prolegomena*. Kant supported the absolute theory of space.[3] In particular he thought that the relational theory could not do justice to the difference between a left hand and a right hand. Suppose that the universe consisted of only one human hand. It seemed to Kant that this hand would have to be either a left hand or a right hand, but how could this difference be defined? To any relation between the positions of points on a left hand there corresponds a relation between the positions of the corresponding points on a right hand. The tip of the thumb of the right hand is a certain distance from the tip of the first finger. The angle from thumb to finger is of about half a right angle, and is anticlockwise in sense if we are looking at the palm. Precisely the same spatial relations obtain if we consider the distance from the tip of the thumb to the tip of the first finger of an extended left hand. The angle between them is the same, though its sense now is clockwise. However, this difference of sense could not itself be defined and so could not be used to tell whether the hand was a left or right one, supposing this hand to be the only thing in the universe. In fact the question of whether the hand is a left or a right one is the very same question as what constitutes a clockwise or anticlockwise rotation. If you can decide one you can decide the other. "Left" and "right" are in practice defined ostensively, but they could be defined explicitly in terms of the bodily characteristics of human beings, together with the notions of clockwise and anticlockwise. If you look at the front of a man and see him put one of his hands on his head, and then rotate that hand in a clockwise direction so that his arm sticks out horizontally, then he is pointing to his *left*. Conversely you could easily define "left" in terms of "clockwise." It is easy to see, therefore, why Kant held that if the universe consisted simply of one human hand, you could not define the difference between "left" and "right." He thought, nevertheless, that the hand would have to be a left one or a right one, and he concluded that spatial properties have to be grasped by intuition and elude definition in terms of relations. However the relationalist could simply reply to Kant that if the universe consisted of only one human hand then the question of whether it was a left one or a right one would be meaningless.

It is indeed an important question, which has come to the fore in modern physics, whether—and if so, how far—the distinction between left and right, or clockwise and anticlockwise, has an absolute significance in nature. A recent discovery in physics, that parity is not conserved, suggests that the distinction between clockwise and anticlockwise rotations has an absolute significance in nature, in the sense that if we could get into telegraphic radio communication with denizens of a distant nebula we could explain to them our notions of left and right.[4] This would not be so, however, if the distant nebula was made of anti-matter (anti-protons where we have protons, positrons where we have electrons). In that case their "left" would be our "right" and we should have to fix arbitrarily whose hand was a "left" one and whose was "anti-left."

In Kant's later *Critique of Pure Reason* we find that according to his developed view space was neither a stuff nor a set of objective relations between things, but rather something subjective with which in thought we clothe nonspatial "things in themselves." (He had a similar view about time.) Kant was here partly motivated by the first of his "antinomies," which is also given in our selection. However, recent developments in mathematics have enabled us to resolve Kant's antinomies about space and time. On this point see the selection from Russell in this volume. Kant's antinomy about time also depends on the confused notion of our advance through time, which I shall criticize later in this Introduction. If the universe journeys through time, there must be a first event, for all journeys must begin somewhere. But if the notion of a journey through time is rejected, then there is no more objection to an infinite series of past events than there is to an infinite series of future ones.

Kant's philosophy was much influenced by Newtonian mechanics, and it will be seen from our selection that Newton himself came down on the side of absolute space and time. It is important to remember, however, that these metaphysical views of his are strictly irrelevant to his dynamics. What is essential in Newtonian dynamics is not the notion of absolute space but that of an inertial system, that is, a system of axes relative to which a free particle travels with uniform velocity in a straight line. Mach in particular set himself to show that we can perfectly well analyze the notion of an inertial system from the point of view of a relational theory of space. According to Mach, an inertial system is determined by the general distribution of matter in the universe. This principle, known as "Mach's principle" is still a bone of contention in cosmological theorizing. The principle clearly invites, though it does not compel, a relational theory of space, such as was held by Mach himself.

Euclidean and Non-Euclidean Space

There was one of his axioms which Euclid regarded as less intuitive than the rest. This is the axiom which is in fact equivalent to the so-called axiom of parallels, which it will be more convenient to discuss here. The axiom of parallels states that if $A B$ is a straight line produced indefinitely in both directions, and if C is a point not on $A B$, then there is one and only one straight line through C and in the same plane as $A B$ which does not intersect $A B$. Many attempts were made to deduce the axiom of parallels from other more evident ones. Thus in the seventeenth and eighteenth centuries Saccheri and Lambert each tried to prove the axiom by a *reductio ad absurdum* method. The idea was to assume the falsity of the axiom of parallels and, with the aid of the other axioms, to deduce a contradiction. This attempt was not successful, and in fact Saccheri and Lambert succeeded in proving a number of quite respectable theorems of non-Euclidean geometry. However, they were not bold enough to make this interpretation of what they were doing. Bolyai and Lobachevsky, in the first half of the nineteenth century, were (independently) the first writers who consciously published a system of non-Euclidean geometry in which the axiom of parallels is denied. They replaced the axiom of parallels by the postulate that more than one parallel can be drawn.

This geometry is called "hyperbolic." On the other hand we can deny that there is any parallel at all, in which case we get "elliptic" geometry. However if we take the latter course we have to make one or two minor adjustments to the other axioms. In elliptic geometry space is finite but unbounded, whereas in Euclidean and hyperbolic geometries it is infinite. Klein and others have shown that hyperbolic and elliptic geometry are consistent if Euclidean gometry is. It is also easy to show that all of these geometries are consistent if the theory of the real number continuum is consistent. A priori, therefore, the non-Euclidean geometries are every bit as respectable as Euclid's geometry is. Mathematicians had unfortunately at first to battle against baseless philosophical objections raised by the followers of Kant, who thought that they had an intuition that space is Euclidean.

The relation between Euclidean, hyperbolic, and elliptic geometry was profoundly illuminated through the work of Arthur Cayley and Felix Klein. They showed that the three geometries could be regarded as specializations of a more general sort of geometry: projective geometry. However, to do so they had to make use of so-called "imaginary" points (that is, coordinates range over the complex number domain, not just that of the real numbers), and the unified treatment should not lead us to minimize the differences between the geometries as they occur in physics. In physics we must confine ourselves to "real" points, and topological differences between the spaces become important. A topological property is one which persists however much a space is deformed and stretched. For example, a cube and a sphere are topologically alike, since with suitable bending and stretching one can be turned into the other. On the other hand, the infinite plane and the surface of a sphere are not topologically alike, nor is the surface of a sphere and that of a torus. It is tempting to suppose that the physicist could use either Euclidean or non-Euclidean geometry as he pleases, just as the cartographer can use either a globe or, say, a Mercator projection on a plane sheet of paper. Notice, however, that if you portray the globe on flat paper you cannot show the North and South poles on your map, and similarly the track of a ship will on occasion disappear over one edge of the paper and reappear over the opposite edge. Something has been lost when we forsake the globe for plane paper, and this is because of the topological difference between the surface of a sphere and a plane surface. Reichen-

bach has rightly stressed the importance of topological differences between various spaces, and has pointed out that we get
causal anomalies in our physics if we use a geometry with
the wrong topological properties. Suppose that the universe is,
as many cosmologists have believed, finite but unbounded, so
that space is the three-dimensional analogue of the surface of
a sphere. We can also take it that the galaxies are approximately evenly spaced throughout this space. If we tried to
map this space as a Euclidean or hyperbolic one we would
get a curious situation. Galaxies would get further and further
apart from one another as their distances increased from some
point of origin. The choice of the "wrong" geometry would
show itself by the fact that some one galaxy would be singled
out in that it would be at a minimal distance from all others.
Even more obviously, the "wrongness" of our geometry would
show itself in the emergence of causal anomalies. For example, a light ray which circumnavigated the elliptic universe
would appear in the hyperbolic or Euclidean representation
as a light ray disappearing to infinity in one direction and
mysteriously reappearing from infinity in the opposite direction. Such causal anomalies suggest that a geometry may give
a simply false picture of the world, quite apart from pragmatic
considerations of the simplicity with which it can be used to
express physical theory. Moreover Adolf Grünbaum has
shown that, once *criteria of congruence* are agreed on, then
the choice of a geometry becomes a completely empirical
matter.[5]

There is here no need to enter further into the relationship
between geometry and physics, since this topic is fully discussed in the selections from Robertson and Reichenbach.
The dialogue by Eddington is also of great interest in this
connection, though at one place he suggests a form of conventionalism about geometry which is open to objection and
which has been criticized by Grünbaum.[6]

Space-Time and Relativity

It is a consequence of Einstein's special theory of relativity
that the notions of space and time must be fused together into
the notion of space-time. This is shown in a particularly
elegant and instructive way by Minkowski, in the paper included in this volume. It is therefore impossible fully to discuss

either space or time in isolation from the other. In this section of the Introduction I shall say something about relativity and the concept of space-time, and how this bears on our philosophical problems. I shall leave to a later section of the Introduction problems which are more closely connected with time in so far as it can be isolated from space-time.

So far in this Introduction we have been using the word "space" as the name of a continuant. That is, we have allowed ourselves to think in terms of "continuing to occupy space," "moving from one part of space to another," and of parts of space beginning to be occupied, ceasing to be occupied, and staying occupied or unoccupied. We have therefore been thinking of space as something which endures through time. This is without prejudice to the question of whether or not it is to be regarded as a concrete entity. Nations and political constitutions are continuants in the relevant sense. Now there is also a different, timeless, use of the word "space." When a geometer tells us that he is studying a space of three dimensions we take it that he is studying a timeless entity. For example if he were to consider the proposition that a sphere changed into an ellipsoid we should consider that he was no longer thinking within the language of solid geometry. In this timeless—or perhaps we had better now say "tenseless"—sense, we can consider a four-dimensional space, three of whose dimensions correspond to space in the ordinary sense of this word, and one of whose dimensions is taken to be a time dimension. An instantaneous state of space, in the ordinary or "continuant" sense of this word, is given by a three-dimensional cross section of this four-dimensional space-time.

The only velocity in nature which seems to have cosmic significance is the velocity of light. Let us therefore choose our units of length and of period of time in such a way that the velocity of light is unity. On this scale one second of time is equal to 186,300 miles, and so the dimensions of a man or a star, throughout even a small part of his or its lifetime, are such that we can think of the man or star as a very elongated four-dimensional worm. Of course if you do not wish to think of yourself as a worm, and wish to give yourself, relatively speaking, more thickness, you can do so by taking a different ratio between the unit of time and that of space. The one we have chosen, which makes the velocity of light equal to unity, makes the mathematics more elegant. Any other choice would appear arbitrary, reflecting for example that we use "feet" as opposed to "meters," or vice versa, and "seconds" as opposed to, say, Martian "years."

Thinking, therefore, of material objects as very long space-time worms, then two stars which are in uniform velocity with respect to one another and which are not accelerated with reference to our frame of reference, will appear in the space-time picture as two very long straight worms inclined at a certain small angle to one another. An observer on each star would regard himself as at rest. He would therefore naturally take his world-line (the line along which his star, as a four-dimensional "worm," lies) as his time axis. It would be natural to suppose that he takes his space axes as, in a certain sense, at right angles to this line. In other words, observers on stars whose world-lines are inclined to one another will slice the four-dimensional cake at different angles: they will regard different sets of events as simultaneous to one another. This consideration makes the relativity of simultaneity seem entirely natural. Minkowski, in the paper included in this volume, shows that, if we postulate a certain semi-Euclidean geometry for space-time, then the above way of looking at things can be worked out in detail and can be made to explain the Lorentz transformations of the special theory of relativity. He shows that the Lorentz transformations can be understood simply as a rotation of axes in space-time. It is true that in diagrams such as the one on p. 300 the rotation looks rather odd—as $O\,t$ moves to the right $O\,x$ goes up instead of down. But this is simply because in the diagram a semi-Euclidean space is being drawn on a Euclidean piece of paper.

We must not forget that space-time is a space in the mathematical sense of the word. Quite clearly it cannot be space in the sense of something which endures through time. This is sometimes half-forgotten in popular expositions of relativity. It is sometimes said, for example, that a light signal is propagated from one part of space-time to another. What should be said is that the light signal lies (tenselessly) along a line between these two regions of space-time. Moritz Schlick has expressed this point well when he says (see p. 293 of this volume): "One may not, for example, say that a point traverses its world-line; or that the three-dimensional section which represents the momentary state of the actual present, wanders along the time-axis through the four-dimensional world. For a wandering of this kind would have to take place in time; and time is already represented within the model and cannot be introduced again from outside." And if there can be no change in space-time, neither can there be any staying the same. As Schlick points out, it is an error to claim that the Minkowski world is static: it neither changes nor stays the same. Changes and stayings the same can both of course be *represented* within the world picture, for example a changing velocity by a curved line and a constant velocity by a straight line.

The tenseless way of talking which is appropriate to the four-dimensional space-time world seems to suggest to some people that some sort of fatalism must be true, and that the future is already somehow "laid up." This, however, is a confusion, for the "is" in "is already laid up" is a tensed one and suggests that the *future* exists *now*, which is absurd. The events of the future, like those of the past, certainly *exist*, in the sense in which this verb is used tenselessly, but of course they do not exist *now*. Nor does the four-dimensional picture imply determinism. It is quite neutral between determinism and indeterminism. The issue between determinism and indeterminism can be put quite easily in the language of space-time. It is as follows: From a complete knowledge of a certain three-dimensional (spacelike) slice of space-time together with a knowledge of the laws of nature, could the properties of later (and indeed earlier) slices of space-time be deduced? For present purposes let us be agnostic as to the answer to this question.

The power of the Minkowski representation can be illustrated if we take a brief look at the so-called clock paradox of special relativity. It is a familiar consequence of the special theory of relativity that if two inertial systems of axes A and B are in relative motion with respect to one another, then processes in B will, when measured relatively to A, run slower than they do when measured in B itself. Similarly processes in A, when measured relative to B, go slower then they do when measured in A. This phenomenon is called the "time dilation." So far there is complete symmetry between A and B. The clock paradox has been thought to arise as follows. Suppose that there are two twins, Peter and Paul, and that while Peter stays at rest in some inertial frame of reference, Paul is shot off in a rocket with a high constant velocity. When he gets to a distant star he is decelerated and re-accelerated so as to come back to meet Peter again. If Paul's velocity of travel is sufficiently great, then owing to the time dilation he will be considerably younger than his twin brother Peter, when he returns. Critics of special relativity object that there is a paradox here, since Paul could regard himself at rest and Peter as in motion, and so, looking at it this way, Peter would have to be younger when they meet. But Peter and Paul cannot *both* be younger. What they forget is that on account of Paul's deceleration and re-acceleration at the distant star he is not throughout in one particular inertial frame but during the journey he changes from one inertial frame of reference to another. Peter does not. So the alleged symmetry between the two twins does not exist. A Minkowski diagram vividly shows this lack of symmetry. In this diagram[7] $A B$ is Peter's world line. $A C$ is Paul's on the outward journey. $C B$ is Paul's on the return journey. And it

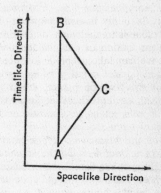

is clear that $A B$ is not equal to $A C + C B$. If space-time were Euclidean, then as every schoolboy knows, $A B$ would be less than $A C + C B$. However, owing to the semi-Euclidean structure of space-time it turns out that $A B$ is greater than $A C + C B$. So it is quite evident that Paul will be younger than Peter when they meet again. There is of course a lot more to be said, but the above may be sufficient to indicate, even to those readers who are not yet conversant with relativity, the way in which the geometrical approach of Minkowski may make something which at first appears counterintuitive look entirely natural.

Absolute and Relational Theories of Space and Time

It is sometimes thought that the acceptance of the special and general theories of relativity forces us to settle the controversy between the absolute and relational theories of space in favor of the relational theory. Now it is certainly true that the theory of relativity does force us to give up the view of space, *taken quite apart from time*, as something absolute, but it is important to realize that it does not force us to give up an absolute theory of *space-time*. Quine points this out (see pp. 376-377). Certainly it is true that any instantaneous state of space corresponds to a slice of space-time which is different according to the motion of the frame of reference. However, the fact that we can take slices at different angles through a sausage does not force us to give up an absolute theory of sausages. We may conclude that the special theory of relativity is neutral so far as the controversy between absolutists and relationalists is concerned, so long as we shift the object of this controversy from the notions of space and time taken separately to that of space-time. The four-dimensional array of the Minkowski world can be given either interpretation.

Nor does the *general* theory of relativity force us toward a relational theory of space-time. It is true that according to the general theory of relativity we can take space-time in a small region as flat or curved according to whether we take one or another set of mutually accelerated axes. Such adjustments can be made only in limited regions of space-time: a "smoothing out" in one small region is made at the expense of great curvature elsewhere. (Compare trying to smooth out a carpet which does not fit properly.) Perhaps it would be safest to adopt the position that the general theory of relativity is neutral between the two philosophical theories, but in his first selection in this volume Adolf Grünbaum argues that Einstein's general theory of relativity cannot in fact be freed from the notion of absolute space-time. Moreover a very natural metaphysical interpretation of general relativity is that a piece of matter simply *is* a region of special curvature of space-time. Space-time, on this view, would be a modern version of Descartes' extension. By taking a space of variable curvature we get over the obvious objection to Descartes' theory that it cannot differentiate between more or less densely occupied regions of space, or between matter and the void. Nevertheless, it is impossible at present to take such a metaphysical position with any confidence. This is because quantum mechanics is a particle physics, and it is not at present possible to reconcile the field theory of general relativity with it. We must wait to see whether a unified field theory is able to absorb quantum mechanics or whether quantum mechanics is able to absorb general relativity. On this issue will probably depend the result of the controversy between absolute and relational theories of space and time, that is, whether particles are to be thought of as singularities in space-time or whether space-time is to be elucidated in terms of relations between particles.

The issue between absolute and relational theories of space can be sharpened into one of ontology, in the sense in which Quine has used this word. Quine has shown how to reconstruct scientific language in such a way that we mention only material objects, classes of these, classes of classes of these, and so on. (Using a well-known device, he defines relations in terms of classes.) His program is in line with the view that a particle physics will win the day. In the section on "Geometrical Objects" which is included in this volume he elucidates geometrical points in terms of relations between material particles. But his program could easily be modified if the unified field theory should win the day. In this case a Quinean ontology would consist of points of space-time, classes of these, classes of classes of these, and so on.

It is also possible that science will develop in such a way that the simple notion of space as a system of relations between particles as well as that of space as absolute will have to be given up. It is just possible that we shall come to regard space and time as statistical properties on the macroscopic level only—just as, for example, temperature is a statistical property on the macrolevel, which has no meaning in microphysics. In this case the particles of microphysics will be related only by relations which are not spatio-temporal, and so these particles will bear a remarkable likeness to Kant's "things-in-themselves." This suggestion is remarkably hard to grasp, and it would be premature to discuss it here. However, it is beginning to turn up in the literature, and it is perhaps worth drawing the attention of philosophers to it.[8]

Time

Time has always seemed peculiarly enigmatic to philosophers. Even more than space it has emotional significance for human beings: old age and death lie in wait for us, and time seems the great enemy. Sometimes, again, it appears as the healer of old wounds. All this is, of course, to speak in metaphor, and these metaphors get their power from an interesting illusion: the illusion that time flows, or alternatively that we advance through it. That there is something wrong with this notion should be evident from the passage from Schlick to which I alluded earlier in this Introduction (p. 12), but some readers may be helped if I go into the matter again without bringing in the notion of space-time.

The events in our life, it seems, recede further and further into the past. The future is continually coming to meet us, and then moves away into the past. Alternatively, we advance from the past into the future. Now there is something inconsistent in this idea. *How fast* do we advance through time, or *how fast* does time flow? (See Broad, pp. 330.) In what units is the rate of time flow to be measured? Seconds per— what? Not seconds per seconds surely! It is true that Grünbaum has objected (see p. 420 of this volume) to this argument on the grounds that this is a *metrical* refutation. According to him, the notion of the flux of time is a *qualitative* conception without any metrical ingredients. As I understand him, he holds that there is a flux of time, in the sense that we are aware of the events in our experience, and we are aware that any such act of awareness is itself an event over and above the set of events of which we were aware in any previous such act. For example, our stock of memories is continually being augmented. I would not wish to deny that Grünbaum is extremely perceptive in thus characterizing the "flux of time," and if this is all that is meant by a "flux of time" then there is indeed such a flux. I would prefer, however, to regard the facts to which he draws attention as facts which explain the *illusion* that there is a flux, in the sense of motion *through* time. I think that Grünbaum is too kind to Weyl when he quotes the latter's remarks about our consciousness "crawling up the world-line of our body." Grünbaum regards Weyl as talking in metaphor here, but if so it seems to me a highly misleading metaphor.

We get the idea that future events come to meet us and then recede into the past partly because we misunderstand the notions of "past," "present," "future," and "now." As Quine and Goodman both show (see pp. 372-373 and pp. 356-365), words like "past," "present," "future," "now," and the tenses of verbs are expressions which refer indirectly to their own utterance. Quine and Goodman help to show in what way these linguistic devices are dispensable. Reichenbach, in his *Elements of Symbolic Logic*, has given a very thorough analysis of these expressions. He calls them "token-reflexive."[9] A token is a particular utterance or inscription (that is, a sound or mark on paper), and a token-reflexive utterance is one which refers to itself. Reichenbach shows that words like "I," "you," "here," "now," "past," "present," and "future" are all token-reflexive, that is, can all be defined in terms of the clearly token-reflexive expression "this token." Thus "I" means "the person uttering this token," and "you" means "the person addressed by this token," and "here" means "near this token," "now" means "simultaneous with this token," "past" means "earlier than this token," and "future" means "later than this token." Tenses are token-reflexive also: thus "Caesar crossed the Rubicon" could be translated "Caesar crosses (tenseless present) the Rubicon earlier than this utterance." Goodman and Quine do not carry out the analysis of tenses in as much detail as Reichenbach, since they are concerned with remodeling language for scientific purposes rather than with giving an analysis of our colloquial language. It is important to stress that Reichenbach does his analysis in terms of "this token." Sometimes critics of this sort of approach unfairly take the analysis of "past," say, to be "earlier than *now*," and if this were so, of course we should not have accomplished much. Or they say that "this token" means "the token which is now." But we must reply that Reichenbach takes the notion of "this token" or token-reflexivity as primitive, and that "now" is defined in terms of it, not vice versa. Thus "future" means "later than this utterance." Every time it is used, therefore, "future" refers to a different utterance. Now, it is no paradox that an event can be later than one human utterance and earlier than another. And this is about all that is legitimately meant when we say that the future flows into the past.

On p. 144 of our selection from him Bergson wishes us to place ourselves "by an effort of intuition, in the concrete flow of duration." There is, we can now see, no such thing as the concrete flow of duration. We get the idea that there is such a thing only because we think of the present or the now as moving through time. The token-reflexiveness of the notion of "present" or "now" has misled us. The present does not move, because there is no such entity as the present. Those who think of the present, or of our span of consciousness, as moving through time, are trapped by a misunderstanding of the platitude that if a person says that something is simultaneous with his utterance on different occasions, then he refers on these different occasions to different times. On p. 143 Bergson says that "duration disintegrates into a powder of moments, none of which endures, each being an instanteity" and ". . . the unity that binds the moments together, this cannot endure either. . . ." So the mobility of duration seems to be lost. In reply I should simply agree with Bergson that instants do not endure. Nor, I should add, do finite spans of time either. *Things* endure, which means simply that (thinking four-dimensionally) they have some thickness in a timelike direction. Things also move and change: that is, thinking four-dimensionally again, their world lines are not parallel to world lines of things taken to be at rest, and their three-dimensional cross sections vary from point to point along the time axis. To talk of *moments* or *spans* of time changing is nonsense, and that this sort of "movement" cannot be grasped intellectually is not surprising. It cannot be grasped by intuition either, for whether or not there is such a thing as "intuition," we can be quite sure that there is no such thing as "the concrete flow of duration."

One way of avoiding the confusions into which we are liable to fall when we think about time is to seek out analogies between space and time. This has been done by Richard Taylor, both in the article included in this volume and in others (see Bibliography). Taylor has shown that if we think clearly enough we will see that in most cases space and time are perfectly analogous to one another. I now wish to pass on, however, to a respect in which the world does not seem to be symmetrical as between its spatial and temporal aspects.

Why is the universe temporally asymmetrical in a way in which it is not markedly spatially asymmetrical? On a large scale the universe is much the same in whatever direction we look. But in the time direction there does seem to be an important difference between earlier and later. There are traces of earlier events and not of future ones. From fossils, photographs, footmarks, and so forth, we can learn about the past in a way in which we cannot learn about the future. We know what animals lived on earth millions of years ago and of the history of Greece and Rome, but we are in the dark about so relatively small a segment as the next hundred years in human history.[10] We can of course predict eclipses and the like, just as we can retrodict them, but there is nothing like a future analogue of a trace. Grünbaum is concerned with this problem. He rightly rejects the verbalistic solution that traces are by definition of the past, not of the future. This does not remove the problem of why there are not future analogues of traces. Uncles are by definition male, but there is no difficulty in conceiving of a female analogue of an uncle; it is an aunt. Grünbaum shows the close connection between the physical concept of entropy and the temporal asymmetry of the world.

In our cosmic epoch at least, as we have seen, the universe is asymmetrical temporally in a way in which it is not so spatially. Some readers might think that this has some connection with free will and human choice. "We can change the future," they may say, "but not the past."[11] I suggest that this is a confusion. It makes no more sense to talk of changing the future than it does to talk of changing the past. Suppose that I decide to change the future, by having coffee for breakfast tomorrow instead of my usual tea. Have I changed the future? No. For coffee for breakfast *was* the future. It has been objected to me that the above argument is perhaps misleading. For, it has been said, there is quite clearly a sense in which I can change the future and not the past, and this is because my acts of will determine the future and not the past—I cannot undo what has been done. Now I do not wish to deny that we can causally affect the future and not the past, and indeed this causal directionality of time is part of the problem of the "direction of time." Nevertheless I would reiterate that the fact that our present actions determine the future would be most misleadingly expressed or described by saying that we can change the future. A man can change his trousers, his club, or his job. Perhaps he may even change the course of world history or the state of scientific thought. But one thing that he cannot change is the future, since whatever he brings about *is* the future, and nothing else is, or ever was.

Some comments are in order about the selection from Broad. He rejects the notion of time flow but holds fast to a notion of absolute becoming, in terms of which he analyzes the notions of past, present, and future. It is of interest that in an earlier paper, published in Hastings' *Encyclopaedia of Religion and Ethics*, he adopted what is very near to Reichenbach's account of past, present, and future in terms of token-reflexive expressions. Indeed on pp. 334-336 Broad considers such an analysis of "past," "present," and "future" in terms of tenseless verbs and the notions of "earlier," "later," and "simultaneous." He complains, however, on p. 337 that this "leaves altogether out of account the transitory aspect of time." He says that on this theory there is no question of events "becoming" or "passing away." "In some sense of 'is'," he says, "there 'is' timelessly or sempiternally all that there ever has been or will be. . . ." Here Broad has incorrectly stated the theory. According to the tenseless way of speaking events do not exist either timelessly or sempiternally. They neither exist at no time nor do they exist at every time. They exist at some time or other, and the verb "exist" here really is meant to be tenseless and not to be the present tense. (Nor is it meant to be some sort of "sempiternal" tense.) Furthermore, the transitory aspect of time must be totally rejected, for it is meaningless to talk, as Broad does, of events "becoming" or "passing away." *Things* become, *events* happen. And things do not just become, they become something or other. For example a man's hair becomes grey or a traffic light becomes green. To say that the becoming green of the traffic light *becomes* is absurd, unless what is meant is simply that the becoming green of the traffic light *happens* or *occurs*. But if this is what is meant then the theory which Broad is criticizing does not deny it. Events certainly happen, and this verb "happen" can perfectly well be made tenseless. Events happen (tenselessly) later than this utterance, simultaneously with this utterance, or earlier than this utterance. On p. 333 Broad correctly argues that in any ordinary sense it is incorrect to say that events become. Anything which, for example, becomes hot, he says "must be a more or less persistent substance." But then he explains the "absolute becoming" of events as "becoming present." This, however, is to treat presentness as a changing property of events, and we must simply deny that this can be done. On the analysis which I have advocated, the sentence "Event *E* was future, is present, and will become past" is an extremely misleading one. If it means anything at all it means: "Event *E* is (tenselessly) later than some utterance earlier than this utterance, and is (tenselessly) simultaneous with this utterance, and is (tenselessly) earlier than some utterance later than this utterance." It is evident that it is highly misleading to think of pastness, presentness, and futurity as properties (even as relational properties) of events.

It is clear, then, that Broad has himself already provided the tools for rejecting the notions of absolute becoming and the transitory aspect of time. In spite of the occasional confusions which I have noted, he is so clear-headed that it is hard to understand why he rejected his earlier theory. This has indeed puzzled his able commentator C. W. K. Mundle.[12]

I wish to commend Findlay's article to the reader as a first-rate attempt to think about time clearly and without mysticism. That I do not agree with everything in it will be evident from my approach in this Introduction, but it was Findlay's article which first made me think that time could be discussed rationally, and without the original inspiration of Findlay's article it is quite possible that this volume, at least under its present editorship, might never have come to exist.

NOTES

[1] From *Albert Einstein: Philosopher-Scientist*, P. A. Schilpp, ed. (2d ed.; La Salle, Ill.: Open Court, 1951), 679.

[2] Leibniz's view cannot be stated consistently, since it is also true that he did not believe in relations, but only in subjects and (one place) predicates. See *The Philosophy of Leibniz* by Bertrand Russell (London: Allen and Unwin, 1900).

[3] As something "phenomenally" real. He denied that space is "noumenally" real. We need not here concern ourselves with this peculiarly Kantian distinction.

[4] See the article by O. R. Frisch, "Parity Not Conserved, a New Twist to Physics?" *Universities Quarterly* (England), 11 (1957), 235-244.

[5] See his "Law and Convention in Physical Theory," in H. Feigl and G. Maxwell, eds., *Current Issues in the Philosophy of Science* (New York: Holt, Rinehart and Winston, 1961).

[6] *Ibid.*

[7] Diagrams similar to this one but with a little more detail in them can be seen in a note by W. H. McCrea in *Discovery, 18* (February, 1957), 56-58, and in G. J. Whitrow's *Natural Philosophy of Time* (London and Edinburgh: Nelson, 1961), 232. If the diagram shown here were drawn more accurately in detail it would show rounded curves at *A, B,* and *C,* since changes of velocity cannot be instantaneous. However, these curved parts of the lines can be neglected (and hence the problem can be treated easily within the *special* theory of relativity), since, as McCrea has pointed out, if we consider a sufficiently long journey the proportion of the journey during which there is acceleration or deceleration can be made as small as we please. See McCrea's notes in *Nature, 167* (1951), 680 and *177* (1956), 782.

[8] See E. J. Zimmerman, "The Macroscopic Nature of Space-Time," *American Journal of Physics, 30* (1962), 97-105 and Banesh Hoffmann, *The Strange Story of the Quantum* (New York: Dover, 1959), 197-8.

[9] H. Reichenbach, *Elements of Symbolic Logic* (New York: The Macmillan Company, 1947), sec. 50-51.

[10] See Christian Ehrenfels, *Cosmogony* (New York: Philosophical Library, 1948), Chap. 1.

[11] See, for example, H. Reichenbach, *The Direction of Time* (University of California Press, 1957), 21.

[12] For Mundle's excellent discussion of Broad's changing theories about time see his paper, "Broad's Views about Time," in P. A. Schilpp, ed., *The Philosophy of C. D. Broad* (La Salle, Ill.: Open Court, 1959).

I

*Space and Time
in the History
of Philosophy*

THE CONCEPT OF SPACE
IN ANTIQUITY

MAX JAMMER

Reprinted from *Concepts of Space*, Harvard University Press, Cambridge, Mass., 1954. This selection consists of Chapter 1 of Jammer's book with the first two paragraphs omitted.

Space as a subject of philosophical inquiry appears very early in Greek philosophy. According to Aristotle, numbers were accredited with a kind of spatiality by the Pythagoreans: "The Pythagoreans, too, asserted the existence of the void and declared that it enters into the heavens out of the limitless breath—regarding the heavens as breathing the very vacancy—which vacancy 'distinguishes' natural objects, as constituting a kind of separation and division between things next to each other, its prime seat being in numbers, since it is this void that delimits their nature."[1] Spatial vacancies were necessary to guarantee the discreteness of individual numbers in the Pythagorean geometrization of number. Space here has not yet any physical implications apart from serving as the limiting agent between different bodies. In early Pythagorean philosophy this kind of "space" is still called *pneuma apeiron* and only occasionally *kenon* (void). The concept of space is still confounded with that of matter. As J. Burnet says: "The Pythagoreans, or some of them, certainly identified 'air' with the void. This is the beginning, but no more than the beginning, of the conception of abstract space or extension."[2] Only later on is this confusion cleared up by Xutus and Philolaus.[3] In Simplicius we find that Archytas, the Pythagorean, already had a clear understanding of this abstract notion, since, as related by Eudemus, he asked whether it would be possible at the end of the world to stretch out one's hand or not.[4] Unfortunately, Archytas' work on the nature of space is lost except for a few fragments to be found in Simplicius' *Commentaries*,[5] according to which Archytas composed a book on our subject. Archytas distinguishes between place (*topos*), or space, and matter. Space differs from matter and is independent of it. Every body occupies some place, and cannot exist unless its place exists. "Since what is moved is moved into a certain place and doing and suffering are motions, it is plain that place, in which what is done and suffered exists, is the first of things. Since everything which is moved is moved into a certain place, it is plain that the place where the thing moving or being moved shall be, must exist first. Perhaps it is the first of all beings, since everything that exists is in a place and cannot exist without a place. If place has existence in itself and is independent of bodies, then, as Archytas seems to mean, place determines the volume of bodies."[6] A characteristic property of space is that all things are in it, but it is never in something else; its surroundings are the infinite void itself. Apart from this metaphysical property, space has the physical property of setting frontiers or limits to bodies in it and of preventing these bodies from becoming indefinitely large or small. It is also owing to this constraining power of space that the universe as a whole occupies a finite space. To Archytas, space is therefore not some pure extension, lacking all qualities or force, but is rather a kind of primordial atmosphere, endowed with pressure and tension and bounded by the infinite void.

The function of the void, or of space, in the atomism of Democritus is too well known to need any elaboration here. But it is of interest to note that according to Democritus infinity of space is not only inherent in the concept itself,[7] but may be deduced from the infinite number of atoms in existence, since these, although indivisible, have a certain magnitude and extension, even if they are not perceptible to our senses. Democritus himself seems not to have attributed weight to the atoms but to have assumed that as a result of constant collisions among themselves they were in motion in infinite space. It was only later, when an explanation of the cause of their motion was sought, that his disciples introduced weight as the cause of the "up and down" movements (Epicurus). If Aristotle says that Democritus' atoms differed in weight according to their size, one has to assume—in modern words—that it was not gravitational force but "force of impact" that was implied. This point is of some importance for our point of view, since it shows that in the first atomistic conception of physical reality space was conceived as an empty extension without any influence on the motion of matter.

However, there still remains one question to be asked: Was space conceived by the atomists of antiquity as an unbounded extension, permeated by all bodies and permeating all bodies, or was it only the sum total of all the *diastemata*, the intervals that separate atom from atom and body from body, assuring their discreteness and possibility of motion? The stress laid time and again by the atomists on the existence of the void was directed against the school of Parmenides and Melissus, according to whom the universe was a compact plenum, one continuous unchanging whole. "Nor is there anything empty," says Melissus, "for the empty is nothing and that which is nothing cannot be." Against such argument Leucippus and Democritus maintained the existence of the void as a logical conclusion of the assumption of the atomistic structure of reality. But here the void or the empty means clearly unoccupied space. The universe is the full and the empty. Space, in this sense, is complementary to matter and is bounded by matter; matter and space are mutually exclusive. This interpretation gains additional weight if we note that the term "the empty" (*kenon*) was used often as synonymous with the word "space"; the term "the empty" obviously implies only the unoccupied space. Additional evidence is furnished by Leucippus' explicit use of the adjective "porous" (*manon*) for the description of the structure of space, which indicates that he had in mind the intervals between particles of matter and not unbounded space. Although Epicurus' recurrent description of the universe as "body and void" seems also to confirm this interpretation, we find in Lucretius, who bases himself on Epicurus, a different view. In general, Lucretius' complete and coherent scheme of atomistic natural philosophy is the best representation of Epicurean views. As far as the problem of space is concerned, Lucretius emphasizes in the first book of *De rerum natura* the maxim: "All nature then, as it exists, by itself, is founded on two things: there are bodies and there is void in which these bodies are placed and through which they move about."[8]

Here we find, in contrast to the early Greek atomism, a clear and explicit expression of the idea that bodies are placed in the void, in space. With Lucretius, therefore, space becomes an infinite receptacle for bodies. Lucretius' proof for the unboundedness of space, resembling Archytas' argument mentioned earlier,[9] runs as follows: "Now since we must admit that there is nothing outside the sum, it has no outside, and therefore is without end and limit. And it matters not in which of its regions you take your stand; so invariably, whatever position any one has taken up, he leaves the universe just as infinite as before in all directions. Again, if for the moment all existing space be held to be bounded, supposing a man runs forward to its outside borders and stands on the utmost verge and then throws a winged javelin, do you choose that when hurled with vigorous force it shall advance to the point to which it has been sent and fly to a distance, or do you decide that something can get in its way and stop it? For you must admit and adopt one of the two suppositions; either of which shuts you out from all escape and compels you to grant that the universe stretches without end."[10]

This argument, and in particular the idea of a man placed at the supposed boundary of space stretching out his hand or throwing a spear, is a recurrent idea in the history of natural philosophy. In fact, an illustration of this kind is to be expected. We find it in Richard of Middleton's writings in the fourteenth century (perhaps with reference to Simplicius' *Physics* 108a), still before the rediscovery of the *De rerum natura* in 1418 by Poggio. We also find it as late as in Locke's *Essay concerning human understanding* (1690), where the question is asked "whether if God placed a man at the extremity of corporeal beings, he could not stretch his hand beyond his body."[11]

Lucretius adduces a further argument for the infinitude of space which reveals an important physical aspect of the atomistic theory: If space were not infinite, he claims, all matter would have sunk in the course of past eternity in a mass to the bottom[12] of space and nothing would exist any more. This remark shows clearly that Lucretius, in the wake of Epicurus, conceived space as endowed with an objectively distinguished direction, the vertical. It is in this direction in which the atoms are racing through space in parallel lines. According to Epicurus and Lucretius, space, though homogeneous, is not isotropic.

Although the idea of a continuous homogeneous and isotropic space, as we see, seems to have been too abstract even for the theoretically minded atomists, it has been justly pointed out that their conception of the noncorporeal existence of a void introduced a new conception of reality.[13] Indeed, it is a strange coincidence that the very founders of the great materialistic school in antiquity had to be "the first to say distinctly that a thing might be real without being a body."

The first clear idea of space and matter as belonging to different categories is to be found in Gorgias.[14] Gorgias first proves that space cannot be infinite. For if the existent were infinite, it would be nowhere. For were it anywhere, that wherein it would be, would be different from it, and therefore the existent, encompassed by something, ceases to be infinite; for the encompassing is larger than the encompassed, and nothing can be larger than the infinite; therefore the infinite is not anywhere. Nor on the other hand, can it be encompassed by itself. For in that case, that wherein it is found would be identical with that which is found therein, and the existent would become two things at a time, space and matter; but this is impossible. The impossibility of the existence of the infinite excludes the possibility of infinite space.

Plato, who, according to Aristotle, was not satisfied, as his predecessors were, with the mere statement of the existence of space, but "attempted to tell us what it is,"[15] develops his theory of space mainly in *Timaeus*. The upshot of the rather obscure exposition of this dialogue, as interpreted by Aristotle,[16] and in modern times by E. Zeller,[17] is that matter—at least in one sense of the word—has to be identified with empty space. Although "Platonic matter" was sometimes held to be a kind of body lacking all quality (Stoics, Plutarch, Hegel) or to be the mere possibility of corporeality (Chalcidius, Neoplatonists), critical analysis seems to show that Plato intended to identify the world of physical bodies with the world of geometric forms. A physical body is merely a part of space limited by geometric surfaces containing nothing but empty space.[18] With Plato physics becomes geometry, just as with the Pythagoreans it became arithmetic. Stereometric similarity becomes the ordering principle in the formation of macroscopic bodies. "Now the Nurse of Becoming, being made watery and fiery and receiving the characters of earth and air, and qualified by all the other affections that go with these, had every sort of diverse appearance to the sight; but because it was filled with powers that were neither alike nor evenly balanced, there was no equipoise in any region of it; but it was everywhere swayed unevenly and shaken by these things, and by its motion shook them in turn. And they, being thus moved, were perpetually being separated and carried in different directions; just as when things are shaken and winnowed by means of winnowing-baskets and other instruments for cleaning corn, the dense and heavy things go one way, while the rare and light are carried to another place and settle there. In the same way at that time the four kinds were shaken by the Recipient, which itself was in motion like an instrument for shaking, and it separated the most unlike kinds farthest apart from one another, and thrust most alike closest together; whereby the different kinds came to have different regions, even before the ordered whole consisting of them came to be."[19] Physical coherence, or, if one likes, chemical affinity, is the outcome of stereometric formation in empty space, which itself is the undifferentiated material substrate, the raw material for the Demiurgus. The shaking and the winnowing process characterizes space with a certain stratification and anisotropy which is manifested physically in the difference between the layers of the elements. Geometric structure is the final cause of what has been called "selective gravitation," where like attracts like.

In accordance with certain ideas expressed by the Pythagorean Philolaus,[20] Plato conceived the elements as endowed with definite spatial structures: to water he assigned the spatial structure of an icosahedron, to air of an octahedron, to fire of a pyramid, and to earth of a cube.[21] Earth, in Plato's view, owing to its cubical form, is the most immovable of the four, having the most stable bases. It is only natural, therefore, that this element is found at the center of the universe; like a nucleus it is embedded in layers of the other elements of space according to their increasing movability. The varieties of the four elements and their gravitational behavior are due to differences in their form and size, or, in the final analysis, are due to differences in form and size of the elementary triangles of which their plane surfaces are formed. As much as matter is reduced to space, physics is reduced to geometry.

This identification of space and matter, or, in the words of later pseudo-Platonic teachings, of tridimensionality and matter, had a great influence on physical thought during the Middle Ages. For although Aristotle's *Organon* was the standard text in logic, Plato's *Timaeus* was succeeded by Aristotle's *Physics* only in the middle of the twelfth century. It is perhaps not wrong to assume that the obscure and vague language of the *Timaeus* contributed to preventing the concept of space from becoming a subject of strict mathematical research. Greek mathematics disregards the geometry of space. Plato himself, for whom solid bodies and their geometry were of fundamental importance in the formulation of his philosophy, lamented the neglecting of this branch of mathematics. In the *Republic*[22] he apologizes for failing to discuss solid geometry when listing the essential subjects for instruction. So we read:

> GLAUCON: It is true that they possess an extraordinary attractiveness and charm. But explain more clearly what you were just speaking of. The investigation of plane surfaces, I presume, you took to be geometry?
>
> SOCRATES: Yes.
>
> GLAUCON: And then at first you took astronomy next and then you drew back.
>
> SOCRATES: Yes, for in my haste to be done I was making less speed. For while the next thing in order is the study of the third dimension or solids, I passed it over because of our absurd neglect to investigate it, and mentioned next after geometry, astronomy, which deals with the movements of solids.[23]

Aristotle's theory of space is expounded chiefly in his *Categories* and, what is of greater relevance for our purpose, in his *Physics*. In the *Categories*, Aristotle begins his short discussion with the remark that quantity is either discrete or continuous. "Space," belonging to the category of quantity, is a continuous quantity. "For the parts of a solid occupy a certain space, and these have a common boundary; it follows that the parts of space also, which are occupied by the parts of the solid, have the same common boundary as the parts of the solid. Thus, not only time, but space also, is a continuous quantity, for its parts have a common boundary."[24] "Space" here is conceived as the sum total of all places occupied by bodies, and "place" (*topos*), conversely, is conceived as that part of space whose limits coincide with the limits of the occupying body.[25]

In the *Physics* Aristotle uses exclusively the term "place" (*topos*), so that strictly speaking the *Physics* does not advance a theory of space at all, but only a theory of place or a theory of positions in space. However, since the Platonic and Democritian conceptions of space are unacceptable to the Aristotelian system of thought, and since the notion of empty space is incompatible with his physics, Aristotle develops only a theory of positions in space, with the exclusion of the rejected conception of general space.

For our purpose, Aristotle's theory of places is of greatest pertinence not only because of its important implications for physics, but also because it was the most decisive stage for the further development of space theories. In our treatment we shall adhere as much as possible to Aristotle's original terminology and use the term "place."

In Book IV of the *Physics* Aristotle develops on an axiomatic basis a deductive theory of the characteristics of place. Place is an accident, having real existence, but not independent existence in the sense of a substantial being. Aristotle's four primary assumptions regarding our concept are as follows: "(1) That the place of a thing is no part or factor of the thing itself, but is that which embraces it; (2) that the immediate or 'proper' place of a thing is neither smaller nor greater than the thing itself; (3) that the place where the thing is can be quitted by it, and is therefore separable from it; and lastly (4), that any and every place implies and involves the correlatives of 'above' and 'below,' and that all the elemental substances have a natural tendency to move towards their own special places, or to rest in them when there—such movement being 'upward' or 'downward,' and such rest 'above' or 'below,' "[26] It is this last assumption that makes space a carrier of qualitative differences and furnishes thereby the metaphysical foundation of the mechanics of "natural" motion. Starting from these assumptions, Aristotle proceeds by a lucid process of logical elimination[27] to his famous definition of "place" as the adjacent boundary of the containing body. By this definition the concept became immune to all the criticisms that were designed to show the logical inconsistency of former definitions, as, for instance, Zeno's famous epicheirema (Everything is in place; this means that it is in something; but if place is something, then place itself is in something, etc.). In fact, this "nest of superimposed places" is mentioned as an argument against the existence of a kind of dimensional entity—distinct from the body that has shifted away when the encircled content is taken out and changed again and again, while the encircling content remains unchanged.

Further, this "replacement" of the content of a vessel by another content reveals that place is something different from its changing contents and so proves the reality of space. Of great importance from our point of view is a passage in Aristotle's *Physics* in which space is likened (using a modern expression) to a field of force: "Moreover the trends of the physical elements (fire, earth, and the rest) show not only that locality or place is a reality but also that it exerts an active influence; for fire and earth are borne, the one upwards and the other downwards, if unimpeded, each towards its own 'place,' and these terms—'up' and 'down' I mean, and the rest of the six dimensional directions—indicate subdivisions or distinct classes of positions or places in general."[28]

The dynamical field structure, inherent in space, is conditioned by the geometric structure of space as a whole. Space, as defined by Aristotle, namely, as the inner boundary of the containing receptacle, is, so to speak, a reference system which generally is of very limited scope. The place of the sailor is in the boat, the boat itself is in the river, and the river is in the river bed. This last receptacle is at rest relative to the earth and therefore also to the universe as a whole, according to contemporary cosmology. For astronomy, with its moving spheres, the reference system has to be generalized still further, leading to the finite space of the universe limited by the interior boundary of the outermost sphere, which itself is not contained in any further receptacle. This universal space, of spherical symmetry, has as its center the center of the earth, to which heavy bodies move under the dynamic influence intrinsic to space. It is natural for us, who have read Mach and Einstein, to raise the question whether the geometric aspect of this dynamical "field structure" depends on the distribution of matter in space or is completely independent of mass. Aristotle anticipated this question and tried to show that the dynamics of natural motion depends on spatial conditions only.

It might be asked, since the center of both (i.e., the earth and the universe) is the same point, in which capacity the natural motion of heavy bodies, or parts of the earth, is directed towards it; whether as center of the universe or of the earth. But it must be towards the center of the universe that they move, seeing that light bodies like fire, whose motion is contrary to that of the heavy, move to the extremity of the region which surrounds the center. It so happens that the earth and the universe have the same center, for the heavy bodies do move also towards the center of the earth, yet only incidentally, because it has its center at the center of the universe.[29]

This description is suggestive of the electrostatic field that exists between a small charged sphere enclosed by another sphere at a different potential. As is well known, the field itself may be nonspherically symmetric, as in the case of an excentric position of the inner sphere, which corresponds to the earth when shifted from the center of the universe, although the lines of force leave the surface of the enclosed body in a normal direction. To Aristotle, such a distortion seemed to be absurd; his world is a world of order and symmetry.

The directional tendencies of the elemental particles are possible only because of the difference in the conditions of the place in which they move from the conditions of the place to which they move. It is clear, therefore, that it is not a kind of buoyancy (corresponding to Archimedes' principle) that causes the motion of heavy or light bodies. For in this case the dynamical field structure would be mass-dependent. However, although these tendencies are independent of the distribution of mass, they are dependent on the very existence of matter. A void, conceived by Aristotle as the privation of all conceivable properties, cannot by its very definition be something differentiated directionally. It is well known how Aristotle exploited this argument in his repudiation of the void.

In conformity with the rejection of a vacuum, Aristotle insists repeatedly that the containing body has to be everywhere in contact with the contained. Polemizing against the Pythagorean doctrine of spatial vacancies, Aristotle offers a psychological explanation of the origin of such "gap" theories. "Because the encircled content may be taken out and changed again and again, while the encircling continent remains unchanged—as when water passes out of a vessel—the imagination pictures a kind of dimensional entity left there, distinct from the body that has shifted away."[30] But, he holds, to suppose that this "interval" is the place or space of the contained would inescapably lead to serious inconsistencies. He argues that on the basis of such a "gap" theory "place" would have to change its "place" and an ascending series of orders of spaces would be involved. Thus, when carrying a vessel of water from one place to another, one has to carry about also the "interval" and a transport of space in space is implied. His second objection is based on the assertion that transporting a vessel full of water means changing the place of the whole but not the places of its parts. According to Simplicius, Aristotle's line of thought seems to have used the following *reductio ad absurdum*: On the basis of an "interval" theory every part of water has to have its own place, since a transport of a vessel of water is accompanied by a rotation or wave disturbance of the liquid, which is possible only if the parts can shift from one interval to another. However, matter is indefinitely divisible and the number of such intervals must consequently be unlimited even for the smallest quantity of water. It follows that the volume, the sum total of all these intervals, being a sum of an infinite series, is infinitely great.

While expounding the inadequacy of these "interval" theories, Aristotle, on his part, ignores the fact that his very insistence on the all-over contact of the two distinct surfaces of the container and the contained must necessarily lead to a serious inconsistency between his own space theory on the one hand and his cosmology and theology on the other. For if the interior concave surface of the sphere of one planet is everywhere in contact with the convex surface of the sphere of another, then obviously the "fifth body," the substance of which the heavens are made, is not continuous, a conclusion that is contrary to the results of his cosmological doctrines as presented in *De caelo*.[31] Simplicius, who noticed this inconsistency, tried to avoid it by maintaining that all celestial spheres extend to one common center which coincides with the center of the earth. But obviously, Simplicius' solution of the problem is not only a theory *ad hoc*, but is also incompatible with the principles of Aristotelian physics, which explicitly rejects the interpenetrability of different bodies.[32]

It should be noted that Aristotle's remarks in the *Categories* indicate a different way of attacking the problem of space. Here space seems to be some kind of continuous extension; it is given no strict definition and, what is more important for our point of view, it has no physical implications for Aristotle's natural philosophy or that of his successors.

It is evident that space as an accident of matter is, according to Aristotle, finite, matter being itself finite.

Space, here, means the sum total of all places. The idea of a finite physical space, thus understood, is not as absurd today as it must have appeared fifty years ago, when physics acknowledged solely the conception of an infinite Euclidean space and when a finite material universe could but be conceived as an island, so to speak, in the infinite ocean of space. It is perhaps not wholly unjustified to suggest a comparison between the notion of physical space in Aristotle's cosmology and the notion of Einstein's "spherical space" as expounded in early relativistic cosmology. In both theories a question of what is "outside" finite space is nonsensical. Furthermore, the idea of "geodesic lines," determined by the geometry of space, and their importance for the description of the paths of material particles or light rays, suggest a certain analogy to the notion of "natural places" and the paths leading to them. The difference is, of course, that in Einstein's theory the geometry of space itself is a function of the mass-energy distribution in accord with the famous field equations, and is not Euclidean but Riemannian.

Although until the fourteenth century Aristotle's and Plato's conceptions were the prototypes, with only minor changes, of all theories of space, yet these conceptions were the object of constant attack, mostly on metaphysical grounds. Aristotle's pupil Theophrastus criticizes the master's theory[33] and speaks of the possibility of a motion of space, of the incomprehensibility of the universe as not being in space, and comes to the conclusion that space is no entity in itself but only an ordering relation that holds between bodies and determines their relative positions.[34] Like a biologist who dissects an animal and considers one organ in relation to another, so Theophrastus views space as a system of interconnected relations.

Concerned as we are with the problem of space in its implications for physics, we may disregard the few original contributions of the Epicureans, Skeptics, and other schools. We should, however, mention in this connection the important deviation of the Stoics from the traditional Aristotelian conception of the cosmos. Continuity, which for Aristotle was a purely geometric property of coherent matter, became with the Stoa a physical principle, an agent responsible for the propagation of physical processes through space. It is by this inner connection, manifested as a tension (*tonos*) in its active state, that distant parts of the universe are able to influence each other, thereby turning the cosmos into one field of action. The void, being incorporeal and therefore lacking all continuity, necessarily precludes all sense perception and so cannot exist inside the world.[35]

This elaboration of the Aristotelian idea of tendencies permeating the continuous plenum is an important generalization in two respects; in the variety of phenomena envisaged, and in their extension beyond the sublunar world (for instance, Posidonius' discovery of the moon's "influence" on the tides, which was regarded as an ostensible proof of the reality of this transmitting agent connecting even celestial with mundane phenomena [Chrysippus]). The range of activity of the propagating tensions is the whole material universe (*holon*) as distinct from the "all" (*pan*).

In order to explain this important distinction we have to refer to the changed definition of "space." In general, the Stoics accepted not Aristotle's definition of space, as the containing surface of the encircling body, but his discarded alternative, that is, the dimensional extension lying between the points of the containing surface. This change enabled the Stoics to maintain the existence of a void outside the material universe, whereas the material universe was conceived as an island of continuous matter surrounded by an infinite void. Needless to say, this infinite void lacked all qualities and differentiations, and, being thus completely indeterminate, it could not act in any way on the matter surrounded by it.[36] Hence the position of bodies was not determined by any properties of the void, but by their own nature. With no reason to move as a whole, the material world rests immovable in the infinite void. To the Stoics it made no sense to speak of the center of the "all"; on the other hand, the center of the material universe was a clear concept, cosmologically and physically well founded. Criticizing this doctrine, the Peripatetics raised the following question: If the material world is really surrounded by an infinite void, why does it not become dissipated and lost in the course of time?[37] The answer is now clear: the different parts of the material world are connected, not, as Aristotle thought, by an exterior continent, an upper sphere which forces the parts to stay together, like samples in a box, but by an internal cohesion (*hexis*), which is only another aspect of the tension mentioned before. It is this binding force that holds the world together, and the void without force of its own can do nothing to loosen it. In the void there is no "up" or "down" or any other direction or dimension.[38] In other words, it is isotropic, bare of any qualities whatever. As for space in the material universe, the Stoics adhere to the traditional Aristotelian doctrine. It was owing chiefly to these Stoic controversies that the problem of space could no longer be considered as one simple question, but had to appear under the form of two different considerations: space and void.

As we have tried to show in this chapter, space was conceived by classical Greek philosophy and science at first as something inhomogeneous because of its local geometric variance (as with Plato), and later as something anisotropic owing to directional differentiation in the substratum (Aristotle). It is perhaps not too conjectural to assume that these doctrines concerning the nature of space account for the failure of mathematics, especially geometry, to deal with space as a subject of scientific inquiry. Perhaps this is the reason why Greek geometry was so much confined to the plane. It may be objected that "space" according to Aristotle is "the adjacent boundary of the containing body" and so by its very definition is only of a two-dimensional character. But this objection ignores a clear passage in the *Physics*[39] and another passage in *De caelo*.[40] As Euclid's *Elements* show, the science of solid geometry was developed only to a small extent and mostly confined to the mensuration of solid bodies, which is at least one reason why even the *termini technici* of solid geometry, compared with those of plane geometry, were so little standardized. The idea of coördinates in the plane seems to go back to pre-Greek sources, the ancient Egyptian hieroglyphic symbol for "district" (*hesp*) being a grid (plane rectangular coördinate system). It would therefore be only natural to expect some reference to spatial coördinates in Greek mathematics. But in the whole history of Greek mathematics no such reference is found. Longitude (*mēkos*) and latitude (*platos*) as spherical coördinates on the celestial sphere or on the earth's surface were obviously used by Eratosthenes, Hipparchus, Marinus of Tyre, and Ptolemy, being the ideal two-dimensional system for concentric spheres in Aristotle's world of spherical symmetry. Simplicius mentions in his commentary on the first book of Aristotle's *De caelo* that Ptolemy composed an essay *On Extension* (*Peri diastaseon*) in which he demonstrated that bodies can have three dimensions. Moritz Cantor refers to this passage and says: "Bei der Unbestimmtheit dieser Angabe müssen wir allerdings dahin gestellt sein lassen, ob man glauben will, es seien in jener Schrift Gedanken enthalten gewesen, welche dem Begriffe von Raumkoordinaten nahe kommen."[41] So our assertion of the absence of spatial coördinates in Greek mathematics may stand. The use of a three-dimensional coördinate system, and in particular of a rectangular spatial coördinate system, was not thought reasonable until the seventeenth century (Descartes, Frans van Schooten, Lahire, and Jean Bernoulli), when the concept of space had undergone a radical change. Undoubtedly, Greek mathematics dealt with three-dimensional objects; Euclid himself, as related by Proclus, saw perhaps in the construction and investigation of the Platonic bodies the final aim of his *Elements*.[42] Yet space, as adopted in mechanics or in astronomy, had never been geometrized in Greek science. For how could Euclidean space, with its homogeneous and infinite lines and planes, possibly fit into the finite and anisotropic Aristotelian universe?

NOTES

[1] Aristotle, *Metaphysics*, 1080 b 33.

[2] J. Burnet, *Early Greek philosophy* (London, 1914), p. 51.

[3] P. Tannery, *Revue philosophique* 20 (1885), 389.

[4] Simplicius, *Physics*, 108 a.

[5] In *Aristotelis categorias commentarium* (ed. Carolus Kalbfleisch; Berolini (1907), p. 13.

[6] *Ibid.*, p. 357.

[7] Aristotle, *De caelo*, III, 2, 300 b.

[8] T. Lucreti Cari, *De rerum natura* (trans. by Munro; Cambridge, 1886), vol. 3, p. 23. The original Latin text is:
 . . . nam corpora sunt et inane,
 haec in quo sita sunt et qua diversa moventur.
     —Liber I, 420.

[9] See p. 28.

[10] Reference 8.

[11] John Locke, *An Essay concerning human understanding*, book II, 13, 21; see, for example, the edition by A. S. Pringle-Pattison (Clarendon Press, Oxford, 1950), p. 102.

[12] *Ad imum.* Liber I, 987.

[13] J. Burnet, *Early Greek philosophy* (London, ed. 3, 1920), p. 389.

[14] *Sexti Empirici opera*, "Adversus dogmaticos" (ed. H. Mutschmann; Leipzig, 1912–14), vol. 2, p. 17.

[15] Aristotle, *Physics*, 209 b.

[16] *Ibid.*, 203 a, 209 b.

[17] E. Zeller, *Die Philosophie der Griechen* (Leipzig, 1869–1879), vol. 2.

[18] Plato, *Timaeus*, 55 ff.

[19] *Ibid.*, 52 d; F. M. Cornford, *Plato's cosmology* (Harcourt, Brace, New York, 1937), p. 198.

[20] Zeller, *op. cit.*, vol. 1, p. 376.

[21] Plato, *Timaeus*, 56.

[22] Plato, *Republic*, 528.

[23] Plato, *Republic*, book VII, trans. by P. Shorey (Loeb Classical Library; Harvard University Press, Cambridge, 1946), vol. 2, p. 179.

[24] Aristotle, *Categories*, 5 a, 8-14. See Richard McKeon, *The Basic works of Aristotle* (Random House, New York, 1941), p. 15.

[25] For this interpretation, see Pierre Duhem, *Le systeme du monde* (Paris, 1913–1917), vol. 1, p. 197.

[26] Aristotle, *Physics*, 211 a, trans. by P. H. Wicksteed and F. M. Cornford (Loeb Classical Library; Harvard University Press, Cambridge, 1929), vol. 1, p. 303.

[27] *Ibid.*, 211 b.

[28] *Ibid.*, 208 b; Loeb edition, p. 279.

[29] Aristotle, *De caelo*, II, 14, 296 b; Loeb edition, p. 243.

[30] Aristotle, *Physics*, IV, 211 b 15; Loeb edition, p. 309.

[31] Aristotle, *De caelo*, I, 3, 270 a *et seq.*, Loeb edition, p. 21.

[32] Aristotle, *Physics*, IV, 209 a 7; Loeb edition, p. 283.

[33] Simplicius, *Physics*, 141.

[34] *Ibid.*, 141, 149.

[35] Cleomedes, "De motu circulari corporum caelestium libri duo," in J. ab Arnim, ed., *Stoicorum veterum fragmenta*, II, 546 (Leipzig, 1905), p. 172.

[36] *Ibid.*, II, 173, 176 (pp. 49, 51).

[37] *Ibid.*, II, 540.

[38] *Ibid.*, 557.

[39] Aristotle, *Physics*, 209 a 4-6.

[40] Aristotle, *De caelo*, 268 a 7-10.

[41] M. Cantor, *Vorlesungen über Geschichte der Mathematik* (Leipzig, 1880), vol. 1, p. 357.

[42] Procli Diadochi, *In primum Euclidis elementorum librum commentarii* (Leipzig, 1873), p. 64.

ARISTOTLE
AND THE SEA BATTLE

DE INTERPRETATIONE, CHAPTER IX

G. E. M. ANSCOMBE

A revised version, with some omissions and additions, of an article published in *Mind*, Volume 65, 1956.

1   For what is and for what has come about, then, it is
necessary that affirmation, or negation, should be true or
false; and for universals universally quantified it is always
necessary that one should be true, the other false; and for
singulars too, as has been said; while for universals not
universally quantified it is not necessary. These have been
discussed.[1]

For what is and for what has come about: he has in fact not
mentioned these, except to say that a verb or a tense—*sc.*
other than the present, which he regards as the verb *par
excellence*—must be part of any proposition.
it is necessary: given an *antiphasis* about the present or past,
the affirmative proposition must be true or false; and similarly
for the negative. An *antiphasis* is a pair of propositions in
which the same predicate is in one affirmed, in the other
denied, of the same subject. Note that Aristotle has not the
idea of the negation of a proposition, with the negation sign
outside the whole proposition; that was (I believe) invented
by the Stoics. What Aristotle says in this sentence is ambiguous; that this is deliberate can be seen by the contrast with
the next sentence. The ambiguity—between necessarily having
a truth-value, and having a necessary truth-value—is first
sustained, and then resolved at the end of the chapter.
for universals universally quantified: he does not mean, as this
place by itself would suggest, that of "All men are white" and
"No men are white" one must be true and the other false. But
that if you take "All men are white" and "No men are white"
and construct the antiphasis of which each is a side, namely,

"All men are white—Not all men are white" and "No men are white—Some man is white," then one side of each antiphasis must be true, and the other side must be false.

for singulars too, as has been said: sc. of "Socrates is white—Socrates is not white" one side is necessarily true, the other necessarily false. (This is what a modern reader cannot take in; but see the "Elucidation.")

for universals not universally quantified: his example rendered literally is "man is white—man is not white." From his remarks I infer that these would be correctly rendered "men are . . ." For, he says, men are beautiful, and they are also not beautiful, for they are ugly too, and if they are ugly they are not beautiful. I believe that we (nowadays) are not interested in these unquantified propositions.

These have been discussed: i.e. in the immediately preceding chapters, by which my explanations can be verified.

2   But for what is singular and future it isn't like this. For if every affirmation and negation is true or false, then it is also necessary for everything to be the case or not be the case. So if one man says something will be, and another says not, clearly it is necessary for one of them to be speaking truly, if every affirmation and nega-
3 tion is true or false. For both will not hold at once on such conditions. For if it is true to say that something is white or is not white, its being white or not white is necessary, and if it is white or not white, it is true to say or deny it. And if it is not the case, then it is false, and if it is false, it is not the case; so that it is necessary as regards either the affirmation or the negation that it is true or false.

singular and future: sc. there will be a relevant discussion tonight; this experiment will result in the mixture's turning green; you will be sent down before the end of term.

it isn't like this: namely, that the propositions (or their negations) must be true or false. Throughout this paragraph the ambiguity is carefully preserved and concealed.

it is also necessary for everything to be the case or not be the case: the Greek "or" is, like the English, ambiguous between being exclusive and being non-exclusive. Here it is exclusive, as will appear; hence the "or" in the conditional "if every affirmation and negation is true or false" is also exclusive, and to point this he says "every affirmation and negation," not, as in (1) "every affirmation or negation"; that "or" was non-exclusive.

For both will not hold at once on such conditions: namely, on the conditions that every affirmation is true or false. This condition is not a universal one; it does not apply to the unquantified propositions, though if the "or" is non-exclusive it does. But if the conditions hold, then just one of the two speakers must be speaking the truth.

it is true to say or deny it: ἦν is the common philosophical imperfect.

4   So nothing is or comes about by chance or 'whichever happens.' Nor will it be or not be, but everything of necessity and not 'whichever happens.' For either someone saying something or someone denying it will be right. For it would either be happening or not happening accordingly. For whichever happens is not more thus or not thus than it is going to be.

'whichever happens': the Greek phrase suggests both "as it may be" and "as it turns out." "As the case may be" would have been a good translation if it could have stood as a subject of a sentence. The 'scare-quotes' are mine; Aristotle is not overtly discussing the *expression* "whichever happens."

is not more thus or not thus than it is going to be: as the Greek for "or" and for "than" are the same, it is so far as I know a matter of understanding the argument whether you translate as here, or (as is more usual) *e.g.*: "isn't or (*sc.* and) isn't going to be rather thus than not thus." But this does not make good sense. Aristotle is arguing: "We say 'whichever happens' or 'as the case may be' about the present as well as about the future; but you don't think the present indeterminate, so why say the future is?" Or rather (as he is not talking about the expression): "Whatever happens *will* be just as determinately thus or not thus as it *is*."

5   Further, if something is white now, it was true earlier to say it was going to be white, so that it was always true to say of any of the things that have come about: "it is, or will be." But if it was always true to say: "it is, or will be," then: impossible for that not to be or be going to be. But if it is impossible for something not to come about, then it is unable not to come about. But if something is unable not to come about it is

6 necessary for it to come about. Therefore it is necessary
that everything that is going to be should come about.
So nothing will be 'whichever happens' or by chance.
For if by chance, then not by necessity.

*But if it is impossible for something not to come about, then it
is unable not to come about:* the reader who works through
to the end and understands the solution will examine the
dialectic to see where it should be challenged. It will turn out
that the point is here, in spite of the equivalence of the two
Greek expressions. It is impossible for the thing not to come
about, *i.e.*, necessary that it should come about, by *necessitas
consequentiae*, which does not confer the character of necessity on what does come about. A 'necessary consequence' of
what is true need not be necessary.

Still, it is not open to us, either, to say that neither is
true, as: that it neither will be nor will not be. For
7 firstly, the affirmation being false the negation will not be
true, and this being false the affirmation won't be true.
And besides, if it is true to say that something is big and
white, both must hold. And if they are going to hold
tomorrow, they must hold tomorrow. And if something
is neither going to be nor not going to be tomorrow,
'whichever happens' won't be. Take a sea-battle, for
example: it would have to be the case that a sea-battle
neither came about nor didn't come about tomorrow.

Still, it is not open to us, either, to say that neither is true: and
yet Aristotle is often supposed to have adopted this as the
solution.
For firstly: this goes against what he has shown at the end of
(3): "if it is false, it does not hold." So much, however, is
obvious, and so this is not a very strong objection if we are
willing to try whether neither is true. What follows is stronger.
And if they are going to hold tomorrow: from here to the end
of the paragraph the argument is: if it is the case that something will be, then it will be the case that it is. In more detail:
you say, or deny, two things about the future. If what you
say is true, then when the time comes you must be able to say
those two things in the present or past tenses.
'whichever happens' won't be: i.e. 'whichever happens' *won't*
happen.

8   These are the queer things about it. And there is
     more of the sort, if it is necessary that for every affirma-
     tion and negation, whether for universals universally
     quantified or for singulars, one of the opposites should be
     true and one false, that there is no 'whichever happens'
     about what comes about, but that everything is and
     comes about of necessity. So that there would be no
     need to deliberate or take trouble, e.g.: "if we do this,
9   this will happen, if not, not." For there is nothing to
     prevent its being said by one man and denied by another
     ten thousand years ahead that this will happen, so that
     whichever of the two was then true to say will of neces-
     sity happen. And indeed it makes no difference even if
     people have said the opposite things or not; for clearly
     this is how things are, even if there isn't one man saying
     something and another denying it; nor is it its having
     been asserted or denied that makes it going to be or not,
     nor its having been ten thousand years ahead or at any
10   time you like. So if in the whole of time it held that the
     one was the truth, then it was necessary that this came
     about, and for everything that has been it always held,
     so that it came about by necessity. For if anyone has
     truly said that something will be, then it can't not happen.
     And it was always true to say of what comes about: it
     will be.

These are the queer things about it. And: I have diverged from
the usual punctuation, which leads to the rendering: "These
and similar strange things result, if . . ." This seems illogical.
e.g.: often rendered "since": "since if we do this, this will
happen, if not, not." This does not appear to me to make
good sense. The Oxford translator sits on the fence here.
So if in the whole of time it held: one must beware of supposing that Aristotle thinks the conclusion stated in the
apodosis of this sentence follows from the condition. It only
follows if the previous arguments are sound. He is going to
reject the conclusion, but there is no reason to think that he
rejects the condition: on the contrary.

11   Now if this is impossible! For we see that things
     that are going to be take their start from deliberating
     and from acting, and equally that there is in general a
     possibility of being and not being in things that are not

always actual. In them, both are open, both being and
not being, and so also both becoming and not becoming.
12 And plenty of things are obviously like this; for example,
this coat is capable of getting cut up, and it won't get
cut up but will wear out first. And equally it is capable
of not getting cut up, for its getting worn out first would
not have occurred if it had not been capable of not
13 getting cut up. So this applies too to all other processes
that are spoken of in terms of this kind of possibility.
So it is clear that not everything is or comes about of
necessity, but with some things 'whichever happens,' and
the affirmation is not true rather than the negation; and
with other things one is true rather and for the most part,
but still it is open for either to happen, and the other not.

take their start: literally: "there is a starting point of things
that are going to be." The word also means "principle." A
human being is a prime mover (in the engineer's sense), but
one that works by deliberating. As if a calculating machine
not merely worked, but was, in part, precisely *qua* calculating,
a prime mover. But Aristotle's approach is not that of someone enquiring into human nature, but into causes of events
and observing that among them is this one.
acting: he means human action, which is defined in terms of
deliberation; see *Nichomachean Ethics,* VI, 1139: there he
repeats the word "ἀρχή": "ἡ τοιαυτη ἀρχη ἀνθρωπος": *the* cause
of this sort is man. An animal too, or a plant, is a prime
mover. Hence his thought is not that there are *new* starting
points constantly coming into existence; that would not matter.
It is first of all the nature of deliberation that makes him think
that the fact of human action proves the dialectic must be
wrong. I cannot pursue this here; though I should like to
enter a warning against the idea (which may present itself):
"the nature of deliberation presupposes freedom of the will
as a condition." That is not an Aristotelian idea.
things that are not always actual: things that are always actual
are the sun, moon, planets and stars. Aristotle thought that
what these do is necessary. The general possibility that he
speaks of is of course a condition required if deliberation and
'action' are to be possible. If what the typewriter is going to
do is necessary, I cannot do anything else with the typewriter.
Not that this is Aristotle's ground for speaking of the general
possibility. That is shown in his consideration about the coat:

he implicitly claims that the assumption that the coat *will* be worn out does not conflict with our knowledge that it can be cut up. He is surely right. We know a vast number of possibilities of this sort.

in terms of this kind of possibility: I take it that we have here the starting point for the development of Aristotle's notion of potentiality. The sentence confirms my view of the point where he would say the dialectic went wrong.

with other things one is true rather and for the most part: as we should say: more probable.

14 The existence of what is when it is, and the non-existence of what isn't when it isn't, is necessary. But still, for everything that is to be is not necessary, nor for everything that isn't not to be. For it isn't the same: for everything that is to be of necessity when it is, and: for it simply to be of necessity. And the same for what isn't. And the same reasoning applies to the antiphasis. For it is necessary that everything should be or not, and should be going to be or not. But it is not the case, separately

15 speaking, that either of the sides is necessary. I mean, *e.g.* that it is necessary that there will be a sea-battle tomorrow or not, but that it is not necessary that there should be a sea-battle tomorrow, nor that it should not happen. But for it to come about or not is necessary. So that since propositions are true as the facts go, it is clear that where things are such as to allow of 'whichever happens' and of opposites, this must hold for the antiphasis too.

The existence of what is when it is . . . is necessary: i.e. it cannot be otherwise. A modern gloss, which Aristotle could not object to, and without which it is not possible for a modern person to understand his argument, is: and cannot be shown to be otherwise. It will by now have become very clear to a reader that the implications of "necessary" in this passage are not what he is used to. But see the "Elucidation," below.

simply to be of necessity: there is a temptation to recognise what we are used to under the title "logical necessity" in this phrase. But I believe that Aristotle thought the heavenly bodies and their movements were necessary in this sense. On the other hand, he seems to have ascribed something like logical necessity to them. . . .

But it is not the case, separately speaking, that either of the sides is necessary: the ambiguity of the opening "it is necessary that an affirmation (or negation) should be true or false" is here resolved. And we learn that when Aristotle said that, he meant that if p is a statement about the present or the past, then either p is necessary or *not-p* is necessary. But this means that in order to ascribe necessity to certain propositions (the ones, namely, that are not 'simply' necessary) we have to be informed about particular facts. So, one may ask, what has this necessity got to do with logic? Aristotle, however, states no facts, past, present, or future. (I do in what follows; I hope this will not prove misleading: the purpose is only didactic.) His results could perhaps be summarised as follows: we use indices $_p$ and $_f$ to the propositional sign to indicate present and past time references on the one hand, and future time reference on the other. Then for all p, p vel *not-p* is necessary (this covers the unquantified propositions too) and p_p is necessary vel $(not\text{-}p)_p$ is necessary; but it is not the case that for all p, p_f is necessary vel $(not\text{-}p)_f$ is necessary.

This is how it is for what is not always existent or not
16 always non-existent. For such things it is necessary that
a side of the antiphasis should be true or false, but not
this one or that one, but whichever happens; and that
one should be true rather than the other; but that does
not mean that it is true, or false. So it is clear that it is
not necessary for every affirmation and negation that this
one of the opposites should be true and that one false; for
it does not hold for what does not exist but is capable of
being or not being; but it is as we have said.

whichever happens: sc.: it is a matter of whichever happens.
that one should be true rather than the other: cf. "rather and for the most part" above; note that this is governed by "it is necessary"; I infer that Aristotle thought that correct statements of probability were true propositions.
but that does not mean: ἤδη, logical, not temporal;[2] ἤδη works rather like the German "schon" (only here of course it would be "noch nicht"). ἤδη in a non-temporal sense is, like οὐκέτι, frequent in Greek literature. English translators of philosophical texts usually either neglect to translate it or mistranslate it. For examples, see *Theaetetus* 201e4, *Physics*, 187a36, *De Interpretatione* 16a8, *Metaphysics* 1006a16. Bonitz gives some more examples.

An Elucidation of the Foregoing from a Modern Point of View

A. The Vice Chancellor will either be run over next week or not. And therefore either he will be run over next week or he will not. Please understand that I was *not* repeating myself!

B. I think I understand what you were trying to do; but I am afraid you were repeating yourself and, what is more, you cannot fail to do so.

A. Can't fail to do so? Well, listen to this: The Vice Chancellor *is* going to be run over next week.

B. Then I am going to the police as soon as I can.

A. You will only be making a fool of yourself. It's not true.

B. Then why did you say it?

A. I was merely trying to make a point: namely, that I have succeeded in saying something true about the future.

B. What have you said about the future that is true?

A. I don't know: but this I do know, that I have said something true; and I know that it was either when I told you the Vice Chancellor would be run over, or on the other hand when I said he wouldn't.

B. I am sorry, but that is no more than to say that Either he will or he won't be run over. Have you given me any information about the future? Don't tell me you have, with one of these two remarks, for that is to tell me nothing, just because the two remarks together cover all the possibilities. If what you tell me is an Either/Or and it embraces all possibilities, you tell me nothing.

A. Can an Either/Or be true except by the truth of *one* of its components? I seem to remember Quine speaking of Aristotle's "fantasy," that "it is true that either *p* or *q*" is not a sufficient condition for "Either it is true that *p* or it is true that *q*." Now I will put it like this: Aristotle seems to think that the truth of a truth-functional expression is independent of the truth values of the component propositions.

B. But that is a howler! The 'truth' of Either *p* or *not-p* is determined, as you know very well, by its truth value's being *T* for all possible combinations of the truth *possibilities* of its components; that is why its 'truth' gives no information. Having set out the full truth-table and discovered that for all possibilities you get T in the final column, you need make no enquiry to affirm the truth of p or not-p—any enquiry would be comic. If on the other hand you tell me p or not-q (q being different from p) you do give me some information, for certain truth-combinations are excluded. There is therefore the possibility of enquiring whether your information is correct. And that I do by discovering which of the truth-possibilities is fulfilled; and if one of the combinations of truth-possibilities which is a truth-condition for p or not-q is fulfilled, then I discover that your information is correct. But to tell me "It will rain, or it won't," is not to tell me of any truth-possibility that it is—or, if you like, will be, satisfied. Now will you actually tell me something about the future?

A. Very well. Either you are sitting in that chair or it will not rain tomorrow.

B. I agree, that is true, because I am sitting in this chair. But still I have been told nothing about the future, because since I know I am sitting in this chair I know what I have been told is true whether it rains tomorrow or not—*i.e.* for all truth possibilities of "It will rain tomorrow." But do you mind repeating your information?

A. Either you are sitting in that chair or it will not rain tomorrow.

B. (*Having stood up*). I am glad to be told it will be fine— but is it certain? Do you get it from the meteorologists? I have heard that they are sometimes wrong.

A. But surely we are talking about truth, not certainty or knowledge.

B. Yes, and I am asking whether your information—which I agree is information this time—is true.

A. I can't tell you till some time tomorrow; perhaps not till midnight. But whatever I tell you then will have been so now—I mean if I tell you then "True," that means not just that it will be true then but that it was true now.

B. But I thought it was the great point against Aristotle that "is true" was timeless.

A. Yes—well, what I mean is that if I tell you—as I shall be able to—"True" tomorrow—I mean *if* I am able to, of course—why, then it will have been, I mean is now correct to say it is true.

B. I understand you. If it is going to rain tomorrow it is true that it is going to rain tomorrow. I should be enormously surprised if Aristotle were to deny this.

A. But Aristotle says it isn't true that it is going to rain tomorrow!

B. I did not read a single prediction in what Aristotle said. He only implied that it didn't have to be true that it will rain tomorrow, *i.e.* it doesn't have to rain tomorrow.

A. What? Even if it is going to rain tomorrow?

B. Oh, of course, if it is going to rain tomorrow, then it necessarily will rain tomorrow: (if *p* then *p*) is necessary. But is it going to?

A. I told you, I can't say, not for certain. But *why* does that matter?

B. Can't you say anything for certain about tomorrow?

A. I am going to Blackwell's tomorrow.

B. And that is certain?

A. Yes, I am absolutely determined to go. (Partly because of this argument: it is a point of honour with me to go, now.)

B. Good. I fully believe you. At least, I believe you as fully as I can. But do I—or you—know you will go? Can nothing stop you?

A. Of course lots of things can stop me—anything from a change of mind to death or some other catastrophe.

B. Then you aren't necessarily going to Blackwell's?

A. Of course not.

B. Are you necessarily here now?

A. I don't understand you.

B. Could it turn out that this proposition that you, N.N., are in All Souls today, May 7th, 1954, is untrue? Or is this certain?

A. No, it is quite certain—my reason for saying so is that if you cared to suggest any test, which could turn out one way or the other, I can't see any reason to trust the test if, situated as I am, I have any doubt that I am here. I don't mean I can't imagine doubting it; but I can't imagine anything that would make it doubtful.

B. Then what is true about the present and the past is *necessarily* true?

A. Haven't you passed from certainty to truth?

B. Do you mean to tell me that something can be certain

without being true? And isn't what is true about the present and the past quite necessary?

A. What does "necessary" mean here, since it obviously doesn't mean that these are what we call necessary propositions?

B. I mean that nothing whatever could make what is certain untrue. Not: if it is true, it is necessary, but: since it is certainly true it is necessary. Now if you can show me that anything about the future is so certain that nothing could falsify it, then (perhaps) I shall agree that it is necessarily true that that thing will happen.

A. Well: the sun will rise tomorrow.

B. That is so certain that nothing could falsify it?

A. Yes.

B. Not even: the sun's not rising tomorrow?

A. But this is absurd! When I say it is certain I am here, am I saying it wouldn't falsify it for me not to be here? But I am here, and the sun will rise tomorrow.

B. Well, let me try again: Could anything that can happen make it untrue that you are here? If not, I go on to ask: Could anything that can happen make it untrue that the sun rises tomorrow?

A. No.

B. If we continued in darkness, the appearance of the night being continued for the rest of our lives, all the same the sun will have risen; and so on?

A. But that can't happen.

B. Is that as certain as that you are here now?

A. I won't say. But what does Aristotle mean when he says that one part of the antiphasis in necessarily true (or false) when it is the present or the past that was in question? Right at the beginning, when I said "The Vice Chancellor will either be run over or not, therefore either he will be run over or he will not" you said that I was repeating myself and could not fail to be repeating myself. And then you referred to the truth-table-tautological account of that proposition. But does not precisely the same point apply to what Aristotle says about "Either *p* or *not-p*" when *p* is a proposition about the present or the past?

B. You could have avoided repeating yourself if you had said "The Vice Chancellor will either be run over or not, therefore either it is necessary that he should be run over or it is necessary that he should not be run over." But as you would have been disinclined to say that—seeing no possible meaning for an ascription of necessity except what we are used to call "logical necessity"—you could not avoid repeating yourself.

Thus Aristotle's point (as we should put it) is that "Either *p* or *not-p*" is always necessary, and this necessity is what we are familiar with. But—and this is from our point of view the right way to put it, for this is a novelty to us—that when *p* describes a present or past situation, then either *p* is necessarily true, or *not-p* is necessarily true; and here "necessarily true" has a sense which is unfamiliar to us. In this sense I say it is necessarily true that there was not—or necessarily false that there was—a big civil war raging in England from 1850 to 1870; necessarily true that there is a University in Oxford; and so on. But "necessarily true" is not simply the same as "true"; for while it may be true that there will be rain tomorrow, it is not necessarily true. As everyone would say: there may be or may not. We also say this about things which we don't know about the past and the present. The question presents itself to us then in this form: does "may" express mere ignorance on our part in both cases?

Suppose I say to someone: "In ten years' time you will have a son; and when he is ten years old he will be killed by a tyrant." Clearly this is something that may be true and may not. But equally clearly there is no way of finding out (unless indeed you say that waiting and seeing is finding out; but it is not finding out that it will happen, only that it does happen).

Now if I really said this to someone, she would either be awestruck or think me dotty; and she would be quite right. For such a prediction is a prophecy. Now suppose that what I say comes true. The whole set of circumstances—the prophecy together with its fulfilment—is a miracle; and one's theoretical attitude (if one has one at all) to the supposition of such an occurrence ought to be exactly the same as one's theoretical attitude to the supposition that one knew of someone's rising from the dead and so on.

As Newman remarks, a miracle ought not to be a silly trivial kind of thing—*e.g.* if my spoon gets up one day and dances a jig on my plate, divides into several pieces and then joins up again, it qualifies ill as a miracle, though it qualifies perfectly well for philosophical discussion of naturally impossible but imaginable occurrences. Similarly if one were discussing impossible predictions one would take such an example as the following: Every day I receive a letter from someone giving an accurate account of my actions and experiences from the time of posting to the time I received the letter. And whatever I do (I do random, absurd actions for example, to see if he will still have written a true account) the letter records it. Now, since we are dealing in what can be imagined and therefore can be supposed to happen, we must settle whether this would be knowledge of the future: its certainty would surely be a proof that what I did I did necessarily.

It is interesting to note that Wittgenstein agrees with Aristotle about this problem, in the *Tractatus*. "The freedom of the will consists in the fact that future actions cannot be known. The connexion of knowing and the known is that of logical necessity. 'A knows that *p*' is senseless, if *p* is a tautology." We are therefore presented with the logical necessariness of what is known's being true, together with the logical non-necessity of the kind of things that are known. The "logical necessity" of which he speaks in the remark on knowledge is thus not just truth-table necessariness. It is the unfamiliar necessariness of which Aristotle also speaks. "A knows that *p*" makes sense only if *p* describes a fact about the past or present; so it comes out in Wittgenstein, and in Aristotle: past and present facts are necessary. (In more detail, by the *Tractatus* account: if A knows *p*, for some *q* (if *q* then *p*) is a tautology, and *q* expresses a fact that A is "acquainted" with.)

Then this letter about my actions would not have been knowledge even if what it said was always right. However often and invariably it was verified, it would still not be certain, because the facts could go against it.

However, Aristotle's considerations aren't about knowledge and certainty, and *are* about necessity and its corresponding possibility. They are probably the more difficult to understand nowadays because—in a way that is not obvious—they supply an instance of his willingness to deny what it does not make sense to assert. For the affirmation of the necessity of *p* is equivalent to the denial of the possibility of *not-p*. But the possibility in question relates only to the future; hence by some current conceptions the negation of such possibility also relates only to the future. *E.g.* "This plaster can be painted for the next eight hours and not after that, since by then it will have set too hard." Neither the affirmation nor the negation of this sort of possibility can be straightforwardly constructed in relation to the past. "It can(not) be painted yesterday" demands emendation, perhaps to "It may (can't) have been painted yesterday," perhaps to "It could (not) have been painted yesterday." The contingency of the past is that something was possible, not that it is possible. (That is not the same as saying that it will be possible, for the possibility may be extinguished.) Now Aristotle, who readily uses "No men are numbers" as a premise, would pass from the denial that possibility of this sort holds in relation to the past (or present) to the assertion that the past and present are necessary.

NOTES

1 This and following passages from Aristotle are in Miss Anscombe's own translation.—ED.

2 I am indebted to Miss M. Hartley of Somerville College for pointing this out to me.—Anscombe.

QUESTIONS
ABOUT TIME

St. Augustine

From *Confessions*, Book Eleven, reprinted as translated and
edited by Albert C. Outler in Volume VII of the Library of
Christian Classics, Westminster Press and SCM Press, Philadelphia and London, 1955.

Chapter XIV

17. There was no time, therefore, when thou hadst not
made anything, because thou hadst made time itself. And
there are no times that are coeternal with thee, because thou
dost abide forever; but if times should abide, they would not
be times.

For what is time? Who can easily and briefly explain it?
Who can even comprehend it in thought or put the answer
into words? Yet is it not true that in conversation we refer to
nothing more familiarly or knowingly than time? And surely
we understand it when we speak of it; we understand it also
when we hear another speak of it.

What, then, is time? If no one asks me, I know what it is.
If I wish to explain it to him who asks me, I do not know.
Yet I say with confidence that I know that if nothing passed
away, there would be no past time; and if nothing were still
coming, there would be no future time; and if there were
nothing at all, there would be no present time.

But, then, how is it that there are the two times, past and
future, when even the past is now no longer and the future is
now not yet? But if the present were always present, and did
not pass into past time, it obviously would not be time but
eternity. If, then, time present—if it be time—comes into
existence only because it passes into time past, how can we
say that even this *is*, since the cause of its being is that it will

cease to be? Thus, can we not truly say that time *is* only as it
tends toward nonbeing?

CHAPTER XV

18. And yet we speak of a long time and a short time; but
never speak this way except of time past and future. We call
a hundred years ago, for example, a long time past. In like
manner, we should call a hundred years hence a long time to
come. But we call ten days ago a short time past; and ten
days hence a short time to come. But in what sense is something long or short that is nonexistent? For the past is not
now, and the future is not yet. Therefore, let us not say, "It *is*
long"; instead, let us say of the past, "It *was* long," and of the
future, "It *will be* long." And yet, O Lord, my Light, shall
not thy truth make mockery of man even here? For that long
time past: was it long when it was already past, or when it
was still present? For it might have been long when there was
a period that could be long, but when it was past, it no longer
was. In that case, that which was not at all could not be long.
Let us not, therefore, say, "Time past was long," for we shall
not discover what it was that was long because, since it is past,
it no longer exists. Rather, let us say that "time *present* was
long, because when it was present it *was* long." For then it had
not yet passed on so as not to be, and therefore it still was in a
state that could be called long. But after it passed, it ceased
to be long simply because it ceased to be.

19. Let us, therefore, O human soul, see whether present
time can be long, for it has been given you to feel and measure
the periods of time. How, then, will you answer me?

Is a hundred years when present a long time? But, first, see
whether a hundred years can be present at once. For if the
first year in the century is current, then it is present time, and
the other ninety and nine are still future. Therefore, they are
not yet. But, then, if the second year is current, one year is
aleady past, the second present, and all the rest are future. And
thus, if we fix on any middle year of this century as present,
those before it are past, those after it are future. Therefore,
a hundred years cannot be present all at once.

Let us see, then, whether the year that is now current can
be present. For if its first month is current, then the rest are

future; if the second, the first is already past, and the remainder are not yet. Therefore, the current year is not present all at once. And if it is not present as a whole, then the year is not present. For it takes twelve months to make the year, from which each individual month which is current is itself present one at a time, but the rest are either past or future.

20. Thus it comes out that time present, which we found was the only time that could be called "long," has been cut down to the space of scarcely a single day. But let us examine even that, for one day is never present as a whole. For it is made up of twenty-four hours, divided between night and day. The first of these hours has the rest of them as future, and the last of them has the rest as past; but any of those between has those that preceded it as past and those that succeed it as future. And that one hour itself passes away in fleeting fractions. The part of it that has fled is past; what remains is still future. If any fraction of time be conceived that cannot now be divided even into the most minute momentary point, this alone is what we may call time present. But this flies so rapidly from future to past that it cannot be extended by any delay. For if it is extended, it is then divided into past and future. But the present has no extension[1] whatever.

Where, therefore, is that time which we may call "long"? Is it future? Actually we do not say of the future, "It is long," for it has not yet come to be, so as to be long. Instead, we say, "It will be long." *When* will it be? For since it is future, it will not be long, for what may be long is not yet. It will be long only when it passes from the future which is not as yet, and will have begun to be present, so that there can be something that may be long. But in that case, time present cries aloud, in the words we have already heard, that it cannot be "long."

CHAPTER XVI

21. And yet, O Lord, we do perceive intervals of time, and we compare them with each other, and we say that some are longer and others are shorter. We even measure how much longer or shorter this time may be than that time. And we say that this time is twice as long, or three times as long, while

this other time is only just as long as that other. But we measure the passage of time when we measure the intervals of perception. But who can measure times past which now are no longer, or times future which are not yet—unless perhaps someone will dare to say that what does not exist can be measured? Therefore, while time is passing, it can be perceived and measured; but when it is past, it cannot, since it is not.

CHAPTER XVII

22. I am seeking the truth, O Father; I am not affirming it. O my God, direct and rule me.

Who is there who will tell me that there are not three times—as we learned when boys and as we have also taught boys—time past, time present, and time future? Who can say that there is only time present because the other two do not exist? Or do they also exist; but when, from the future, time becomes present, it proceeds from some secret place; and when, from times present, it becomes past, it recedes into some secret place? For where have those men who have foretold the future seen the things foretold, if then they were not yet existing? For what does not exist cannot be seen. And those who tell of things past could not speak of them as if they were true, if they did not see them in their minds. These things could in no way be discerned if they did not exist. There are therefore times future and times past.[2]

CHAPTER XVIII

23. Give me leave, O Lord, to seek still further. O my Hope, let not my purpose be confounded. For if there are times past and future, I wish to know where they are. But if I have not yet succeeded in this, I still know that wherever they are, they are not there as future or past, but as present. For if they are there as future, they are there as "not yet"; if they are there as past, they are there as "no longer." Wherever they are and whatever they are they exist therefore only as present. Although we tell of past things as true, they

are drawn out of the memory—not the things themselves, which have already passed, but words constructed from the images of the perceptions which were formed in the mind, like footprints in their passage through the senses. My childhood, for instance, which is no longer, still exists in time past, which does not now exist. But when I call to mind its image and speak of it, I see it in the present because it is still in my memory. Whether there is a similar explanation for the foretelling of future events—that is, of the images of things which are not yet seen as if they were already existing—I confess, O my God, I do not know. But this I certainly do know: that we generally think ahead about our future actions, and this premeditation is in time present; but that the action which we premeditate is not yet, because it is still future. When we shall have started the action and have begun to do what we were premeditating, then that action will be in time present, because then it is no longer in time future.

24. Whatever may be the manner of this secret foreseeing of future things, nothing can be seen except what exists. But what exists now is not future, but present. When, therefore, they say that future events are seen, it is not the events themselves, for they do not exist as yet (that is, they are still in time future), but perhaps, instead, their causes and their signs are seen, which already do exist. Therefore, to those already beholding these causes and signs, they are not future, but present, and from them future things are predicted because they are conceived in the mind. These conceptions, however, exist *now*, and those who predict those things see these conceptions before them in time present.

Let me take an example from the vast multitude and variety of such things. I see the dawn; I predict that the sun is about to rise. What I see is in time present, what I predict is in time future—not that the sun is future, for it already exists; but its rising is future, because it is not yet. Yet I could not predict even its rising, unless I had an image of it in my mind; as, indeed, I do even now as I speak. But that dawn which I see in the sky is not the rising of the sun (though it does precede it), nor is it a conception in my mind. These two[3] are seen in time present, in order that the event which is in time future may be predicted.

Future events, therefore, are not yet. And if they are not yet, they do not exist. And if they do not exist, they cannot be seen at all, but they can be predicted from things present, which now are and are seen. . . .

CHAPTER XXI

27. I have said, then, that we measure periods of time as they pass so that we can say that this time is twice as long as that one or that this is just as long as that, and so on for the other fractions of time which we can count by measuring.

So, then, as I was saying, we measure periods of time as they pass. And if anyone asks me, "How do you know this?" I can answer: "I know because we measure. We could not measure things that do not exist, and things past and future do not exist." But how do we measure present time since it has no extension? It is measured while it passes, but when it has passed it is not measured; for then there is nothing that could be measured. But whence, and how, and whither does it pass while it is being measured? Whence, but from the future? Which way, save through the present? Whither, but into the past? Therefore, from what is not yet, through what has no length, it passes into what is now no longer. But what do we measure, unless it is a time of some length? For we cannot speak of single, and double, and triple, and equal, and all the other ways in which we speak of time, except in terms of the length of the periods of time. But in what "length," then, do we measure passing time? Is it in the future, from which it passes over? But what does not yet exist cannot be measured. Or, is it in the present, through which it passes? But what has no length we cannot measure. Or is it in the past into which it passes? But what is no longer we cannot measure.

CHAPTER XXII

28. My soul burns ardently to understand this most intricate enigma. O Lord my God, O good Father, I beseech thee through Christ, do not close off these things, both the familiar and the obscure, from my desire. Do not bar it from entering into them; but let their light dawn by thy enlightening mercy,

O Lord. Of whom shall I inquire about these things? And to whom shall I confess my ignorance of them with greater profit than to thee, to whom these studies of mine (ardently longing to understand thy Scriptures) are not a bore? Give me what I love, for I do love it; and this thou hast given me. O Father, who truly knowest how to give good gifts to thy children, give this to me. Grant it, since I have undertaken to understand it, and hard labor is my lot until thou openest it. I beseech thee, through Christ and in his name, the Holy of Holies, let no man interrupt me. "For I have believed, and therefore do I speak."[4] This is my hope; for this I live: that I may contemplate the joys of my Lord.[5] Behold, thou hast made my days grow old, and they pass away—and how I do not know.

We speak of this time and that time, and these times and those times: "How long ago since he said this?" "How long ago since he did this?" "How long ago since I saw that?" "This syllable is twice as long as that single short syllable." These words we say and hear, and we are understood and we understand. They are quite commonplace and ordinary, and still the meaning of these very same things lies deeply hid and its discovery is still to come.

CHAPTER XXIII

29. I once heard a learned man say that the motions of the sun, moon, and stars constituted time; and I did not agree. For why should not the motions of all bodies constitute time? What if the lights of heaven should cease, and a potter's wheel still turn round: would there be no time by which we might measure those rotations and say either that it turned at equal intervals, or, if it moved now more slowly and now more quickly, that some rotations were longer and others shorter? And while we were saying this, would we not also be speaking in time? Or would there not be in our words some syllables that were long and others short, because the first took a longer time to sound, and the others a shorter time? O God, grant men to see in a small thing the notions that are common to all things,[6] both great and small. Both the stars and the lights of heaven are "for signs and seasons, and for days and years."[7] This is doubtless the case, but just as I should not say that the circuit of that wooden wheel was a day,

neither would that learned man say that there was, therefore, no time.

30. I thirst to know the power and the nature of time, by which we measure the motions of bodies, and say, for example, that this motion is twice as long as that. For I ask, since the word "day" refers not only to the length of time that the sun is above the earth (which separates day from night), but also refers to the sun's entire circuit from east all the way around to east—on account of which we can say, "So many days have passed" (the nights being included when we say, "So many days," and their lengths not counted separately)—since, then, the day is ended by the motion of the sun and by his passage from east to east, I ask whether the motion itself is the day, or whether the day is the period in which that motion is completed; or both? For if the sun's passage is the day, then there would be a day even if the sun should finish his course in as short a period as an hour. If the motion itself is the day, then it would not be a day if from one sunrise to another there were a period no longer than an hour. But the sun would have to go round twenty-four times to make just one day. If it is both, then that could not be called a day if the sun ran his entire course in the period of an hour; nor would it be a day if, while the sun stood still, as much time passed as the sun usually covered during his whole course, from morning to morning. I shall, therefore, not ask any more what it is that is called a day, but rather what time is, for it is by time that we measure the circuit of the sun, and would be able to say that it was finished in half the period of time that it customarily takes if it were completed in a period of only twelve hours. If, then, we compare these periods, we could call one of them a single and the other a double period, as if the sun might run his course from east to east sometimes in a single period and sometimes in a double period.

Let no man tell me, therefore, that the motions of the heavenly bodies constitute time. For when the sun stood still at the prayer of a certain man in order that he might gain his victory in battle, the sun stood still but time went on. For in as long a span of time as was sufficient the battle was fought and ended.[8]

I see, then, that time is a certain kind of extension. But do I see it, or do I only seem to? Thou, O Light and Truth, wilt show me.

CHAPTER XXIV

31. Dost thou command that I should agree if anyone says
that time is "the motion of a body"? Thou dost not so command. For I hear that no body is moved but in time; this thou
tellest me. But that the motion of a body itself is time I do not
hear; thou dost not say so. For when a body is moved, I
measure by time how long it was moving from the time when
it began to be moved until it stopped. And if I did not see
when it began to be moved, and if it continued to move so
that I could not see when it stopped, I could not measure the
movement, except from the time when I began to see it until
I stopped. But if I look at it for a long time, I can affirm only
that the time is long but not how long it may be. This is
because when we say, "How long?" we are speaking comparatively as: "This is as long as that," or, "This is twice as long
as that"; or other such similar ratios. But if we were able to
observe the point in space where and from which the body,
which is moved, comes and the point to which it is moved; or
if we can observe its parts moving as in a wheel, we can say
how long the movement of the body took or the movement
of its parts from this place to that. Since, therefore, the motion of a body is one thing, and the norm by which we measure how long it takes is another thing, we cannot see which
of these two is to be called time. For, although a body is
sometimes moved and sometimes stands still, we measure not
only its motion but also its rest as well; and both by time!
Thus we say, "It stood still as long as it moved," or, "It stood
still twice or three times as long as it moved"—or any other
ratio which our measuring has either determined or imagined,
either roughly or precisely, according to our custom. Therefore, time is not the motion of a body.

CHAPTER XXV

32. And I confess to thee, O Lord, that I am still ignorant
as to what time is. And again I confess to thee, O Lord, that
I know that I am speaking all these things in time, and that I

have already spoken of time a long time, and that "very long" is not long except when measured by the duration of time. How, then, do I know this, when I do not know what time is? Or, is it possible that I do not know how I can express what I do know? Alas for me! I do not even know the extent of my own ignorance. Behold, O my God, in thy presence I do not lie. As my heart is, so I speak. Thou shalt light my candle; thou, O Lord my God, wilt enlighten my darkness.[9]

CHAPTER XXVI

33. Does not my soul most truly confess to thee that I do measure intervals of time? But what is it that I thus measure, O my God, and how is it that I do not know what I measure? I measure the motion of a body by time, but the time itself I do not measure. But, truly, could I measure the motion of a body—how long it takes, how long it is in motion from this place to that—unless I could measure the time in which it is moving?

How, then, do I measure this time itself? Do we measure a longer time by a shorter time, as we measure the length of a crossbeam in terms of cubits?[10] Thus, we can say that the length of a long syllable is measured by the length of a short syllable and thus say that the long syllable is double. So also we measure the length of poems by the length of the lines, and the length of the line by the length of the feet, and the length of the feet by the length of the syllable, and the length of the long syllables by the length of the short ones. We do not measure by pages—for in that way we would measure space rather than time—but when we speak the words as they pass by we say: "It is a long stanza, because it is made up of so many verses; they are long verses because they consist of so many feet; they are long feet because they extend over so many syllables; this is a long syllable because it is twice the length of a short one."

But no certain measure of time is obtained this way; since it is possible that if a shorter verse is pronounced slowly, it may take up more time than a longer one if it is pronounced hurriedly. The same would hold for a stanza, or a foot, or a syllable. From this it appears to me that time is nothing other than extendedness;[11] but extendedness of what I do not know. This is a marvel to me. The extendedness may be of the mind itself. For what is it I measure, I ask thee, O my God, when I say either, roughly, "This time is longer than that," or, more precisely, "This is *twice* as long as that." I know that I am measuring time. But I am not measuring the future, for it is not yet; and I am not measuring the present because it is extended by no length; and I am not measuring the past because it no longer is. What is it, therefore, that I am measuring? Is it time in its passage, but not time past [*praetereuntia tempora, non praeterita*]? This is what I have been saying.

Chapter XXVII

34. Press on, O my mind, and attend with all your power. God is our Helper: "it is he that hath made us and not we ourselves."[12] Give heed where the truth begins to dawn.[13] Suppose now that a bodily voice begins to sound, and continues to sound—on and on—and then ceases. Now there is silence. The voice is past, and there is no longer a sound. It was future before it sounded, and could not be measured because it was not yet; and now it cannot be measured because it is no longer. Therefore, while it was sounding, it might have been measured because then there was something that could be measured. But even then it did not stand still, for it was in motion and was passing away. Could it, on that account, be any more readily measured? For while it was passing away, it was being extended into some interval of time in which it might be measured, since the present has no length. Supposing, though, that it might have been measured—then also suppose that another voice had begun to sound and is still sounding without any interruption to break its continued flow. We can measure it only while it is sounding, for when it has ceased to sound it will be already past and there will not be anything there that can be measured. Let us measure it exactly; and let us say how much it is. But while it is sounding, it cannot be measured except from the instant when it began to sound, down to the final moment when it left off. For we measure the time interval itself from some beginning point to some end. This is why a voice that has not yet ended

cannot be measured, so that one could say how long or how briefly it will continue. Nor can it be said to be equal to another voice or single or double in comparison to it or anything like this. But when it is ended, it is no longer. How, therefore, may it be measured? And yet we measure times; not those which are not yet, nor those which no longer are, nor those which are stretched out by some delay, nor those which have no limit. Therefore, we measure neither times future nor times past, nor times present, nor times passing by; and yet we do measure times.

35. *Deus Creator omnium:*[14] this verse of eight syllables alternates between short and long syllables. The four short ones—that is, the first, third, fifth, and seventh—are single in relation to the four long ones—that is, the second, fourth, sixth, and eighth. Each of the long ones is double the length of each of the short ones. I affirm this and report it, and common sense perceives that this indeed is the case. By common sense, then, I measure a long syllable by a short one, and I find that it is twice as long. But when one sounds after another, if the first be short and the latter long, how can I hold the short one and how can I apply it to the long one as a measure, so that I can discover that the long one is twice as long, when, in fact, the long one does not begin to sound until the short one leaves off sounding? That same long syllable I do not measure as present, since I cannot measure it until it is ended; but its ending is its passing away.

What is it, then, that I can measure? Where is the short syllable by which I measure? Where is the long one that I am measuring? Both have sounded, have flown away, have passed on, and are no longer. And still I measure, and I confidently answer—as far as a trained ear can be trusted—that this syllable is single and that syllable double. And I could not do this unless they both had passed and were ended. Therefore I do not measure them, for they do not exist any more. But I measure something in my memory which remains fixed.

36. It is in you, O mind of mine, that I measure the periods of time. Do not shout me down that it exists [objectively]; do not overwhelm yourself with the turbulent flood of your impressions. In you, as I have said, I measure the periods of time. I measure as time present the impression that things make on you as they pass by and what remains after they have passed by—I do not measure the things themselves which

have passed by and left their impression on you. This is what I measure when I measure periods of time. Either, then, these are the periods of time or else I do not measure time at all.

What are we doing when we measure silence, and say that this silence has lasted as long as that voice lasts? Do we not project our thought to the measure of a sound, as if it were then sounding, so that we can say something concerning the intervals of silence in a given span of time? For, even when both the voice and the tongue are still, we review—in thought —poems and verses, and discourse of various kinds of various measures of motions, and we specify their time spans—how long this is in relation to that—just as if we were speaking them aloud. If anyone wishes to utter a prolonged sound, and if, in forethought, he has decided how long it should be, that man has already in silence gone through a span of time, and committed his sound to memory. Thus he begins to speak and his voice sounds until it reaches the predetermined end. It has truly sounded and will go on sounding. But what is already finished has already sounded and what remains will still sound. Thus it passes on, until the present intention carries the future over into the past. The past increases by the diminution of the future until by the consumption of all the future all is past.[15]

Chapter XXVIII

37. But how is the future diminished or consumed when it does not yet exist? Or how does the past, which exists no longer, increase, unless it is that in the mind in which all this happens there are three functions? For the mind expects, it attends, and it remembers; so that what it expects passes into what it remembers by way of what it attends to. Who denies that future things do not exist as yet? But still there is already in the mind the expectation of things still future. And who denies that past things now exist no longer? Still there is in the mind the memory of things past. Who denies that time present has no length, since it passes away in a moment? Yet, our attention has a continuity and it is through this that what is present may proceed to become absent. Therefore, future time, which is nonexistent, is not long; but "a long future" is "a long expectation of the future." Nor is time past,

which is now no longer, long; a "long past" is "a long memory of the past."

38. I am about to repeat a psalm that I know. Before I begin, my attention encompasses the whole, but once I have begun, as much of it as becomes past while I speak is still stretched out in my memory. The span of my action is divided between my memory, which contains what I have repeated, and my expectation, which contains what I am about to repeat. Yet my attention is continually present with me, and through it what was future is carried over so that it becomes past. The more this is done and repeated, the more the memory is enlarged—and expectation is shortened—until the whole expectation is exhausted. Then the whole action is ended and passed into memory. And what takes place in the entire psalm takes place also in each individual part of it and in each individual syllable. This also holds in the even longer action of which that psalm is only a portion. The same holds in the whole life of man, of which all the actions of men are parts. The same holds in the whole age of the sons of men, of which all the lives of men are parts.

NOTES

[1] *Spatium,* which means extension either in space or time.

[2] I am indebted to Mr. Kenneth Friedman for pointing out a slip in the originally published version of the translation here, and to Professor Outler for authorizing the correction.—ED.

[3] The breaking light and the image of the rising sun.

[4] Cf. Ps. 116:10.

[5] Cf. Matt. 25:21, 23.

[6] *Communes notitias,* the universal principles of "common sense." This idea became a basic category in scholastic epistemology.

[7] Gen. 1:14.

[8] Cf. Josh. 10:12-14.

[9] Cf. Ps. 18:28.

[10] *Cubitum,* literally the distance between the elbow and the tip of the middle finger; in the imperial system of weights and measures it was 17.5 inches.

[11] *Distentionem,* "spread-out-ness"; cf. Descartes' notion of *res extensae,* and *its* relation to time.

[12] Ps. 100:3.

[13] Here Augustine begins to summarize his own answers to the questions he has raised in his analysis of time.

[14] The same hymn of Ambrose quoted above, Bk. IX, Ch. XII, 32, and analyzed again in *De musica*, VI, 2:2.

[15] This theory of time is worth comparing with its most notable restatement in modern poetry, in T. S. Eliot's *Four Quartets* and especially "Burnt Norton."

SPACE

AND MATTER

RENÉ DESCARTES

Reprinted from sections 4–21 of Part II of Descartes' *Principles of Philosophy,* in the volume *Descartes' Philosophical Writings,* translated and edited by Elizabeth Anscombe and Peter Thomas Geach, Thomas Nelson & Sons, Ltd., Edinburgh and London, 1959.

IV. . . . The nature of matter, or of body considered in general, does not consist in its being a thing that has hardness or weight, or colour, or any other sensible property, but simply in its being a thing that has extension in length, breadth, and depth. For as regards hardness, our sensation tells us no more than that the parts of a hard body resist the movement of our hands when they encounter it; if, whenever our hands moved in a given direction, all the bodies lying that way were always to retreat with the same speed as our hands approached, we should never have any sensation of hardness. Now it is inconceivable that, if bodies did retreat in this way, they would thereby lose their nature as bodies; so this nature cannot consist in hardness. By the same reasoning it may be shown that weight, colour, and all other such sensible qualities of corporeal matter can be removed from body while it itself remains in its entirety; so it follows that its real nature depends upon none of them.

V. There remain, however, two possible reasons for doubting whether the real nature of body consists merely in extension. First, many people hold that various bodies can be rarefied and condensed, so that when rarefied they have more extension than when condensed; some people are indeed so subtle as to distinguish the substance of a body from its quantity, and even the quantity from its extension. Secondly, if we conceive a place to contain nothing but extension in length, breadth, and depth, we do not usually say there is a body there; we just say there is space there—empty space, which almost everyone is convinced is mere nonentity.

VI. But as regards rarefaction and condensation, anybody who thinks attentively, and will admit only what he clearly perceives, will hold that all that happens here is change of shape. What I mean is this: rarefied bodies are those that have many gaps between their particles, which are occupied by other bodies; and increase of density results merely from these particles approaching one another, so as to diminish these gaps or altogether obliterate them. In the latter case the body becomes so dense that it is contradictory to suppose it could be any denser. A body is, however, not of less extent in this case than when it occupies a greater space through the separation of its particles; for the extension comprised in the pores or gaps that remain between its particles must be assigned not to it but to the other bodies, whatever they may be, that fill the gaps. It is just as when we see a sponge swollen with water or some other fluid; we do not think its several parts have any greater extension than when it is squeezed dry; we just think that its pores are open wider, so that it is spread over a bigger space.

VII. I really cannot see the motive of people who choose to say that rarefaction happens by an increase of quantity, rather than explain it by this example of the sponge. Of course when air or water is rarefied we cannot see any pores growing bigger, nor yet any new body coming to fill them up; but it is irrational to invent something unintelligible as a merely verbal account of rarefaction, rather than infer from rare-faction that there are pores or gaps that grow bigger, that there is some new body that comes and fills them up, although we do not perceive this body by any of our senses. For there is no compelling reason to believe that all bodies that exist must affect our senses. Again, we can very easily see how rarefaction can come about in this way, but not how it could in any other way. Finally, it is a flat contradiction that any-thing should be increased by new quantity, new extension, without a simultaneous addition of new extended substance, that is new body. No increment of extension or quantity is conceivable without an increment of substance to which this quantity and extension shall belong. This will be made clearer by what follows.

VIII. Quantity differs from the extended substance, not in actuality, but only as regards our way of conceiving them; just as number does from what is numbered. We may consider the entire nature of the corporeal substance that in fact occupies a space of ten feet, without attending to the magnitude *ten feet*; for this nature is conceived as being just the same in any given part of the space as in the whole space. Conversely, the number ten, and similarly the continuous quantity *ten feet*, may be conceived without our attending to this definite substance. The concept of the number ten is just the same whether it is referred to this ten-foot magnitude or to anything else; and although the continuous quantity *ten feet* cannot be conceived apart from some substance whose quantity it shall be, it can be conceived without this definite substance. But in actuality it is not possible to subtract the least bit of the quantity or extension without likewise removing just as much of the substance; or conversely, to remove never so little of the substance without subtracting just as much of the quantity or extension.

IX. People may speak otherwise, but I do not think they have any conception other than this. When they distinguish substance from extension or quantity, either they mean nothing by the term substance; or they simply have a confused notion of an incorporeal substance, which they falsely attach to corporeal substance; the genuine notion of corporeal substance falls for them under extension, which, however, they call an accident. Thus what they express in words is quite different from what they grasp in their minds.

X. A space, or intrinsic place, does not differ in actuality from the body that occupies it; the difference lies simply in our ordinary ways of thinking. In reality the extension in length, breadth, and depth that constitutes the space is absolutely the same as that which constitutes the body. The difference lies in this: when we consider the extension as belonging to the body, we regard it as something individual, so that there is a new extension in the place as often as there is a new body; but when we consider the extension as belonging to the space, we are ascribing to it only a generic identity, so that when a new body comes to occupy the space, the extension of the space is deemed not to be a new extension, but to be just the same as before. (So long, that is, as it still has the same size and shape, and keeps the same position relatively to certain external bodies that we use to determine the space.)

XI. It is easy to see that it is the same extension that essentially constitutes (*naturam . . . constituit*) a body and a space; that there is no more difference here than there is between the essence (*natura*) of a genus or species and the essence (*natura*) of the individual. We have only to attend to our idea of some body, e.g. a stone, and remove from it whatever we know is not entailed by the very nature of body. We first reject hardness; for if the stone is melted, or divided into a very fine powder, it will lose this quality without ceasing to be a body. Again, we reject colour; we have often seen stones so transparent as to be colourless. We reject heaviness; fire is extremely light, but none the less conceived as a body. Finally, we reject coldness and heat and all other such qualities; either they are not what we are considering in thinking of the stone, or at least their changing does not mean that the stone is regarded as having lost the nature of a body. We may now observe that absolutely no element of our idea remains, except extension in length, breadth, and depth. Now this is just what is implied in the idea of space; not merely of a space occupied by bodies, but even of a so-called vacuum.

XII. There is, however, a conceptual difference. When a stone is removed from the space or place where it is, we think of its extension as being likewise removed; for we are then regarding the extension as something individual, and inseparable from the stone. At the same time, we regard the extension of the place where the stone was as something persisting and identical, although the place of the stone is now occupied by wood, water, air, or any other body, or is believed to be empty; for now we are considering the extension as a general property, and it is deemed to be "the same" extension in stone, wood, water, or any other body (or even in a vacuum, if such there be) so long as it still has the same shape and size, and keeps the same position relatively to the external bodies that determine this space.

XIII. The terms *place* and *space* do not signify something different from the body that is said to be in a place; they merely mean its size, shape, and position relative to other bodies. To determine the position we have to look to some other bodies, regarded as unmoving; and we may say—relatively to different sets of bodies—that the same thing is simultaneously changing and not changing its place. E.g. when a ship is sailing at sea, a man sitting in the poop remains in one place relatively to the parts of the ship, for he keeps in one position among these; and yet he is continually changing his place relatively to the shore, for he is continually receding from one shore and approaching the other. Again, if we conceive of the Earth as moving, and as travelling from West to

East exactly as far as the ship travels from East to West in the same time, we shall again say that the man sitting in the poop "is not changing his place"; we shall now be deriving the determination of place from some unmoving points in the heavens. But we may well end by thinking that no such genuinely unmoving points are to be found in the universe . . .; and in that case we shall conclude that no object has a permanent place except by the determination of our thought (*cogitatione*).

XIV. The terms *place* and *space* differ in that *place* signifies position more expressly than size or shape, and these features, conversely, are rather what we have in mind when we speak of space. We often say that one body takes the "place" of another, even if it has not exactly the same size or shape; but we then say it does not occupy the same space. On the other hand, when the position is changed, we say the place is changed, even if the body keeps the same shape and size. When we say an object is "in" a place we are merely thinking of its occupying a position relatively to other objects; when we add that it "fills" the place or space, we are also thinking of it as having a definite size and shape.

XV. Thus we always take *a space* to mean an extension in length, breadth, and depth. Place is considered sometimes as intrinsic to the object that is in a place, and sometimes as extrinsic to it. Intrinsic place is just the same as space; extrinsic place may be taken to mean the surface immediately surrounding the body that is in the place. It should be noticed that *surface* here does not mean a part of the surrounding body, but only the common boundary of the surrounding and surrounded bodies, which is a mere aspect of them; at least, what is meant is the surface as a common property, which is not part of one body rather than the other, and is deemed to be always "the same" so long as it keeps the same size and shape. For even if the body, and the surface of the body, surrounding a given object, should completely change, yet the object so surrounded is not considered as changing its place, provided that it meanwhile retains the same position relatively to the bodies that are taken as unmoving. E.g. if we suppose that a ship is equally impelled in one direction by the flow of the river and in the opposite direction by the wind, so that its position relative to the banks is unchanged, it will readily be held to be staying in the same place, although the surface surrounding it is entirely changed.

XVI. The impossibility of a vacuum in the philosophical sense—a place in which there is absolutely no substance—is obvious from the fact that the extension of a space or intrinsic place is in no way different from the extension of a body. For the extension of a body in length, breadth and depth justifies us in concluding that it is a substance, since it is wholly contradictory that there should be extension that is the extension of nothing; and we must draw the same conclusion about the supposedly empty space—viz., that since there is extension there, there must necessarily be substance there as well.

XVII. In common speech the term *empty* usually means, not a place or space where there is no object at all, but simply a place where there is no object such as we think there ought to be. Since e.g. a jug is made to hold water, it is called "empty" when it is only full of air. A fish-pond "has nothing in it," although there is plenty of water in it, if there are no fish. A ship fitted out to carry merchandise is "empty" if it is loaded only with sand to break the force of the wind. Finally, a space containing nothing sensible is "empty," even if it is full of created and self-subsistent matter; for we ordinarily consider only such things as our senses attain to. If, then, we neglect the proper meaning of the terms *empty* and *nothing*, and suppose that when we call a space "empty" it contains, not just nothing that is sensible, but no object at all, we shall be falling into the same error as though we inferred, from our way of calling a jug that contains only air an "empty" jug, that the air contained in it is not a substantial reality.

XVIII. Almost all of us have fallen into this error at an early age. We could discern no necessary connexion between a vessel and the body it contains; so we thought there was nothing to prevent a body's being removed from the vessel it fills without any other taking its place; that at any rate God could bring this about. To correct this error, we must reflect that, whereas there is no connexion between a vessel and this or that particular body contained in it, there is a very close, and absolutely necessary, connexion between the concave shape of the vessel and the general concept of the extension that must be contained in that concavity. It is no less contradictory than to think of a mountain without a valley, if we conceive that there can be this concavity without extension contained in it, or that there can be this extension without a substance whose extension it shall be; for, as I have often said, there can be no extension that is extension of nothing. It may be asked what would happen if God removed all the body contained in a vessel, and allowed no other body to come and take the place of what was removed. The answer must be that in that case the sides of the vessel would *ipso facto* be in contact; for when there is nothing between two bodies, they must necessarily touch each other. It is manifestly contradictory for them to be apart, or to have a distance between them, while at the same time the distance is nothing; for any distance is an aspect (*modus*) of extension, and thus cannot exist without an extended substance.

XIX. We have thus seen that the nature of corporeal substance consists in its being something extended (*res extensa*), and that its extension is none other than is commonly ascribed to a space however "empty." From this we readily see that it is impossible for any part of matter to occupy more space at one time than at another; thus rarefaction is not possible except in the way already explained. And, again, there can be no more matter (corporeal substance) in a vessel filled with lead, gold, or some other such body, as heavy and solid (*duro*) as you will, than there is when it just contains air and is considered "empty." The quantity of a piece of matter depends not on its heaviness or solidity (*duritie*), but simply on its extension; and in a given vessel this is constant.

XX. We see also the impossibility of atoms—pieces of matter that are by their nature indivisible. If they exist, they must necessarily be extended, however small they are imagined to be; so we can still divide any one of them in thought (*cogitatione*) into two or more smaller ones, and thus we can recognise their divisibility. There is nothing we can divide in thought but we can see to *be* divisible; if we were to judge that it was indivisible, our judgment would go against what we knew. Even if we imagined a Divine decree that some particle of matter could not be divided into smaller ones, it would not be properly speaking indivisible. Even if God made it not to be divisible by any creatures, he could not take away his own power of dividing it; for it is quite impossible for God to diminish his own power. . . . So, speaking absolutely, it will still be divisible, being such by its very nature.

XXI. We see, furthermore, that this world—the totality of corporeal substance—has no limits to its extension. Wherever we imagine the boundaries to be, there is always the possibility, not merely of imagining further space indefinitely extended,[1] but also of seeing that this imagination is true to fact—that such space actually exists. And hence there must also be indefinitely extended[2] corporeal substance contained in this space. For, as has already been abundantly shown, the idea of the extension that we conceive any given space to have is identical with the idea of corporeal substance.

NOTES

[1] Cf. *Principia*, I. xxvi-vii.—Tr.
[2] *Op. cit.*

ABSOLUTE
SPACE AND TIME

ISAAC NEWTON

This selection consists of the Scholium to the Definitions of Newton's *Mathematical Principles of Natural Philosophy*. Reprinted from *Sir Isaac Newton's Mathematical Principles of Natural Philosophy and His System of the World*, Florian Cajori edition, University of California Press, Berkeley, Calif., 1934.

Hitherto I have laid down the definitions of such words as are less known, and explained the sense in which I would have them to be understood in the following discourse. I do not define time, space, place, and motion, as being well known to all. Only I must observe, that the common people conceive those quantities under no other notions but from the relation they bear to sensible objects. And thence arise certain prejudices, for the removing of which it will be convenient to distinguish them into absolute and relative, true and apparent, mathematical and common.

I. Absolute, true, and mathematical time, of itself, and from its own nature, flows equably without relation to anything external, and by another name is called duration: relative, apparent, and common time, is some sensible and external (whether accurate or unequable) measure of duration by the means of motion, which is commonly used instead of true time; such as an hour, a day, a month, a year.

II. Absolute space, in its own nature, without relation to anything external, remains always similar and immovable. Relative space is some movable dimension or measure of the absolute spaces; which our senses determine by its position to bodies; and which is commonly taken for immovable space; such is the dimension of a subterraneous, an aerial, or celestial space, determined by its position in respect of the earth. Absolute and relative space are the same in figure and magnitude; but they do not remain always numerically the same. For if the earth, for instance, moves, a space of our air, which relatively and in respect of the earth remains always the same, will at one time be one part of the absolute space into which the air passes; at another time it will be another part of the same, and so, absolutely understood, it will be continually changed.

III. Place is a part of space which a body takes up, and is according to the space, either absolute or relative. I say, a part of space; not the situation, nor the external surface of the body. For the places of equal solids are always equal; but their surfaces, by reason of their dissimilar figures, are often unequal. Positions properly have no quantity, nor are they so much the places themselves, as the properties of places. The motion of the whole is the same with the sum of the motions of the parts; that is, the translation of the whole, out of its place, is the same thing with the sum of the translations of the parts out of their places; and therefore the place of the whole is the same as the sum of the places of the parts, and for that reason, it is internal, and in the whole body.

IV. Absolute motion is the translation of a body from one absolute place into another; and relative motion, the translation from one relative place into another. Thus in a ship under sail, the relative place of a body is that part of the ship which the body possesses; or that part of the cavity which the body fills, and which therefore moves together with the ship: and relative rest is the continuance of the body in the same part of the ship, or of its cavity. But real, absolute rest, is the continuance of the body in the same part of that immovable space, in which the ship itself, its cavity, and all that it contains, is moved. Wherefore, if the earth is really at rest, the body, which relatively rests in the ship, will really and absolutely move with the same velocity which the ship has on the earth. But if the earth also moves, the true and absolute motion of the body will arise, partly from the true motion of the earth, in immovable space, partly from the relative motion of the ship on the earth; and if the body moves also relatively in the ship, its true motion will arise, partly from the true motion of the earth, in immovable space, and partly from the relative motions as well of the ship on the earth, as of the body in the ship; and from these relative motions will arise the relative motion of the body on the earth. As if that part of the earth, where the ship is, was truly moved towards the east, with a velocity of 10010 parts; while the ship itself, with a fresh gale, and full sails, is carried towards the west, with a velocity expressed by 10 of those parts; but a sailor walks in the ship towards the east, with 1 part of the said velocity; then the sailor will be moved truly in immovable space towards the east, with a velocity of 10001 parts, and relatively on the earth towards the west, with a velocity of 9 of those parts.

Absolute time, in astronomy, is distinguished from relative, by the equation or correction of the apparent time. For the natural days are truly unequal, though they are commonly considered as equal, and used for a measure of time; astronomers correct this inequality that they may measure the celestial motions by a more accurate time. It may be, that there is no such thing as an equable motion, whereby time may be accurately measured. All motions may be accelerated and retarded, but the flowing of absolute time is not liable to any change. The duration or perseverance of the existence of things remains the same, whether the motions are swift or slow, or none at all: and therefore this duration ought to be distinguished from what are only sensible measures thereof; and from which we deduce it, by means of the astronomical equation. The necessity of this equation, for determining the times of a phenomenon, is evinced as well from the experiments of the pendulum clock, as by eclipses of the satellites of Jupiter.

As the order of the parts of time is immutable, so also is the order of the parts of space. Suppose those parts to be moved out of their places, and they will be moved (if the expression may be allowed) out of themselves. For times and spaces are, as it were, the places as well of themselves as of all other things. All things are placed in time as to order of succession; and in space as to order of situation. It is from their essence or nature that they are places; and that the primary places of things should be movable, is absurd. These are therefore the absolute places; and translations out of those places, are the only absolute motions.

But because the parts of space cannot be seen, or distinguished from one another by our senses, therefore in their stead we use sensible measures of them. For from the positions and distances of things from any body considered as immovable, we define all places; and then with respect to such places, we estimate all motions, considering bodies as transferred from some of those places into others. And so, instead of absolute places and motions, we use relative ones; and that without any inconvenience in common affairs; but in philosophical disquisitions, we ought to abstract from our senses, and consider things themselves, distinct from what are only sensible measures of them. For it may be that there is no body really at rest, to which the places and motions of others may be referred.

But we may distinguish rest and motion, absolute and relative, one from the other by their properties, causes, and effects. It is a property of rest, that bodies really at rest do rest in respect to one another. And therefore as it is possible, that in the remote regions of the fixed stars, or perhaps far beyond them, there may be some body absolutely at rest; but impossible to know, from the position of bodies to one another in our regions, whether any of these do keep the same position to that remote body, it follows that absolute rest cannot be determined from the position of bodies in our regions.

It is a property of motion, that the parts, which retain given positions to their wholes, do partake of the motions of those wholes. For all the parts of revolving bodies endeavor to recede from the axis of motion; and the impetus of bodies moving forwards arises from the joint impetus of all the parts. Therefore, if surrounding bodies are moved, those that are relatively at rest within them will partake of their motion. Upon which account, the true and absolute motion of a body cannot be determined by the translation of it from those which only seem to rest; for the external bodies ought not only to appear at rest, but to be really at rest. For otherwise, all included bodies, besides their translation from near the surrounding ones, partake likewise of their true motions; and though that translation were not made, they would not be really at rest, but only seem to be so. For the surrounding bodies stand in the like relation to the surrounded as the exterior part of a whole does to the interior, or as the shell does to the kernel; but if the shell moves, the kernel will also move, as being part of the whole, without any removal from near the shell.

A property, near akin to the preceding, is this, that if a place is moved, whatever is placed therein moves along with it; and therefore a body, which is moved from a place in motion, partakes also of the motion of its place. Upon which account, all motions, from places in motion, are no other than parts of entire and absolute motions; and every entire motion is composed of the motion of the body out of its first place, and the motion of this place out of its place; and so on, until we come to some immovable place, as in the before-mentioned example of the sailor. Wherefore, entire and absolute motions can be no otherwise determined than by immovable places; and for that reason I did before refer those absolute motions to immovable places, but relative ones to movable places. Now no other places are immovable but those that, from infinity to infinity, do all retain the same given position one to another; and upon this account must ever remain unmoved; and do thereby constitute immovable space.

The causes by which true and relative motions are distinguished, one from the other, are the forces impressed upon bodies to generate motion. True motion is neither generated nor altered, but by some force impressed upon the body moved; but relative motion may be generated or altered without any force impressed upon the body. For it is sufficient only to impress some force on other bodies with which the former is compared, that by their giving way, that relation may be changed, in which the relative rest or motion of this other body did consist. Again, true motion suffers always some change from any force impressed upon the moving body; but relative motion does not necessarily undergo any change by such forces. For if the same forces are likewise impressed on those other bodies, with which the comparison is made, that the relative position may be preserved, then that condition will be preserved in which the relative motion consists. And therefore any relative motion may be changed when the true motion remains unaltered, and the relative may be preserved when the true suffers some change. Thus, true motion by no means consists in such relations.

The effects which distinguish absolute from relative motion are, the forces of receding from the axis of circular motion. For there are no such forces in a circular motion purely relative, but in a true and absolute circular motion, they are greater or less, according to the quantity of the motion. If a vessel, hung by a long cord, is so often turned about that the cord is strongly twisted, then filled with water, and held at rest together with the water; thereupon, by the sudden action of another force, it is whirled about the contrary way, and while the cord is untwisting itself, the vessel continues for some time in this motion; the surface of the water will at first be plain, as before the vessel began to move; but after that, the vessel, by gradually communicating its motion to the water, will make it begin sensibly to revolve, and recede by little and little from the middle, and ascend to the sides of the vessel, forming itself into a concave figure (as I have experienced), and the swifter the motion becomes, the higher will the water rise, till at last, performing its revolutions in the same times with the vessel, it becomes relatively at rest in it. This ascent of the water shows its endeavor to recede from the axis of its motion; and the true and absolute circular motion of the water, which is here directly contrary to the relative, becomes known, and may be measured by this endeavor. At first, when the relative motion of the water in the vessel was greatest, it produced no endeavor to recede from the axis; the water showed no tendency to the circumference, nor any ascent towards the sides of the vessel, but remained of a plain surface, and therefore its true circular motion had not yet begun. But afterwards, when the relative motion of the water had decreased, the ascent thereof towards the sides of the vessel proved its endeavor to recede from the axis; and this endeavor showed the real circular motion of the water continually increasing, till it had acquired its greatest quantity, when the water rested relatively in the vessel. And therefore this endeavor does not depend upon any translation of the water in respect of the ambient bodies, nor can true circular motion be defined by such translation. There is only one real circular motion of any one revolving body, corresponding to only one power of endeavoring to recede from its axis of motion, as its proper and adequate effect; but relative motions, in one and the same body, are innumerable, according to the various relations it bears to external bodies, and, like other relations, are altogether destitute of any real effect, any otherwise than they may perhaps partake of that one only true motion. And therefore in their system who suppose that our heavens, revolving below the sphere of the fixed stars, carry the planets along with them; the several parts of those heavens, and the planets, which are indeed relatively at rest in their heavens, do yet really move. For they change their position one to another (which never happens to bodies truly at rest), and being carried together with their heavens, partake of their motions, and as parts of revolving wholes, endeavor to recede from the axis of their motions.

Wherefore relative quantities are not the quantities themselves, whose names they bear, but those sensible measures of them (either accurate or inaccurate), which are commonly used instead of the measured quantities themselves. And if the meaning of words is to be determined by their use, then by the names time, space, place, and motion, their [sensible] measures are properly to be understood; and the expression will be unusual, and purely mathematical, if the measured quantities themselves are meant. On this account, those violate the accuracy of language, which ought to be kept precise, who interpret these words for the measured quantities. Nor do those less defile the purity of mathematical and philosophical truths, who confound real quantities with their relations and sensible measures.

It is indeed a matter of great difficulty to discover, and effectually to distinguish, the true motions of particular bodies from the apparent; because the parts of that immovable space, in which those motions are performed, do by no means come under the observation of our senses. Yet the thing is not altogether desperate; for we have some arguments to guide us, partly from the apparent motions, which are the differences of the true motions; partly from the forces, which are the causes and effects of the true motions. For instance, if two globes, kept at a given distance one from the other by means of a cord that connects them, were revolved about their common centre of gravity, we might, from the tension of the cord, discover the endeavor of the globes to recede from the axis of their motion, and from thence we might compute the quantity of their circular motions. And then if any equal forces should be impressed at once on the alternate faces of the globes to augment or diminish their circular motions, from the increase or decrease of the tension of the cord, we might infer the increment or decrement of their motions; and thence would be found on what faces those forces ought to be impressed, that the motions of the globes might be most augmented; that is, we might discover their hindmost faces, or those which, in the circular motion, do follow. But the faces which follow being known, and consequently the opposite ones that precede, we should likewise know the determination of their motions. And thus we might find both the quantity and the determination of this circular motion, even in an immense vacuum, where there was nothing external or sensible with which the globes could be compared. But now, if in that space some remote bodies were placed that kept always a given position one to another, as the fixed stars do in our regions, we could not indeed determine from the relative translation of the globes among those bodies, whether the motion did belong to the globes or to the bodies. But if we observed the cord, and found that its tension was that very tension which the motions of the globes required, we might conclude the motion to be in the globes, and the bodies to be at rest; and then, lastly, from the translation of the globes among the bodies, we should find the determination of their motions. But how we are to obtain the true motions from their causes, effects, and apparent differences, and the converse, shall be explained more at large in the following treatise. For to this end it was that I composed it.

THE RELATIONAL THEORY
OF SPACE AND TIME

GOTTFRIED LEIBNIZ

A selection reprinted from the Third and Fifth Papers in *The Leibniz-Clarke Correspondence*, edited by H. G. Alexander, Manchester University Press, Manchester, England, 1956. Samuel Clarke was an English philosopher and theologian who followed Newton in asserting an absolute theory of space and time.

FROM LEIBNIZ'S THIRD PAPER
(IN ANSWER TO CLARKE'S SECOND REPLY)

. . . As for my own opinion, I have said more than once, that I hold space to be something merely[1] relative, as time is; that I hold it to be an order of coexistences, as time is an order of successions. For space denotes, in terms of possibility, an order of things which exist at the same time, considered as existing together; without enquiring into their manner of existing. And when many things are seen together, one perceives that order of things among themselves.

5. I have many demonstrations, to confute the fancy of those who take space to be a substance, or at least an absolute being. But I shall only use, at the present, one demonstration, which the author here gives me occasion to insist upon. I say then, that if space was an absolute being, there would something happen for which it would be impossible there should be a sufficient reason. Which is against my axiom. And I prove it thus. Space is something absolutely uniform; and, without the things placed in it, one point of space does not absolutely differ in any respect whatsoever from another point of space. Now from hence it follows (supposing space to be something in itself, besides the order of bodies among themselves) that 'tis impossible there should be a reason, why God, preserving the same situations of bodies among themselves, should have

placed them in space after one certain particular manner, and not otherwise; why every thing was not placed the quite contrary way, for instance, by changing East into West. But if space is nothing else, but that order or relation; and is nothing at all without bodies, but the possibility of placing them; then those two states, the one such as it now is, the other supposed to be the quite contrary way, would not at all differ from one another. Their difference therefore is only to be found in our chimerical supposition of the reality of space in itself. But in truth the one would exactly be the same thing as the other, they being absolutely indiscernible; and consequently there is no room to enquire after a reason of the preference of the one to the other.

6. The case is the same with respect to time. Supposing any one should ask, why God did not create every thing a year sooner; and the same person should infer from thence, that God has done something, concerning which 'tis not possible there should be a reason, why he did it so, and not otherwise: the answer is, that his inference would be right, if time was any thing distinct from things existing in time. For it would be impossible there should be any reason, why things should be applied to such particular instants, rather than to others, their succession continuing the same. But then the same argument proves, that instants, consider'd without the things, are nothing at all; and that they consist only in the successive order of things: which order remaining the same, one of the two states, viz. that of a supposed anticipation, would not at all differ, nor could be discerned from, the other which now is. . . .

33. Since space in itself is an ideal thing, like time; space out of the world must needs be imaginary, as the schoolmen themselves have acknowledged. The case is the same with empty space within the world; which I take also to be imaginary, for the reasons before alleged.

34. The author objects against me the vacuum discovered by Mr. Guerike of Magdeburg,[2] which is made by pumping the air out of a receiver; and he pretends that there is truly a perfect vacuum, or a space without matter (at least in part), in that receiver. The Aristotelians and Cartesians, who do not admit a true vacuum, have said in answer to that experiment of Mr. Guerike, as well as to that of Torricellius of Florence (who emptied the air out of a glass-tube by the help of quick-

silver),[3] that there is no vacuum at all in the tube or in the receiver; since glass has small pores, which the beams of light, the effluvia of the load-stone, and other very thin fluids may go through. I am of their opinion: and I think the receiver may be compared to a box full of holes in the water, having fish or other gross bodies shut up in it; which being taken out, their place would nevertheless be filled up with water. There is only this difference; that though water be fluid and more yielding than those gross bodies, yet it is as heavy and massive, if not more, than they: whereas the matter which gets into the receiver in the room of the air is much more subtle. The new sticklers for a vacuum allege in answer to this instance, that it is not the grossness of matter, but its mere quantity, that makes resistance; and consequently that there is of necessity more vacuum, where there is less resistance. They add, that the subtleness of matter has nothing to do here; and that the particles of quicksilver are as subtle and fine as those of water; and yet that quicksilver resists about ten times more. To this I reply, that it is not so much the quantity of matter, as its difficulty of giving place, that makes resistance. For instance, floating timber contains less of heavy matter, than an equal bulk of water does; and yet it makes more resistance to a boat, than the water does.

35. And as for quicksilver; 'tis true, it contains about fourteen times more of heavy matter, than an equal bulk of water does; but it does not follow, that it contains fourteen times more matter absolutely. On the contrary, water contains as much matter; if we include both its own matter, which is heavy; and the extraneous matter void of heaviness, which passes through its pores. . . .

FROM LEIBNIZ'S FIFTH PAPER
(IN ANSWER TO CLARKE'S FOURTH REPLY)

47. I will here show, how men come to form to themselves the notion of space. They consider that many things exist at once and they observe in them a certain order of co-existence, according to which the relation of one thing to another is more or less simple. This order, is their *situation* or distance. When it happens that one of those co-existent things changes its relation to a multitude of others, which do not change

their relation among themselves; and that another thing, newly come, acquires the same relation to the others, as the former had; we then say, it is come into the place of the former; and this change, we call a motion in that body, wherein is the immediate cause of the change. And though many, or even all the co-existent things, should change according to certain known rules of direction and swiftness; yet one may always determine the relation of situation, which every co-existent acquires with respect to every other co-existent; and even that relation which any other co-existent would have to this, or which this would have to any other, if it had not changed, or if it had changed any otherwise. And supposing, or feigning, that among those co-existents, there is a sufficient number of them, which have undergone no change; then we may say, that those which have such a relation to those fixed existents, as others had to them before, have now the *same place* which those others had. And that which comprehends all those places, is called *space*. Which shows, that in order to have an idea of place, and consequently of space, it is sufficient to consider these relations, and the rules of their changes, without needing to fancy any absolute reality out of the things whose situation we consider. And, to give a kind of a definition: *place* is that, which we say is the same to A and, to B, when the relation of the co-existence of B, with C, E, F, G, etc. agrees perfectly with the relation of the co-existence, which A had with the same C, E, F, G, etc. supposing there has been no cause of change in C, E, F, G, etc. It may be said also without entering into any further particularity, that *place* is that, which is the same in different moments to different existent things, when their relations of co-existence with certain other existents, which are supposed to continue fixed from one of those moments to the other, agree entirely together. And *fixed existents* are those, in which there has been no cause of any change of the order of their co-existence with others; or (which is the same thing) in which there has been no motion. Lastly, *space* is that, which results from places taken together. And here it may not be amiss to consider the difference between place, and the relation of situation, which is in the body that fills up the place. For, the place of A and B, is the same; whereas the relation of A to fixed bodies, is not precisely and individually the same, as the relation which B (that comes into its place) will have to the same fixed bodies; but these relations agree only. For, two different subjects, as

A and *B*, cannot have precisely the same individual affection; it being impossible, that the same individual accident should be in two subjects, or pass from one subject to another. But the mind not contented with an agreement, looks for an identity, for something that should be truly the same; and conceives it as being extrinsic to the subjects: and this is what we call *place* and *space*. But this can only be an ideal thing; containing a certain order, wherein the mind conceives the application of relations. In like manner, as the mind can fancy to itself an order made up of genealogical lines, whose bigness would consist only in the number of generations, wherein every person would have his place: and if to this one should add the fiction of a *metempsychosis*, and bring in the same human souls again; the persons in those lines might change place; he who was a father, or a grandfather, might become a son, or a grandson, etc. And yet those genealogical places, lines, and spaces, though they should express real truth, would only be ideal things. I shall allege another example, to show how the mind uses, upon occasion of accidents which are in subjects, to fancy to itself something answerable to those accidents, out of the subjects. The ratio or proportion between two lines *L* and *M*, may be conceived three several ways; as a ratio of the greater *L*, to the lesser *M*; as a ratio of the lesser *M*, to the greater *L*; and lastly, as something abstracted from both, that is, as the ratio between *L* and *M*, without considering which is the antecedent, or which the consequent; which the subject, and which the object. And thus it is, that proportions are considered in music. In the first way of considering them, *L* the greater; in the second, *M* the lesser, is the subject of that accident, which philosophers call relation. But, which of them will be the subject, in the third way of considering them? It cannot be said that both of them, *L* and *M* together, are the subject of such an accident; for if so, we should have an accident in two subjects, with one leg in one, and the other in the other; which is contrary to the notion of accidents. Therefore we must say, that this relation, in this third way of considering it, is indeed out of the subjects; but being neither a substance, nor an accident, it must be a mere ideal thing, the consideration of which is nevertheless useful. To conclude:[4] I have here done much like Euclid, who not being able to make his readers well understand what *ratio* is absolutely in the sense of geometricians; defines what are the *same ratios*. Thus, in like manner, in order to explain what *place* is,

I have been content to define what is the *same place*. Lastly, I observe, that the traces of moveable bodies, which they leave sometimes upon the immoveable ones on which they are moved; have given men occasion to form in their imagination such an idea, as if some trace did still remain, even when there is nothing unmoved. But this is a mere ideal thing, and imports only, that if there was any unmoved thing there, the trace might be marked out upon it. And 'tis this analogy, which makes men fancy places, traces and spaces; though those things consist only in the truth of relations, and not at all in any absolute reality. . . .

49. It cannot be said that [a certain] duration is eternal but [it can be said] that the things which continue always are eternal [gaining always a new duration]. Whatever exists of time and of duration [being successive] perishes continually: and how can a thing exist eternally, which (to speak exactly) does never exist at all? For, how can a thing exist, whereof no part does ever exist? Nothing of time does ever exist, but instants; and an instant is not even itself a part of time. Whoever considers these observations, will easily apprehend that time can only be an ideal thing. And the analogy between time and space, will easily make it appear, that the one is as merely ideal as the other. [But, if in saying that the duration of a thing is eternal, it is only meant that the thing endures eternally, I have nothing to say against it.] . . .

52. In order to prove that space, without bodies, is an absolute reality; the author objected, that a finite material universe might move forward in space. I answered, it does not appear reasonable that the material universe should be finite; and, though we should suppose it to be finite; yet 'tis unreasonable it should have motion any otherwise, than as its parts change their situation among themselves; because such a motion would produce no change that could be observed, and would be without design. 'Tis another thing, when its parts change their situation among themselves; for then there is a motion in space; but it consists in the order of relations which are changed. The author replies now, that the reality of motion does not depend upon being observed; and that a ship may go forward, and yet a man, who is in the ship, may not perceive it. I answer, motion does not indeed depend upon being observed; but it does depend upon being possible to be observed. There is no motion, when there is no change that can be observed. And when there is no change that can

be observed, there is no change at all. The contrary opinion is grounded upon the supposition of a real absolute space, which I have demonstratively confuted by the principle of the want of a sufficient reason of things.

53. I find nothing in the Eighth Definition of the *Mathematical Principles of Nature*, nor in the Scholium belonging to it, that proves, or can prove, the reality of space in itself. However, I grant there is a difference between an absolute true motion of a body, and a mere relative change of its situation with respect to another body. For when the immediate cause of the change is in the body, that body is truly in motion; and then the situation of other bodies, with respect to it, will be changed consequently, though the cause of that change be not in them. 'Tis true that, exactly speaking, there is not any one body, that is perfectly and entirely at rest; but we frame an abstract notion of rest, by considering the thing mathematically. Thus have I left nothing unanswered, of what has been alleged for the absolute reality of space. And I have demonstrated the falsehood of that reality, by a fundamental principle, one of the most certain both in reason and experience; against which, no exception or instance can be alleged. Upon the whole,[5] one may judge from what has been said that I ought not to admit a moveable universe; nor any place out of the material universe.

54. I am not sensible of any objection, but what I think I have sufficiently answered. As for the objection that space and time are quantities, or rather things endowed with quantity; and that situation and order are not so: I answer, that order also has its quantity; there is in it, that which goes before, and that which follows; there is distance or interval. Relative things have their quantity, as well as absolute ones. For instance, ratios or proportions in mathematics, have their quantity, and are measured by logarithms; and yet they are relations. And therefore though time and space consist in relations, yet they have their quantity.

55. As to the question, whether God could have created the world sooner; 'tis necessary here to understand each other rightly. Since I have demonstrated, that time, without things, is nothing else but a mere ideal possibility; 'tis manifest, if any one should say that this same world, which has been actually created, might have been created sooner, without any other change; he would say nothing that is intelligible. For there is no mark or difference, whereby it would be possible to know,

that this world was created sooner. And therefore (as I have already said) to suppose that God created the same world sooner, is supposing a chimerical thing. 'Tis making time a thing absolute, independent upon God; whereas time does only co-exist with creatures, and is only conceived by the order and quantity of their changes. . . .

57. Thus it appears how we are to understand, that God created things at what time he pleased; for this depends upon the things, which he resolved to create. But things being once resolved upon, together with their relations; there remains no longer any choice about the time and the place, which of themselves have nothing in them real, nothing that can distinguish them, nothing that is at all discernible.

58. One cannot therefore say, as the author does here, that the wisdom of God may have good reasons to create this world at such or such a particular time: that particular time, considered without the things, being an impossible fiction; and good reasons for a choice, being not to be found, where every thing is indiscernible.

59. When I speak of *this world*, I mean the whole universe of material and immaterial creatures taken together, from the beginning of things. But if any one mean only the beginning of the material world, and suppose immaterial creatures before it; he would have somewhat more reason for his supposition. For time then being marked by things that existed already, it would be no longer indifferent; and there might be room for choice. And yet indeed, this would be only putting off the difficulty. For, supposing the whole universe of immaterial and material creatures together, to have a beginning; there is no longer any choice about the time, in which God would place that beginning.

60. And therefore, one must not say, as the author does here, that God created things in what particular space, and at what particular time he pleased. For, all time and all spaces being in themselves perfectly uniform and indiscernible from each other, one of them cannot please more than another. . . .

62. I don't say that matter and space are the same thing. I only say, there is no space, where there is no matter; and that space in itself is not an absolute reality. Space and matter differ, as time and motion. However, these things, though different, are inseparable.

63. But yet it does not at all follow, that matter is eternal

and necessary; unless we suppose space to be eternal and necessary: a supposition ill grounded in all respects. . . .

67. The parts of space are not determined and distinguished, but by the things which are in it: and the diversity of things in space, determines God to act differently upon different parts of space. But space without things, has nothing whereby it may be distinguished; and indeed not any thing actual.

68. If God is resolved to place a certain cube of matter at all, he is also resolved in what particular place to put it. But 'tis with respect to other parts of matter; and not with respect to bare space itself, in which there is nothing to distinguish it. . . .

104. I don't say, that space is an order or situation, which makes things capable of being situated: this would be nonsense. Any one needs only consider my own words, and add them to what I said above (Numb. 47), in order to show how the mind comes to form to itself an idea of space, and yet that there need not be any real and absolute being answering to that idea, distinct from the mind, and from all relations. I don't say therefore, that space is an order or situation, but an order of situations; or (an order) according to which, situations are disposed; and that abstract space is that order of situations, when they are conceived as being possible. Space is therefore something [merely]⁶ ideal. But, it seems, the author will not understand me. I have already, in this paper (Numb. 54), answered the objection, that order is not capable of quantity.

105. The author objects here, that time cannot be an order of successive things, because the quantity of time may become greater or less, and yet the order of successions continue the same. I answer; this is not so. For if the time is greater, there will be more successive and like states interposed; and if it be less, there will be fewer; seeing there is no vacuum, nor condensation, nor penetration (if I may so speak), in times, any more than in places.

NOTES

1 "purement."
2 Guericke (1602–86). Inventor of the air pump. He is said to have performed an experiment before the Emperor Ferdinand III

in which he took two hollow copper hemispheres, exhausted the air from them with his pump, and then showed that thirty horses, fifteen pulling on each hemisphere, could not separate them. Leibniz corresponded with Guericke about the air pump in 1671–72 (G.I.193).

[3] Torricelli (1608–47). Pupil of Galileo, and inventor of the barometer. In his most famous experiment he took a long tube closed at one end, filled it with mercury and closing the open end with his finger, inverted it in a basin of mercury. When he removed his finger, the level of mercury in the tube fell to thirty inches above the surface, leaving an apparent vacuum at the top of the tube.

[4] *"au reste."*

[5] *"au reste."*

[6] Clarke's addition.

PLACE, EXTENSION,
AND DURATION

JOHN LOCKE

A selection reprinted from Locke's *An Essay Concerning Human Understanding*, Book II, Chapters 13 and 14.

PLACE AND EXTENSION

7. *Place.* Another idea coming under this head and belonging to this tribe, is that we call "place." As in simple space we consider the relation of distance between any two bodies or points, so in our idea of place we consider the relation of distance betwixt anything and any two or more points, which are considered as keeping the same distance one with another, and so considered as at rest: for when we find anything at the same distance now which it was yesterday from any two or more points, which have not since changed their distance one with another, and with which we then compared it, we say it hath kept the same place; but if it hath sensibly altered its distance with either of those points, we say it hath changed its place; though, vulgarly speaking in the common notion of place, we do not always exactly observe the distance from precise points, but from large portions of sensible objects to which we consider the thing placed to bear relation, and its distance from which we have some reason to observe.

8. Thus a company of chess-men, standing on the same squares of the chess-board where we left them, we say they are all in the *same* place, or unmoved—though perhaps the chess-board hath been in the meantime carried out of one room into another—because we compared them only to the parts of the chess-board which keep the same distance one with another. The chess-board, we also say, is in the same place it was, if it remain in the same part of the cabin, though perhaps the ship which it is in sail all the while; and the ship is said to be in the same place, supposing it kept the same

distance with the parts of the neighbouring land, though perhaps the earth had turned round, and so both chess-men, and board, and ship have every one changed place, in respect of remoter bodies, which have kept the same distance one with another. But yet the distance from certain parts of the board being that which determines the place of the chess-men, and the distance from the fixed parts of the cabin being that which determined the place of the chess-board, and the distance from the fixed parts of the earth that by which we determined the place of the ship; these things may be said properly to be in the same place in those respects; though their distance from some other things, which in this matter we did not consider, being varied, they have undoubtedly changed place in that respect: and we ourselves shall think so when we have occasion to compare them with those other.

9. By this modification of distance we call "place" being made by men, for their common use, that by it they might be able to design the particular position of things, where they had occasion for such designation; men consider and determine of this place by reference to those adjacent things which best served to their present purpose, without considering other things which to another purpose would better determine the place of the same thing. Thus in the chess-board the use of the designation of the place of each chess-man being determined only within that chequered piece of wood, it would cross that purpose to measure it by anything else: but when these very chess-men are put up in a bag, if anyone should ask where the Black King is, it would be proper to determine the place by the parts of the room it was in, and not by the chess-board, there being another use of designing the place it is now in than when in play it was on the chess-board, and so must be determined by other bodies. So, if anyone should ask in what place are the verses which report the story of Nisus and Euryalus, it would be very improper to determine this place by saying, they were in such a part of the earth, or in Bodley's library; but the right designation of the place would be by the parts of Virgil's works and the proper answer would be, that these verses were about the middle of the ninth book of his AEneids, and that they have been always constantly in the same place ever since Virgil was printed: which is true, though the book itself hath moved a thousand times; the use of the idea of place here being to know only in what part of

the book that story is, that so upon occasion we may know where to find it, and have recourse to it for our use.

10. That our idea of place is nothing else but such a relative position of anything as I have before mentioned, I think is plain, and will be easily admitted when we consider that we can have no idea of the place of the universe, though we can of all the parts of it; because beyond that we have not the idea of any fixed, distinct, particular beings, in reference to which we can imagine it to have any relation of distance: but all beyond it is one uniform space or expansion, wherein the mind finds no variety, no marks. For to say that the world is somewhere, means no more than that it does exist: this, though a phrase borrowed from place, signifying only its existence, not location; and when one can find out and frame in his mind clearly and distinctly the place of the universe, he will be able to tell us whether it moves or stands still in the undistinguishable inane of infinite space: though it be true that the word "place" has sometimes a more confused sense, and stands for that space which any body takes up; and so the universe is in a place. The idea therefore of place we have by the same means that we get the idea of space (whereof this is but a particular limited consideration), viz., by our sight and touch, by either of which we receive into our minds the ideas of extension or distance.

11. *Extension and body not the same.* There are some that would persuade us that body and extension are the same thing; who either change the signification of words, which I would not suspect them of, they having so severely condemned the philosophy of others because it hath been too much placed in the uncertain meaning or deceitful obscurity of doubtful or insignificant terms. If therefore, they mean by body and extension the same that other people do, viz., by body, something that is solid and extended, whose parts are separable and movable different ways; and by extension, only the space that lies between the extremities of those solid coherent parts, and which is possessed by them, they confound very different ideas one with another. For I appeal to every man's own thoughts, whether the idea of space be not as distinct from that of solidity, as it is from the idea of scarlet colour? It is true, solidity cannot exist without extension, neither can scarlet colour exist without extension; but this hinders not but that they are distinct ideas. Many ideas

require others as necessary to their existence or conception, which yet are very distinct ideas. Motion can neither be, nor be conceived, without space; and yet motion is not space, nor space motion; space can exist without it, and they are very distinct ideas; and so, I think, are those of space and solidity. Solidity is so inseparable an idea from body, that upon that depends its filling of space, its contact, impulse, and communication of motion upon impulse. And if it be a reason to prove that spirit is different from body, because thinking includes not the idea of extension in it, the same reason will be as valid, I suppose, to prove that space is not body, because it includes not the idea of solidity in it; space and solidity being as distinct ideas as thinking and extension, and as wholly separable in the mind one from another. Body, then, and extension, it is evident, are two distinct ideas. For,

12. First. Extension includes no solidity nor resistance to the motion of body, as body does.

13. Secondly. The parts of pure space are inseparable one from the other; so that the continuity cannot be separated, neither really nor mentally. For I demand of anyone to remove any part of it from another with which it is continued, even so much as in thought. To divide and separate actually, is as I think, by removing the parts one from another, to make two superficies, where before there was a continuity: and to divide mentally, is to make in the mind two superficies, where before there was a continuity, and consider them as removed one from the other; which can only be done in things considered by the mind as capable of being separated; and, by separation, of acquiring new distinct superficies, which they then have not, but are capable of; but neither of these ways of separation, whether real or mental, is, as I think, compatible to pure space.

It is true, a man may consider so much of such a space as is answerable or commensurate to a foot, without considering the rest; which is indeed a partial consideration, but not so much as mental separation or division; since a man can no more mentally divide without considering two superficies separate one from the other, than he can actually divide without making two superficies disjoined one from the other; but a partial consideration is not separating. A man may consider light in the sun without its heat, or mobility in body without its extension, without thinking of their separation. One is only a partial consideration, terminating in one alone; and the other is a consideration of both, as existing separately.

14. Thirdly. The parts of pure space are immovable, which follows from their inseparability; motion being nothing but change of distance between any two things: but this cannot be between parts that are inseparable; which therefore must needs be at perpetual rest one amongst another.

Thus the determined idea of simple space distinguishes it plainly and sufficiently from body, since its parts are inseparable, immovable, and without resistance to the motion of body.

DURATION

1. Duration is fleeting extension. There is another sort of distance or length, the idea whereof we get not from the permanent parts of space, but from the fleeting and perpetually perishing parts of succession: this we call "duration," the simple modes whereof are any different lengths of it whereof we have distinct ideas, as hours, days, years, &c., time, and eternity. . . .

17. *Time is duration set out by measures*. Having thus got the idea of duration, the next thing natural for the mind to do is, to get some measure of this common duration, whereby it might judge of its different lengths, and consider the distinct order wherein several things exist: without which a great part of our knowledge would be confused, and a great part of history be rendered very useless. This consideration of duration, as set out by certain periods, and marked by certain measures or epochs, is that, I think, which most properly we call "time."

TRANSCENDENTAL IDEALITY
OF SPACE AND TIME

IMMANUEL KANT

This selection comprises part of "*The Transcendental Aesthetic*," and "*The First Antinomy*" from the *Critique of Pure Reason*, translated by Norman Kemp Smith, Macmillan and Co., Ltd., London, 1929. References to "A" and "B" are to the original German editions, A being the first and B the second.

THE TRANSCENDENTAL AESTHETIC

Section I: Space

§ 2

METAPHYSICAL EXPOSITION OF THIS CONCEPT[1]

By means of outer sense, a property of our mind, we represent to ourselves objects as outside us, and all without exception in space. In space their shape, magnitude, and relation to one another are determined or determinable. Inner sense, by means of which the mind intuits itself or its inner state, yields indeed no intuition of the soul itself as an object; but there is nevertheless a determinate form [namely, time] in which alone the intuition of inner states is possible, and everything which belongs to inner determinations is therefore represented in relations of time. Time cannot be outwardly intuited, any more than space can be intuited as something in us. What, then, are space and time? Are they real existences? Are they only determinations or relations of things, yet such as would belong to things even if they were not intuited? Or are space and time such that they belong only to the form of intuition, and therefore to the subjective constitution of our mind, apart from which they could not be ascribed to anything whatsoever? In order to obtain light upon these questions, let us first give an exposition of the concept of space.[2] By *exposition*[3] *(expositio)* I mean the clear, though not necessarily exhaustive, representation of that which belongs to a concept: The exposition is *metaphysical* when it contains that which exhibits the concept *as given a priori*.

1. Space is not an empirical concept which has been derived from outer experiences. For in order that certain sensations be referred to something outside me (that is, to something in another region of space from that in which I find myself), and similarly in order that I may be able to represent them as outside and alongside[4] one another, and accordingly as not only different but as in different places, the representation of space must be presupposed. The representation of space cannot, therefore, be empirically obtained from the relations of outer appearance. On the contrary, this outer experience is itself possible at all only through that representation.

2. Space is a necessary *a priori* representation, which underlies all outer intuitions. We can never represent to ourselves the absence of space, though we can quite well think it as empty of objects. It must therefore be regarded as the condition of the possibility of appearances, and not as a determination dependent upon them. It is an *a priori* representation, which necessarily underlies outer appearances.*

3.[5] Space is not a discursive or, as we say, general concept of relations of things in general, but a pure intuition. For, in the first place, we can represent to ourselves only one space; and if we speak of diverse spaces, we mean thereby only parts

---

* [In A there is here inserted the following argument:]

3. The apodeictic certainty of all geometrical propositions and the possibility of their *a priori* construction, is grounded in this *a priori* necessity of space. Were this representation of space a concept acquired *a posteriori*, and derived from outer experience in general, the first principles of mathematical determination would be nothing but perceptions. They would therefore all share in the contingent character of perception; that there should be only one straight line between two points would not be necessary, but only what experience always teaches. What is derived from experience has only comparative universality, namely, that which is obtained through induction. We should therefore only be able to say that, so far as hitherto observed, no space has been found which has more than three dimensions.

of one and the same unique space. Secondly, these parts cannot precede the one all-embracing space, as being, as it were, constituents out of which it can be composed; on the contrary, they can be thought only as *in* it. Space is essentially one; the manifold in it, and therefore the general concept of spaces, depends solely on [the introduction of] limitations. Hence it follows that an *a priori*, and not an empirical, intuition underlies all concepts of space. For kindred reasons, geometrical propositions, that, for instance, in a triangle two sides together are greater than the third, can never be derived from the general concepts of line and triangle, but only from intuition, and this indeed *a priori*, with apodeictic certainty.

4.* Space is represented as an infinite *given* magnitude. Now every concept must be thought as a representation which is contained in an infinite number of different possible representations (as their common character), and which therefore contains these *under* itself; but no concept, as such, can be thought as containing an infinite number of representations *within* itself. It is in this latter way, however, that space is thought; for all the parts of space coexist *ad infinitum*. Consequently, the original representation of space is an *a priori* intuition, not a concept.

§ 3⁶

THE TRANSCENDENTAL EXPOSITION OF THE CONCEPT
OF SPACE

I understand by a transcendental exposition the explanation of a concept, as a principle from which the possibility of other *a priori* synthetic knowledge can be understood. For this purpose it is required (1) that such knowledge does really flow from the given concept, (2) that this knowledge is possible only on the assumption of a given mode of explaining the concept.

Geometry is a science which determines the properties of space synthetically, and yet *a priori*. What, then, must be our

---

* [In A this paragraph runs:]

5. Space is represented as an infinite given magnitude. A general concept of space, which is found alike in a foot and in an ell, cannot determine anything in regard to magnitude. If there were no limitlessness in the progression of intuition, no concept of relations could yield a principle of their infinitude.

representation of space, in order that such knowledge of it may be possible? It must in its origin be intuition; for from a mere concept no propositions can be obtained which go beyond the concept—as happens in geometry . . . Further, this intuition must be *a priori*, that is, it must be found in us prior to any perception of an object, and must therefore be pure, not empirical, intuition. For geometrical propositions are one and all apodeictic, that is, are bound up with the consciousness of their necessity; for instance, that space has only three dimensions. Such propositions cannot be empirical or, in other words, judgments of experience, nor can they be derived from any such judgments . . .

How, then, can there exist in the mind an outer intuition which precedes the objects themselves, and in which the concept of these objects can be determined *a priori*? Manifestly, not otherwise than in so far as the intuition has its seat in the subject only, as the formal character of the subject, in virtue of which, in being affected by objects, it obtains *immediate representation*, that is, *intuition*, of them; and only in so far, therefore, as it is merely the form of outer *sense* in general.

Our explanation is thus the only explanation that makes intelligible the *possibility* of geometry, as a body of *a priori* synthetic knowledge. Any mode of explanation which fails to do this, although it may otherwise seem to be somewhat similar, can by this criterion[7] be distinguished from it with the greatest certainty.

CONCLUSIONS FROM THE ABOVE CONCEPTS

(*a*) Space does not represent any property of things in themselves, nor does it represent them in their relation to one another. That is to say, space does not represent any determination that attaches to the objects themselves, and which remains even when abstraction has been made of all the subjective conditions of intuition. For no determinations, whether absolute or relative, can be intuited prior to the existence of the things to which they belong, and none, therefore, can be intuited *a priori*.

(*b*) Space is nothing but the form of all appearances of outer sense. It is the subjective condition of sensibility, under which alone outer intuition is possible for us. Since, then, the receptivity of the subject, its capacity to be affected by objects, must necessarily precede all intuitions of these objects, it can readily be understood how the form of all appearances can be given prior to all actual perceptions, and so exist in the mind *a priori*, and how, as a pure intuition, in which all objects must be determined, it can contain, prior to all experience, principles which determine the relations of these objects.

It is, therefore, solely from the human standpoint that we can speak of space, of extended things, etc. If we depart from the subjective condition under which alone we can have outer intuition, namely, liability to be affected by objects, the representation of space stands for nothing whatsoever. This predicate can be ascribed to things only, in so far as they appear to us, that is, only to objects of sensibility. The constant form of this receptivity, which we term sensibility, is a necessary condition of all the relations in which objects can be intuited as outside us; and if we abstract from these objects, it is a pure intuition, and bears the name of space. Since we cannot treat the special conditions of sensibility as conditions of the possibility of things, but only of their appearances, we can indeed say that space comprehends all things that appear to us as external, but not all things in themselves, by whatever subject they are intuited, or whether they be intuited or not. For we cannot judge in regard to the intuitions of other thinking beings, whether they are bound by the same conditions as those which limit our intuition and which for us are universally valid. If we add to the concept of the subject of a judgment the limitation under which the judgment is made, the judgment is then unconditionally valid. The proposition, that all things are side by side in space, is valid under[8] the limitation that these things are viewed as objects of our sensible intuition. If, now, I add the condition to the concept, and say that all things, as outer appearances, are side by side in space, the rule is valid universally and without limitation. Our exposition[9] therefore establishes the *reality*, that is, the objective validity, of space in respect of whatever can be presented to us outwardly as object, but also at the same time the *ideality* of space in respect of things when they are considered in themselves through reason, that is, without regard to the constitution of our sensibility. We assert, then, the *empirical reality* of space, as regards all possible outer experience; and yet at the same time we assert its *transcendental ideality*—in other words, that it is nothing at all, immediately we withdraw the above condition, namely, its limitation to possible experience, and so look upon it as something that underlies things in themselves.

With the sole exception of space there is no subjective representation, referring to something *outer*, which could be entitled [at once] objective [and] *a priori*. For* there is no other subjective representation from which we can derive *a priori* synthetic propositions, as we can from intuition in space (§ 3). Strictly speaking, therefore, these other representations have no ideality, although they agree with the representation of space in this respect, that they belong merely to the subjective constitution of our manner of sensibility, for instance, of sight, hearing, touch,[10] as in the case of the sensations of colours, sounds, and heat, which, since they are mere sensations and not intuitions, do not of themselves yield knowledge of any object, least of all any *a priori* knowledge.

The above remark is intended only to guard anyone from supposing that the ideality of space as here asserted can be illustrated by examples so altogether insufficient as colours,

---

* ["For there is . . ." to end of paragraph, substituted in B for the following:]

This subjective condition of all outer appearances cannot, therefore, be compared to any other. The taste of a wine does not belong to the objective determinations of the wine, not even if by the wine as an object we mean the wine as appearance, but to the special constitution of sense in the subject that tastes it. Colours are not properties of the bodies to the intuition of which they are attached, but only modifications of the sense of sight, which is affected in a certain manner by light. Space, on the other hand, as condition of outer objects, necessarily belongs to their appearance or intuition. Taste and colours are not necessary conditions under which alone objects can be for us objects of the senses. They are connected with the appearances only as effects accidentally added by the particular constitution of the sense organs. Accordingly, they are not *a priori* representations, but are grounded in sensation, and, indeed, in the case of taste, even upon feeling[11] (pleasure and pain), as an effect of sensation. Further, no one can have *a priori* a representation of a colour or of any taste; whereas, since space concerns only the pure form of intuition, and therefore involves no sensation whatsoever, and nothing empirical, all kinds and determinations of space can and must be represented *a priori*, if concepts of figures and of their relations are to arise. Through space alone is it possible that things should be outer objects to us.

taste, etc. For these cannot rightly be regarded as properties of things, but only as changes in the subject, changes which may, indeed, be different for different men. In such examples as these, that which originally is itself only appearance, for instance, a rose, is being treated by the empirical understanding as a thing in itself, which, nevertheless, in respect of its colour, can appear differently to every observer. The transcendental concept of appearances in space, on the other hand, is a critical reminder that nothing intuited in space is a thing in itself, that space is not a form inhering in things in themselves as their intrinsic property, that objects in themselves are quite unknown to us, and that what we call outer objects are nothing but mere representations of our sensibility, the form of which is space. The true correlate of sensibility, the thing in itself, is not known, and cannot be known, through these representations; and in experience no question is ever asked in regard to it.

Section II: Time

§ 4

METAPHYSICAL EXPOSITION OF THE CONCEPT OF TIME[12]

1. Time is not an empirical concept that has been derived from any experience. For neither coexistence nor succession would ever come within our perception, if the representation of time were not presupposed as underlying them *a priori*. Only on the presupposition of time can we represent to ourselves a number of things as existing at one and the same time (simultaneously) or at different times (successively).

2. Time is a necessary representation that underlies all intuitions. We cannot, in respect of appearances in general, remove time itself, though we can quite well think time as void of appearances. Time is, therefore, given *a priori*. In it alone is actuality of appearances possible at all. Appearances may, one and all, vanish; but time (as the universal condition of their possibility)[13] cannot itself be removed.

3. The possibility of apodeictic principles concerning the relations of time, or of axioms of time in general, is also grounded upon this *a priori* necessity. Time has only one dimension; different times are not simultaneous but successive (just as different spaces are not successive but simultaneous). These principles cannot be derived from experience, for experience would give neither strict universality nor apodeictic

certainty. We should only be able to say that common experience teaches us that it is so; not that it must be so. These principles are valid as rules under which alone experiences are possible; and they instruct us in regard to[14] the experiences, not by means of them.

4. Time is not a discursive, or what is called a general concept, but a pure form of sensible intuition. Different times are but parts of one and the same time; and the representation which can be given only through a single object is intuition. Moreover, the proposition that different times cannot be simultaneous is not to be derived from a general concept. The proposition is synthetic, and cannot have its origin in concepts alone. It is immediately contained in the intuition and representation of time.

5. The infinitude of time signifies nothing more than that every determinate magnitude of time is possible only through limitations of one single time that underlies it. The original representation, *time*, must therefore be given as unlimited. But when an object is so given that its parts, and every quantity of it, can be determinately represented only through limitation, the whole representation cannot be given through concepts, since they contain only partial representations;[15] on the contrary, such concepts must themselves rest on immediate intuition.

§ 5[16]

THE TRANSCENDENTAL EXPOSITION OF THE CONCEPT OF TIME

I may here refer to No. 3,[17] where, for the sake of brevity, I have placed under the title of metaphysical exposition what is properly transcendental. Here I may add that the concept of alteration,[18] and with it the concept of motion, as alteration of place, is possible only through and in the representation of time; and that if this representation were not an *a priori* (inner) intuition, no concept, no matter what it might be, could render comprehensible the possibility of an alteration, that is, of a combination of contradictorily opposed predicates in one and the same thing in one and the same place. Only in time can two contradictorily opposed predicates meet in one and the same object, namely, *one after the other*. Thus our concept of time explains the possibility of that body of *a priori* synthetic knowledge which is exhibited in the general doctrine of motion, and which is by no means unfruitful.

§ 6[19]

CONCLUSIONS FROM THESE CONCEPTS

(a) Time is not something which exists of itself, or which inheres in things as an objective determination, and it does not, therefore, remain when abstraction is made of all subjective conditions of its intuition. Were it self-subsistent, it would be something which would be actual and yet not an actual object. Were it a determination or order inhering in things themselves, it could not precede the objects as their condition, and be known and intuited *a priori* by means of synthetic propositions. But this[20] last is quite possible if time is nothing but the subjective condition under which alone[21] intuition can take place in us. For that being so, this form of inner intuition can be represented prior to the objects, and therefore *a priori*.

(b) Time is nothing but the form of inner sense, that is, of the intuition of ourselves and of our inner state. It cannot be a determination of outer appearances; it has to do neither with shape nor position, but with the relation of representations in our inner state. And just because this inner intuition yields no shape, we endeavour to make up for this want by analogies. We represent the time-sequence by a line progressing to infinity, in which the manifold constitutes a series of one dimension only; and we reason from the properties of this line to all the properties of time, with this one exception, that while the parts of the line are simultaneous the parts of time are always successive. From this fact also, that all the relations of time allow of being expressed in an outer intuition, it is evident that the representation is itself an intuition.

(c) Time is the formal *a priori* condition of all appearances whatsoever. Space, as the pure form of all *outer* intuition, is so far limited; it serves as the *a priori* condition only of outer appearances. But since all representations, whether they have for their objects outer things or not, belong, in themselves, as determinations of the mind, to our inner state; and since this inner state stands under the formal condition of inner intuition, and so belongs to time, time is an *a priori* condition of all appearance whatsoever. It is the immediate condition of inner appearances (of our souls), and thereby the mediate condition of outer appearances. Just as I can say *a priori* that all outer appearances are in space, and are determined *a priori* in conformity with the relations of space, I can also say, from the principle of inner sense, that all appearances whatsoever, that is, all objects of the senses, are in time, and necessarily stand in time-relations.

If we abstract from *our* mode of inwardly intuiting ourselves—the mode of intuition in terms of which we likewise take up into our faculty of representation[22] all outer intuitions —and so take objects as they may be in themselves, then time is nothing. It has objective validity only in respect of appearances, these being things which we take *as objects of our senses*. It is no longer objective, if we abstract from the sensibility of our intuition, that is, from that mode of representation which is peculiar to us, and speak of *things in general*. Time is therefore a purely subjective condition of our (human) intuition (which is always sensible, that is, so far as we are affected by objects), and in itself, apart from the subject, is nothing. Nevertheless, in respect of all appearances, and therefore of all the things which can enter into our experience, it is necessarily objective. We cannot say that all things are in time, because in this concept of things in general we are abstracting from every mode of their intuition and therefore from that condition under which alone objects can be represented as being in time. If, however, the condition be added to the concept, and we say that all things as appearances, that is, as objects of sensible intuition, are in time, then the proposition has legitimate objective validity and universality *a priori*.

What we are maintaining is, therefore, the *empirical reality* of time, that is, its objective validity in respect of all objects which allow of ever being given to our senses. And since our intuition is always sensible, no object can ever be given to us in experience which does not conform to the condition of time. On the other hand, we deny to time all claim to absolute reality; that is to say, we deny that it belongs to things absolutely, as their condition or property, independently of any reference to the form of our sensible intuition; properties that belong to things in themselves can never be given to us through the senses. This, then, is what constitutes the *transcendental ideality* of time. What we mean by this phrase is that if we abstract from the subjective conditions of sensible intuition, time is nothing, and cannot be ascribed to the objects in themselves (apart from their relation to our intuition) in the way either of subsistence or of inherence. This ideality, like that of space, must not, however, be illustrated by false analogies with sensation,[23] because it is then assumed that the appearance, in which the sensible predicates inhere, itself has objective reality. In the case of time, such objective reality falls entirely away, save in so far as it is merely empirical, that is, save in so far as we regard the object itself merely as appearance. On this subject, the reader may refer to what has been said at the close of the preceding section.[24]

§ 7[25]

ELUCIDATION

Against this theory, which admits the empirical reality of time, but denies its absolute and transcendental reality, I have heard men of intelligence so unanimously voicing an objection, that I must suppose it to occur spontaneously to every reader to whom this way of thinking is unfamiliar. The objection is this. Alterations[26] are real,[27] this being proved by change[28] of our own representations—even if all outer appearances, together with their alterations, be denied. Now alterations are possible only in time, and time is therefore something real. There is no difficulty in meeting this objection. I grant the whole argument. Certainly time is something real, namely, the real form of inner intuition. It has therefore subjective reality in respect of inner experience; that is, I really have the representation of time and of my determinations in it. Time is therefore to be regarded as real, not indeed as object but as the mode of representation of myself as object. If without this condition of sensibility I could intuit myself, or be intuited by another being, the very same determinations which we now represent to ourselves as alterations would yield knowledge into which the representation of time, and therefore also of alteration, would in no way enter. Thus empirical reality has to be allowed to time, as the condition of all our experiences; on our theory, it is only its absolute reality that has to be denied. It is nothing but the form of our inner intuition.[a] If we take away from our inner intuition the peculiar

---

[a] I can indeed say that my representations follow one another; but this is only to say that we are conscious of them as in a time-sequence, that is, in conformity with the form of inner sense. Time is not, therefore, something in itself, nor is it an objective determination inherent in things.

condition of our sensibility, the concept of time likewise vanishes; it does not inhere in the objects, but merely in the subject which intuits them.

But the reason why this objection is so unanimously urged, and that too by those who have nothing very convincing to say against the doctrine of the ideality of space, is this. They have no expectation of being able to prove apodeictically the absolute reality of space; for they are confronted by idealism, which teaches that the reality of outer objects does not allow of strict proof. On the other hand, the reality of the object of our inner sense (the reality of myself and my state) is [they argue] immediately evident through consciousness. The former may be merely an illusion; the latter is, on their view, undeniably something real. What they have failed, however, to recognise is that both are in the same position; in neither case can their reality as representations be questioned, and in both cases they belong only to appearance, which always has two sides, the one by which the object is viewed in and by itself (without regard to the mode of intuiting it—its nature therefore remaining always problematic), the other by which the form of the intuition of this object is taken into account. This form is not to be looked for in the object in itself, but in the subject to which the object appears; nevertheless, it belongs really and necessarily to the appearance of this object.

Time and space are, therefore, two sources of knowledge, from which bodies of *a priori* synthetic knowledge can be derived. (Pure mathematics is a brilliant example of such knowledge, especially as regards space and its relations.) Time and space, taken together, are the pure forms of all sensible intuition, and so are what make *a priori* synthetic propositions possible. But these *a priori* sources of knowledge, being merely conditions of our sensibility, just by this very fact determine their own limits, namely, that they apply to objects only in so far as objects are viewed as appearances, and do not present things as they are in themselves. This is the sole field of their validity; should we pass beyond it, no objective use can be made of them. This ideality[29] of space and time leaves, however, the certainty of empirical knowledge unaffected, for we are equally sure of it, whether these forms necessarily inhere in things in themselves or only in our intuition of them. Those, on the other hand, who maintain the absolute reality of space and time, whether as subsistent[30] or only as inherent, must come into conflict with the principles of experience itself. For if they decide for the former alternative (which is generally the view taken by mathematical students of nature), they have to admit two eternal and infinite self-subsistent[31] non-entities[32] (space and time), which are there (yet without there being anything real) only in order to contain in themselves all that is real. If they adopt the latter alternative (as advocated by certain metaphysical students of nature), and regard space and time as relations of appearances, along side or in succession to one another— relations abstracted from experience, and in this isolation confusedly represented—they are obliged to deny that *a priori* mathematical doctrines have any validity in respect of real things (for instance, in space), or at least to deny their apodeictic certainty. For such certainty is not to be found in the *a posteriori*. On this view, indeed, the *a priori* concepts of space and time are merely creatures of the imagination, whose source must really be sought in experience, the imagination framing out of the relations abstracted from experience something that does indeed contain what is general in these relations, but which cannot exist without the restrictions which nature has attached to them. The former thinkers obtain at least this advantage, that they keep the field of appearances open for mathematical propositions. On the other hand, they have greatly embarrassed themselves by those very conditions [space and time, eternal, infinite, and self-subsistent], when with the understanding they endeavour to go out beyond this field. The latter have indeed an advantage, in that the representations of space and time do not stand in their way if they seek to judge of objects, not as appearances but merely in their relation to the understanding. But since they are unable to appeal to a true and objectively valid *a priori* intuition, they can neither account for the possibility of *a priori* mathematical knowledge, nor bring the propositions of experience into necessary agreement with it. On our theory of the true character of these two original forms of sensibility, both difficulties are removed.

Lastly, transcendental aesthetic cannot contain more than these two elements, space and time. This is evident from the fact that all other concepts belonging to sensibility, even that of motion, in which both elements are united, presuppose something empirical. Motion presupposes the perception of something movable. But in space, considered in itself, there is nothing movable; consequently the movable must be something that is found *in space only through experience*, and must therefore be an empirical datum. For the same reason, transcendental aesthetic cannot count the concept of change among its *a priori* data. Time itself does not change, but only something which is in time. The concept of time thus presupposes the perception of something existing and of the succession of its determinations; that is to say, it presupposes experience.

THE ANTINOMY OF PURE REASON

First Conflict of the Transcendental Ideas

THESIS

The world has a beginning in time, and is also limited as regards space.

ANTITHESIS

The world has no beginning, and no limits in space; it is infinite as regards both time and space.

PROOF

If we assume that the world has no beginning in time, then up to every given moment an eternity has elapsed, and there has passed away in the world an infinite series of successive states of things. Now the infinity of a series consists in the fact that it can never be completed through successive synthesis. It thus follows that it is impossible for an infinite world-series to have passed away, and that a beginning of the world is therefore a necessary condition of the world's existence. This was the first point that called for proof.

PROOF

For let us assume that it has a beginning. Since the beginning is an existence which is preceded by a time in which the thing is not, there must have been a preceding time in which the world was not, *i.e.* an empty time. Now no coming to be of a thing is possible in an empty time, because no part of such a time possesses, as compared with any other, a distinguishing condition of existence rather than of non-existence; and this applies whether the thing is supposed to arise of itself or through some other cause. In

PROOF

As regards the second point let us again assume the opposite, namely, that the world is an infinite given whole of co-existing things. Now the magnitude of a quantum which is not given in intuition[a] as within certain limits, can be thought only through the synthesis of its parts, and the totality of such a quantum only through a synthesis that is brought to completion through repeated addition of unit to unit.[b] In order, therefore, to think, as a whole, the world which fills all spaces, the successive synthesis of the parts of an infinite world must be viewed as completed, that is, an infinite time must be viewed as having elapsed in the enumeration of all co-existing things. This, however, is impossible. An infinite aggregate of actual things cannot therefore be viewed as a given whole, nor consequently as simultane-

PROOF

the world many series of things can, indeed, begin; but the world itself cannot have a beginning, and is therefore infinite in respect of past time.

As regards the second point, let us start by assuming the opposite, namely, that the world in space is finite and limited, and consequently exists in an empty space which is unlimited. Things will therefore not only be related in space but also related to space. Now since the world is an absolute whole beyond which there is no object of intuition, and therefor no correlate with which the world stands in relation, the relation of the world to empty space would be a relation of it to no object. But such a relation, and consequently the limitation of the world by empty space, is nothing. The world cannot, therefore, be limited in space;

---

[a] An indeterminate quantum can be intuited as a whole when it is such that though enclosed within limits we do not require to construct its totality through measurement, that is, through the successive synthesis of its parts. For the limits, in cutting off anything further, themselves determine its completeness.

[b] The concept of totality is in this case simply the representation of the completed synthesis of its parts; for, since we cannot obtain the concept from the intuition of the whole—that being in this case impossible—we can apprehend it only through the synthesis of the parts viewed as carried, at least in idea, to the completion of the infinite.

PROOF

ously given. Th[...]ld is, therefore, as rega[...] exten-sion in space, not infin[...]e, but is enclosed within limits. This was the second point in dispute.

PROOF

that is, it is infinite in respect of extension.[c]

Observations on the First Antinomy

I. ON THE THESIS

In stating these conflicting arguments I have not sought to elaborate sophisms. That is to say, I have not resorted to the method of the special pleader who attempts to take advantage of an opponent's carelessness—freely allowing the appeal to a misunderstood law, in order that he may be in a position to establish his

II. ON THE ANTITHESIS

The proof of the infinitude of the given world-series and of the world-whole, rests upon the fact that, on the contrary assumption, an empty time and an empty space, must constitute the limit of the world. I am aware that attempts have been made to evade this conclusion by arguing that a limit of the world

---

[c] Space is merely the form of outer intuition (formal intuition). It is not a real object which can be outwardly intuited. Space, as prior to all things which determine (occupy or limit) it, or rather which give an empirical intuition in accordance with its form, is, under the name of absolute space, nothing but the mere possibility of outer appearances in so far as they either exist in themselves or can be added to given appearances. Empirical intuition is not, therefore, a composite of appearances and space (of perception and empty intuition). The one is not the correlate of the other in a synthesis; they are connected in one and the same empirical intuition as matter and form of the intuition. If we attempt to set one of these two factors outside the other, space outside all appearances, there arise all sorts of empty determinations of outer intuition, which yet are not possible perceptions. For example, a determination of the relation of the motion (or rest) of the world to infinite empty space is a determination which can never be perceived, and is therefore the predicate of a mere thought-entity.

I. ON THE THESIS

own unrighteous claims by the refutation of that law. Each of the above proofs arises naturally out of the matter in dispute, and no advantage has been taken of the openings afforded by erroneous conclusions arrived at by dogmatists in either party.

I might have made a pretence of establishing the thesis in the usual manner of the dogmatists, by starting from a defective concept of the infinitude of a given magnitude. I might have argued that a magnitude is infinite if a greater than itself, as determined by the multiplicity of given units which it contains, is not possible. Now no multiplicity is the greatest, since one or more units can always be added to it. Consequently an infinite given magnitude, and therefore an infinite world (infinite as regards the elapsed series or as regards extension) is impossible; it must be limited in both respects. Such is the line that my proof might have followed. But the above concept is not adequate to what we mean by an infinite whole. It does not represent *how*

II. ON THE ANTITHESIS

in time and space is quite possible without our having to make the impossible assumption of an absolute time prior to the beginning of the world, or of an absolute space extending beyond the real world. With the latter part of this doctrine, as held by the philosophers of the Leibnizian school, I am entirely satisfied. Space is merely the form of outer intuition; it is not a real object which can be outwardly intuited; it is not a correlate of the appearances, but the form of the appearances themselves. And since space is thus no object but only the form of possible objects, it cannot be regarded as something absolute in itself that determines the existence of things. Things, as appearances, determine space, that is, of all its possible predicates of magnitude and relation they determine this or that particular one to belong to the real. Space, on the other hand, viewed as a self-subsistent something, is nothing real in itself; and cannot, therefore, determine the magnitude or shape of real things. Space, it further follows, whether full or empty,[d]

---

[d] It will be evident that what we here desire to say is that *empty space,* so far as it is *limited by appearances,* that is, empty space *within the world,* is at least not contradictory

I. ON THE THESIS

great it is, and consequently is not the concept of a *maximum*. Through it we think only its relation to any assignable unit in respect to which it is greater than all number. According as the unit chosen is greater or smaller, the infinite would be greater or smaller. Infinitude, however, as it consists solely in the relation to the given unit, would always remain the same. The absolute magnitude of the whole would not, therefore, be known in this way; indeed, the above concept does not really deal with it.

The true transcendental concept of infinitude is this, that the successive synthesis of units required for the enumeration of a quantum can never be completed.[e] Hence it follows with complete certainty that an eternity of actual successive states leading up to a given (the present) moment cannot have elapsed, and that the world must therefore have a beginning.

In the second part of the

II. ON THE ANTITHESIS

may be limited by appearances, but appearances cannot be limited *by an empty space* outside them. This is likewise true of time. But while all this may be granted, it yet cannot be denied that these two nonentities, empty space outside the world and empty time prior to it, have to be assumed if we are to assume a limit to the world in space and in time.

The method of argument which professes to enable us to avoid the above consequence (that of having to assume that if the world has limits in time and space, the infinite void must determine the magnitude in which actual things are to exist) consists in surreptitiously substituting for the sensible world some intelligible world of which we know nothing; for the first beginning (an existence preceded by a time of nonexistence) an existence in general which presupposes no other condition whatsoever; and for the limits of extension boundaries of the world-whole—thus getting rid of

of transcendental principles and may therefore, so far as they are concerned, be admitted. This does not, however, amount to an assertion of its possibility.

[e] This quantum therefore contains a quantity (of given units) which is greater than any number—which is the mathematical concept of the infinite.

I. ON THE THESIS

thesis the difficulty involved in a series that is infinite and yet has elapsed does not arise, since the manifold of a world which is infinite in respect of extension is given as *co-existing*. But if we are to think the totality of such a multiplicity, and yet cannot appeal to limits that of themselves constitute it a totality in intuition, we have to account for a concept which in this case cannot proceed from the whole to the determinate multiplicity of the parts, but which must demonstrate the possibility of a whole by means of the successive synthesis of the parts. Now since this synthesis must constitute a never to be completed series, I cannot think a totality either prior to the synthesis or by means of the synthesis. For the concept of totality is in this case itself the representation of a completed synthesis of the parts. And since this completion is impossible, so likewise is the concept of it.

II. ON THE ANTITHESIS

time and space. But we are here treating only of the *mundus phaenomenon* and its magnitude, and cannot therefore abstract from the aforesaid conditions of sensibility without destroying the very being of that world. If the sensible world is limited, it must necessarily lie in the infinite void. If that void, and consequently space in general as *a priori* condition of the possibility of appearances, be set aside, the entire sensible world vanishes. This world is all that is given us in our problem. The *mundus intelligibilis* is nothing but the general concept of a world in general, in which abstraction is made from all conditions of its intuition, and in reference to which, therefore, no synthetic proposition, either affirmative or negative, can possibly be asserted.

NOTES

¹ ["§ 2" and sub-heading added in B.]
² [In B: *den Begriff des Raumes erörtern* substituted for *zuerst den Raum betrachten*.]
³ ["By *exposition* . . ." to end of sentence added in B.]
⁴ [*und neben* added in B.]
⁵ [In A: "4."]
⁶ [This whole sub-section added in B.]

⁷ [End of the sub-section added in B.]

⁸ [In A: valid only under.]

⁹ [Following the 4th ed. substitution of *Erörterung lehrt* for *Erörterungen lehren.*]

¹⁰ [*Gefühls,* cf. below.]

¹¹ [*Gefühl,* cf. above.]

¹² ["§ 4" and sub-title added in B.]

¹³ [Brackets added in B.]

¹⁴ [Taking the 3rd ed. reading of *von* for *vor.*]

¹⁵ [In A: since in their case the partial representations come first.]

¹⁶ [The whole of sub-section 5 is added in B.]

¹⁷ [I.e. to § 4, No. 3.]

¹⁸ [*Veränderung.* Cf. below, pp. 216-17.]

¹⁹ ["§ 6" added in B.]

²⁰ [Reading, with Grillo, *Dieses* for *Diese.*]

²¹ [Reading, with Erdmann, *allein* for *alle.*]

²² [*in der Vorstellungskraft zu befassen.*]

²³ [*mit den Subreptionen der Empfindung in Vergleichung zu stellen, i.e.* the ideality of time and space must not be confused with the ideality ascribed to sensations.]

²⁴ [Above, p. 110.]

²⁵ ["§ 7" added in B.]

²⁶ [*Veränderungen.*]

²⁷ [*Wirklich* here, as often elsewhere, is used by Kant as the adjective corresponding to the substantive *Realität,* and in such cases it is more suitably translated by 'real' than by 'actual'.]

²⁸ [*Wechsel.*]

²⁹ [Reading, with Laas, Adickes, and Vaihinger, *Idealität* for *Realität.*]

³⁰ [*subsistierend.*]

³¹ [*für sich bestehende.*]

³² [*Undinge.*]

MIRROR IMAGES

Immanuel Kant

This is part of section 13 of Kant's *Prolegomena to Any Future Metaphysics*, translated, with introduction and notes, by P. G. Lucas, Manchester University Press, 1953.

Those who cannot get away from the concept that space and time are real qualities attached to things in themselves, may exercise their sagacity on the following paradox, and when they have tried in vain to solve it, may suspect, free from prejudices at least for a few moments, that the reduction of space and time to mere forms of our sensible intuition may perhaps have some ground.

If two things are completely the same in all points that can be known at all about each separately (in all determinations belonging to quantity and quality), it must follow that each can be replaced by the other in all cases and in all respects, without the exchange causing the slightest recognisable difference. This is in fact the case with plane figures in geometry; but various spherical figures show, notwithstanding this complete inner agreement, an outer relation such that one cannot be replaced by the other. For example two spherical triangles on opposite hemispheres which have an arc of the equator as their common base can be completely equal, in respect of sides as well as angles, so that nothing is found in either, when it is described alone and completely, which does not also appear in the description of the other, and yet one cannot be put in the place of the other (on the opposite hemisphere). Here then is an *inner* difference between the two triangles which no understanding can show to be inner and which only reveals itself through the outer relation in space. But I will quote more usual cases which can be taken from ordinary life.

What can be more like my hand or my ear, and more equal in all points, than its image in the mirror? And yet I cannot put such a hand as is seen in the mirror in the place of its original: for if the original was a right hand, the hand in the mirror is a left hand, and the image of the right ear is a left

ear, which could never serve as a substitute for the other. Here
are no inner differences that any understanding could think;
and yet the differences are inner as far as the senses tell us,
for the left hand cannot be enclosed in the same boundaries
as the right (they cannot be congruent) notwithstanding all
their mutual equality and similarity; the glove of the one hand
cannot be used on the other. What is the solution? These
objects are not representations of the things as they are in
themselves and as a pure[1] understanding would know them,
but sensible intuitions, i.e. appearances, the possibility of
which rests on the relation of certain things, unknown in
themselves, to something else, namely our sensibility. Now
of this sensibility space is the form of outer intuition, and the
inner determination of any space is only possible by determining its outer relation to space as a whole of which it is a
part (the relation to outer sense), i.e. the part is only possible
through the whole, which is never the case with things in
themselves as objects of bare understanding, but can well be
with mere appearances. Hence we cannot make the difference
between similar and equal but yet incongruent things (e.g.
spirals winding opposite ways[2]) intelligible by any single concept, but only by their relation to the right and the left hand,
which bears immediately on intuition.

NOTES

[1] Kant distinguishes the use of "pure" here from its usual use in
"pure intuition," etc., by putting the Latin form *pur* instead of the
usual *rein*. He also writes "bare" (*bloss*) nine lines below.

[2] All the previous translators have noted the memorable rendering of this phrase in Richardson's translation (London, 1819, p.
50): "for instance, snails rolled up contrary to all sense."

NEWTON'S VIEWS

OF TIME, SPACE AND MOTION

ERNST MACH

Reprinted from Chapter II, Section VI, 2–6, of *The Science of Mechanics: A Critical and Historical Account of Its Development* by Ernst Mach, translated by Thomas J. McCormack, as it appears in the sixth edition with revisions through the ninth German edition and a new introduction by Karl Menger (La Salle, Ill., Open Court Publishing Company, 1960).

2. It would appear as though Newton in the remarks here cited[1] still stood under the influence of the mediæval philosophy, as though he had grown unfaithful to his resolves to investigate only actual facts. When we say a thing *A* changes with the time, we mean simply that the conditions that determine a thing *A* depend on the conditions that determine another thing *B*. The vibrations of a pendulum take place *in time* when its excursion *depends* on the position of the earth. Since, however, in the observation of the pendulum, we are not under the necessity of taking into account its dependence on the position of the earth, but may compare it with any other thing (the conditions of which of course also depend on the position of the earth), the illusory notion easily arises that *all* the things with which we compare it are unessential. Nay, we may, in attending to the motion of a pendulum, neglect entirely other external things, and find that for every position of it our thoughts and sensations are different. Time, accordingly, appears to be some particular and independent thing, on the progress of which the position of the pendulum depends, while the things that we resort to for comparison and choose at random appear to play a wholly collateral part. But we must not forget that all things in the world are connected with one another and depend on one another, and that we ourselves and all our thoughts are also a part of nature. It is utterly beyond our power to *measure* the changes of things by *time*. Quite the contrary, time is an

126

abstraction, at which we arrive by means of the changes of things; made because we are not restricted to any one *definite* measure, all being interconnected. A motion is termed uniform in which equal increments of space described correspond to equal increments of space described by some motion with which we form a comparison, as the rotation of the earth. A motion may, with respect to another motion, be uniform. But the question whether a motion is *in itself* uniform, is senseless. With just as little justice, also, may we speak of an "absolute time"—*of a time independent of* change. This absolute time can be measured by comparison with no motion; it has therefore neither a practical nor a scientific value; and no one is justified in saying that he knows aught about it. It is an idle metaphysical conception. . . .

3. Views similar to those concerning time, are developed by Newton with respect to space and motion. We extract here a few passages which characterize his position.[2] . . .

4. It is scarcely necessary to remark that in the reflections here presented Newton has again acted contrary to his expressed intention only to investigate *actual facts*. No one is competent to predicate things about absolute space and absolute motion; they are pure things of thought, pure mental constructs, that cannot be produced in experience. All our principles of mechanics are, as we have shown in detail, experimental knowledge concerning the relative positions and motions of bodies. Even in the provinces in which they are now recognized as valid, they could not, and were not, admitted without previously being subjected to experimental tests. No one is warranted in extending these principles beyond the boundaries of experience. In fact, such an extension is meaningless, as no one possesses the requisite knowledge to make use of it. . . .

Let us look at the matter in detail. When we say that a body K alters its direction and velocity solely through the influence of another body K', we have asserted a conception that it is impossible to come at unless other bodies A, B, C. . . . are present with reference to which the motion of the body K has been estimated. In reality, therefore, we are simply cognizant of a relation of the body K to A, B, C. . . . If now we suddenly neglect A, B, C. . . . and attempt to speak of the deportment of the body K in absolute space, we implicate ourselves in a twofold error. In the first place, we cannot know how K would act in the absence of $A, B, C \dots$; and in the second place, every means would be wanting of forming a judgment of the behavior of K and of putting to the test what we had predicated—which latter therefore would be bereft of all scientific significance.

Two bodies K and K', which gravitate toward each other, impart to each other in the direction of their line of junction accelerations inversely proportional to their masses m, m'. In this proposition is contained, not only a relation of the bodies K and K' to one another, but also a relation of them to other bodies. For the proposition asserts, not only that K and K' suffer with respect to one another the acceleration designated by $x(m + m'/r^2)$, but also that K experiences the acceleration $- x\,m'/r^2$ and K' the acceleration $+ x\,m/r^2$ in the direction of the line of junction; facts which can be ascertained only by the presence of other bodies.

The motion of a body K can only be estimated by reference to other bodies $A, B, C \dots$ But since we always have at our disposal a sufficient number of bodies, that are as respects each other relatively fixed, or only slowly change their positions, we are, in such reference, restricted to no one *definite* body and can alternately leave out of account now this one and now that one. In this way the conviction arose that these bodies are indifferent generally.

It might be, indeed, that the isolated bodies $A, B, C \dots$ play merely a collateral rôle in the determination of the motion of the body K, and that this motion is determined by a *medium* in which K exists. In such a case we should have to substitute this medium for Newton's absolute space. Newton certainly did not entertain this idea. Moreover, it is easily demonstrable that the atmosphere is not this motion-determinative medium. We should, therefore, have to picture to ourselves some other medium, filling, say, all space, with respect to the constitution of which and its kinetic relations to the bodies placed in it we have at present no adequate knowledge. In itself such a state of things would not belong to the impossibilities. It is known, from recent hydrodynamical investigations, that a rigid body experiences resistance in a frictionless fluid only when its velocity *changes*. True, this result is derived theoretically from the notion of inertia; but it might, conversely, also be regarded as the primitive fact from which we have to start. Although, practically, and at present, nothing is to be accomplished with this conception, we might still hope to learn more in the future concerning this hypothetical medium; and from the point of view of science it would be in every respect a more valuable acquisition than the forlorn idea of absolute space. When we reflect that we cannot abolish the isolated bodies A, B, C. . . . , that is, cannot determine by experiment whether the part they play is fundamental or collateral, that hitherto they have been the sole and only competent means of the orientation of motions and of the description of mechanical facts, it will be found expedient provisionally to regard all motions as determined by these bodies.

5. Let us now examine the point on which Newton, apparently with sound reasons, rests his distinction of absolute and relative motion. If the earth is affected with an *absolute* rotation about its axis, centrifugal forces are set up in the earth: it assumes an oblate form, the acceleration of gravity is diminished at the equator, the plane of Foucault's pendulum rotates, and so on. All these phenomena disappear if the earth is at rest and the other heavenly bodies are affected with absolute motion round it, such that the same *relative* rotation is produced. This is, indeed, the case, if we start *ab initio* from the idea of absolute space. But if we take our stand on the basis of facts, we shall find we have knowledge only of *relative* spaces and motions. *Relatively*, not considering the unknown and neglected medium of space, the motions of the universe are the same whether we adopt the Ptolemaic or the Copernican mode of view. Both views are, indeed, equally *correct*; only the latter is more simple and more *practical*. The universe is not *twice* given, with an earth at rest and an earth in motion; but only *once*, with its *relative* motions, alone determinable. It is, accordingly, not permitted us to say how things would be if the earth did not rotate. We may interpret the one case that is given us, in different ways. If, however, we so interpret it that we come into conflict with experience, our interpretation is simply wrong. The principles of mechanics can, indeed, be so conceived, that even for relative rotations centrifugal forces arise.

Newton's experiment with the rotating vessel of water simply informs us, that the relative rotation of the water with respect to the sides of the vessel produces *no* noticeable centrifugal forces, but that such forces *are* produced by its relative rotation with respect to the mass of the earth and the other celestial bodies. No one is competent to say how the experiment would turn out if the sides of the vessel increased in thickness and mass till they were ultimately several leagues thick. The one experiment only lies before us, and our business is, to bring it into accord with the other facts known to us, and not with the arbitrary fictions of our imagination.

6. When Newton examined the principles of mechanics discovered by Galileo, the great value of the simple and precise law of inertia for deductive derivations could not possibly escape him. He could not think of renouncing its help. But the law of inertia, referred in such a naïve way to the earth supposed to be at rest, could not be accepted by him. For, in Newton's case, the rotation of the earth was not a debatable point; it rotated without the least doubt. Galileo's happy discovery could only hold approximately for small times and spaces, during which the rotation did not come into question. Instead of that, Newton's conclusions about planetary motion, referred as they were to the fixed stars, appeared to conform to the law of inertia. Now, in order to have a generally valid system of reference, Newton ventured the fifth corollary of the *Principia* (p. 19 of the first edition). He imagined a momentary terrestrial system of coördinates, for which the law of inertia is valid, held fast in space without any rotation relatively to the fixed stars. Indeed he could, without interfering with its usability, impart to this system any initial position and any uniform translation relatively to the above momentary terrestrial system. The Newtonian laws of force are not altered thereby; only the initial positions and initial velocities—the constants of integration—may alter. By this view Newton gave the *exact* meaning of his hypothetical extension of Galileo's law of inertia. We see that the reduction to absolute space was by no means necessary, for the system of reference is just as relatively determined as in every other case. In spite of his metaphysical liking for the absolute, Newton was correctly led by the *tact of the natural investigator*. This is particularly to be noticed, since, in former editions of this book, it was not sufficiently emphasized. How far and how accurately the conjecture will hold good in future is of course undecided.

The comportment of terrestrial bodies with respect to the earth is reducible to the comportment of the earth with respect to the remote heavenly bodies. If we were to assert that we knew more of moving objects than this their last-mentioned, experimentally-given comportment with respect to the celestial bodies, we should render ourselves culpable of a falsity. When, accordingly, we say, that a body preserves unchanged its direction and velocity *in space*, our assertion is nothing more or less than an abbreviated reference to *the entire universe*. The use of such an abbreviated expression is permitted the original author of the principle, because he knows that as things are no difficulties stand in the way of carrying out its implied directions. But no remedy lies in his power, if difficulties of the kind mentioned present themselves; if, for example, the requisite, relatively fixed bodies are wanting.

NOTES

[1] Mach has just quoted from part of Newton's Scholium. See pp. 81-88 of this volume, especially p. 81.—ED.

[2] See note 1.—ED.

THE UNREALITY
OF SPACE AND TIME

F. H. BRADLEY

Reprinted from Chapter 4 of *Appearance and Reality*, second edition, Clarendon Press, Oxford, England, 1930.

The object of this chapter is far from being an attempt to discuss fully the nature of space or of time. It will content itself with stating our main justification for regarding them as appearance. It will explain why we deny that, in the character which they exhibit, they either have or belong to reality. I will first show this of space.

We have nothing to do here with the psychological origin of the perception. Space may be a product developed from non-spatial elements; and, if so, its production may have great bearing on the question of its true reality. But it is impossible for us to consider this here. For, in the first place, every attempt so to explain its origin has turned out a clear failure.[1] And, in the second place, its reality would not be necessarily affected by the proof of its development. Nothing can be taken as real because, for psychology, it is original; or, again, as unreal, because it is secondary. If it were a legitimate construction from elements that were true, then it might be derived only for our knowledge, and be original in fact. But so long as its attempted derivation is in part obscure and in part illusory, it is better to regard this whole question as irrelevant.

Let us then, taking space or extension simply as it is, inquire whether it contradicts itself. The reader will be acquainted with the difficulties that have arisen from the continuity and the discreteness of space. These necessitate the conclusion that space is endless, while an end is essential to its being. Space cannot come to a final limit, either within itself or on the outside. And yet, so long as it remains something always passing away, internally or beyond itself, it is not space at all. This dilemma has been met often by the ignoring of

132

one aspect, but it has never been, and it will never be, confronted and resolved. And naturally, while it stands, it is the condemnation of space.

I am going to state it here in the form which exhibits, I think, most plainly the root of the contradiction, and also its insolubility. Space is a relation—which it cannot be; and it is a quality or substance—which again it cannot be. It is a peculiar form of the problem which we discussed [earlier] and is a special attempt to combine the irreconcilable. I will set out this puzzle antithetically.

1. Space is not a mere relation. For any space must consist of extended parts, and these parts clearly are spaces. So that, even if we could take our space as a collection, it would be a collection of solids. The relation would join spaces which would not be mere relations. And hence the collection, if taken as a *mere* inter-relation, would not be space. We should be brought to the proposition that space is nothing but a relation of spaces. And this proposition contradicts itself.

Again, from the other side, if any space is taken as a whole, it is evidently more than a relation. It is a thing, or substance, or quality (call it what you please), which is clearly as solid as the parts which it unites. From without, or from within, it is quite as repulsive and as simple as any of its contents. The mere fact that we are driven always to speak of its *parts* should be evidence enough. What could be the *parts* of a relation?

2. But space is nothing but a relation. For, in the first place, any space must consist of parts; and, if the parts are not spaces, the whole is not space. Take then in a space any parts. These, it is assumed, must be solid, but they are obviously extended. If extended, however, they will themselves consist of parts, and these again of further parts, and so on without end. A space, or a part of space, that really means to be solid, is a self-contradiction. Anything extended is a collection, a relation of extendeds, which again are relations of extendeds, and so on indefinitely. The terms are essential to the relation, and the terms do not exist. Searching without end, we never find anything more than relations; and we see that we cannot. Space is essentially a relation of what vanishes into relations, which seek in vain for their terms. It is lengths of lengths of—nothing that we can find.

And, from the outside again, a like conclusion is forced on us. We have seen that space vanishes internally into relations between units which never can exist. But, on the other side, when taken itself as a unit, it passes away into the search for an illusory whole. It is essentially the reference of itself to something else, a process of endless passing beyond actuality. As a whole it *is*, briefly, the relation of itself to a non-existent other. For take space as large and as complete as you possibly can. Still, if it has not definite boundaries, it is not space; and to make it end in a cloud, or in nothing, is mere blindness and *our* mere failure to perceive. A space limited, and yet without space that is outside, is a self-contradiction. But the outside, unfortunately, is compelled likewise to pass beyond itself; and the end cannot be reached. And it is not merely that we fail to perceive, or fail to understand, how this can be otherwise. We perceive and we understand that it cannot be otherwise, at least if space is to be space. We either do not know what space means; and, if so, certainly we cannot say that it is more than appearance. Or else, knowing what we mean by it, we see inherent in that meaning the puzzle we are describing. Space, to be space, must have space outside itself. It for ever disappears into a whole, which proves never to be more than one side of a relation to something beyond. And thus space has neither any solid parts, nor, when taken as one, is it more than the relation of itself to a new self. As it stands, it is *not* space; and, in trying to find space beyond it, we can find only that which passes away into a relation. Space is a relation between terms, which can never be found.

It would not repay us to dwell further on the contradiction which we have exhibited. The reader who has once grasped the principle can deal himself with the details. I will refer merely in passing to a supplementary difficulty. Empty space—space without some quality (visual or muscular) which in itself is more than spatial—is an unreal abstraction. It cannot be said to exist, for the reason that it cannot by itself have any meaning. When a man realizes what he has got in it, he finds that always he has a quality which is more than extension. . . . But, if so, how this quality is to stand to the extension is an insoluble problem. It is a case of "inherence," . . . in principle unintelligible. And, without further delay, I will proceed to consider time. I shall in this chapter confine myself almost entirely to the difficulties caused by the discretion and the continuity of time. With regard to change, I will say something further in the chapter which follows.

Efforts have been made to explain time psychologically—to exhibit, that is to say, its origin from what comes to the mind as timeless. But, for the same reason which seemed conclusive in the case of space, and which here has even greater weight, I shall not consider these attempts. I shall inquire simply as to time's character, and whether, that being as it is, it can belong to reality.

It is usual to consider time under a spatial form. It is taken as a stream, and past and future are regarded as parts of it, which presumably do not coexist, but are often talked of as if they did. Time, apprehended in this way, is open to the objection we have just urged against space. It is a relation—and, on the other side, it is not a relation; and it is, again, incapable of being anything beyond a relation. And the reader who has followed the dilemma which was fatal to space will not require much explanation. If you take time as a relation between units without duration, then the whole time has no duration, and is not time at all. But, if you give duration to the whole time, then at once the units themselves are found to possess it; and they thus cease to be units. Time in fact is "before" and "after" in one; and without this diversity it is not time. But these differences cannot be asserted of the unity; and, on the other hand and failing that, time is helplessly dissolved. Hence they are asserted under a relation. "Before in relation to after" is the character of time; and here the old difficulties about relation and quality recommence. The relation is not a unity, and yet the terms are non-entities, if left apart. Again, to import an independent character into the terms is to make each somehow in itself both before and after. But this brings on a process which dissipates the terms into relations, which, in the end, end in nothing. And to make the relation of time a unit is, first of all, to make it stationary, by destroying within it the diversity of before and after. And, in the second place, this solid unit, existing only by virtue of external relations, is forced to expand. It perishes in ceaseless oscillation, between an empty solidity and a transition beyond itself towards illusory completeness.

And, as with space, the qualitative content—which is not merely temporal, and apart from which the terms related in time would have no character—presents an insoluble problem. How to combine this in unity with the time which it fills, and again how to establish each aspect apart, are both beyond our resources. And time so far, like space, has turned out to be appearance.

But we shall be rightly told that a spatial form is not essential to time, and that, to examine it fairly, we should not force our errors upon it. Let us then attempt to regard time as it stands, and without extraneous additions. We shall only convince ourselves that the root of the old dilemma is not torn up.

If we are to keep to time as it comes, and are to abstain at first from inference and construction, we must confine ourselves, I presume, to time as presented. But presented time must be time present, and we must agree, at least provisionally, not to go beyond the "now." And the question at once before us will be as to the "now's" temporal contents. First, let us ask if they exist. Is the "now" simple and indivisible? We can at once reply in the negative. For time implies before and after, and by consequence diversity; and hence the simple is not time. We are compelled then, so far, to take the present as comprehending diverse aspects.

How many aspects it contains is an interesting question. According to one opinion, in the "now" we can observe both past and future; and, whether these are divided by the present, and, if so, precisely in what sense, admits of further doubt. In another opinion, which I prefer, the future is not presented, but is a product of construction; and the "now" contains merely the process of present turning into past. But here these differences, if indeed they are such, are fortunately irrelevant. All that we require is the admission of some process within the "now."[2]

For any process admitted destroys the "now" from within. Before and after are diverse, and their incompatibility compels us to use a relation between them. Then at once the old wearisome game is played again. The aspects become parts, the "now" consists of "nows," and in the end these "nows" prove undiscoverable. For, as a solid part of time, the "now" does not exist. Pieces of duration may to us appear not to be composite; but a very little reflection lays bare their inherent fraudulence. If they are not duration, they do not contain an after and before, and they have, by themselves, no beginning or end, and are by themselves outside of time. But, if so, time becomes merely the relation between them; and duration is a number of relations of the timeless, themselves also, I suppose, related somehow so as to make one duration. But how a relation is to be a unity, of which these differences are predicable, we have seen is incomprehensible. And, if it fails to be a unity, time is forthwith dissolved. But why should I weary the reader by developing in detail the impossible consequences of either alternative? If he has understood the principle, he is with us; and, otherwise, the uncertain *argumentum ad hominem* would too certainly pass into *argumentum ad nauseam*.

I will, however, instance one result which follows from a denial of time's continuity. Time will in this case fall somehow between the timeless, as *A—C—E*. But the rate of change is not uniform for all events; and, I presume, no one will assert that, when *we* have arrived at *our* apparent units, that sets a limit to actual and possible velocity. Let us suppose then another series of events, which, taken as a whole, coincides in time with *A—C—E*, but contains the six units *a—b—c—d—e—f*. Either then these other relations (those, for example, between *a* and *b*, *c* and *d*) will fall between *A* and *C*, *C* and *E*, and what that can mean I do not know; or else the transition *a—b* will coincide with *A*, which is timeless and contains no possible lapse. And that, so far as I can perceive, contradicts itself outright. But I feel inclined to add that this whole question is less a matter for detailed argument than for understanding in its principle. I doubt if there is any one who has ever grasped this, and then has failed to reach one main result. But there are too many respectable writers whom here one can hardly criticize. They have simply never got to understand.

Thus, if in the time, which we call presented, there exists any lapse, that time is torn by a dilemma, and is condemned to be appearance. But, if the presented is timeless, another destruction awaits us. Time will be the relation of the present to a future and past; and the relation, as we have seen, is not compatible with diversity or unity. Further, the existence, not presented, of future and of past seems ambiguous. But, apart from that, time perishes in the endless process beyond itself. The unit will be for ever its own relation to something beyond, something in the end not discoverable. And this process is forced on it, both by its temporal form, and again by the continuity of its content, which transcends what is given.

Time, like space, has most evidently proved not to be real, but to be a contradictory appearance.

NOTES

[1] I do not mean to say that I consider it to be original. On the contrary, one may have reason to believe something to be secondary, even though one cannot point out its foundation and origin. What has been called "extensity" appears to me (as offered) to involve a confusion. When you know what you mean by it, it seems to turn out to be either spatial at once and downright, or else not spatial at all. It seems useful, in part, only as long as you allow it to be obscure. Does *all* perception of more and less (or all which does not involve degree in the strict sense) imply space, or not? *Any* answer to this question would, I think, dispose of "extensity" as offered. But see *Mind*, iv, pp. 232-5.

[2] On the different meanings of the "present" I have said something in my *Principles of Logic*, pp. 51, ff.

DURATION

AND INTUITION

HENRI BERGSON

Reprinted from *An Introduction to Metaphysics*, translated by T. E. Hulme, Macmillan and Co., Ltd., London, 1913.

There is one reality, at least, which we all seize from within, by intuition and not by simple analysis. It is our own personality in its flowing through time—our self which endures. We may sympathise intellectually with nothing else, but we certainly sympathise with our own selves.

When I direct my attention inward to contemplate my own self (supposed for the moment to be inactive), I perceive at first, as a crust solidified on the surface, all the perceptions which come to it from the material world. These perceptions are clear, distinct, juxtaposed or juxtaposable one with another; they tend to group themselves into objects. Next, I notice the memories which more or less adhere to these perceptions and which serve to interpret them. These memories have been detached, as it were, from the depth of my personality, drawn to the surface by the perceptions which resemble them; they rest on the surface of my mind without being absolutely myself. Lastly, I feel the stir of tendencies and motor habits—a crowd of virtual actions, more or less firmly bound to these perceptions and memories. All these clearly defined elements appear more distinct from me, the more distinct they are from each other. Radiating, as they do, from within outwards, they form collectively the surface of a sphere which tends to grow larger and lose itself in the exterior world. But if I draw myself in from the periphery towards the centre, if I search in the depth of my being that which is most uniformly, most constantly and most enduringly myself, I find an altogether different thing.

There is, beneath these sharply cut crystals and this frozen surface, a continuous flux which is not comparable to any flux I have ever seen. There is a succession of states, each of which announces that which follows and contains that which precedes it. They can, properly speaking, only be said to form multiple states when I have already passed them and turn back to observe their track. Whilst I was experiencing them they were so solidly organised, so profoundly animated with a common life, that I could not have said where any one of them finished or where another commenced. In reality no one of them begins or ends, but all extend into each other.

This inner life may be compared to the unrolling of a coil, for there is no living being who does not feel himself coming gradually to the end of his rôle; and to live is to grow old. But it may just as well be compared to a continual rolling up, like that of a thread on a ball, for our past follows us, it swells incessantly with the present that it picks up on its way; and consciousness means memory.

But actually it is neither an unrolling nor a rolling up, for these two similes evoke the idea of lines and surfaces whose parts are homogeneous and superposable on one another. Now, there are no two identical moments in the life of the same conscious being. Take the simplest sensation, suppose it constant, absorb in it the entire personality: the consciousness which will accompany this sensation cannot remain identical with itself for two consecutive moments, because the second moment always contains, over and above the first, the memory that the first has bequeathed to it. A consciousness which could experience two identical moments would be a consciousness without memory. It would die and be born again continually. In what other way could one represent unconsciousness?

It would be better, then, to use as a comparison the myriad-tinted spectrum, with its insensible gradations leading from one shade to another. A current of feeling which passed along the spectrum, assuming in turn the tint of each of its shades, would experience a series of gradual changes, each of which would announce the one to follow and would sum up those which preceded it. Yet even here the successive shades of the spectrum always remain external one to another. They are juxtaposed; they occupy space. But pure duration, on the contrary, excludes all idea of juxtaposition, reciprocal externality and extension.

Let us, then, rather, imagine an infinitely small elastic body, contracted, if it were possible, to a mathematical point. Let this be drawn out gradually in such a manner that from the point comes a constantly lengthening line. Let us fix our attention not on the line as a line, but on the action by which it is traced. Let us bear in mind that this action, in spite of its duration, is indivisible if accomplished without stopping, that if a stopping point is inserted, we have two actions instead of one, that each of these separate actions is then the indivisible operation of which we speak, and that it is not the moving action itself which is divisible, but, rather, the stationary line it leaves behind it as its track in space. Finally, let us free ourselves from the space which underlies the movement in order to consider only the movement itself, the act of tension or extension; in short, pure mobility. We shall have this time a more faithful image of the development of our self in duration.

However, even this image is incomplete, and indeed, every comparison will be insufficient, because the unrolling of our duration resembles in some of its aspects the unity of an advancing movement and in others the multiplicity of expanding states; and, clearly, no metaphor can express one of these two aspects without sacrificing the other. If I use the comparison of the spectrum with its thousand shades, I have before me a thing already made, whilst duration is continually "in the making." If I think of an elastic which is being stretched, or of a spring which is extended or relaxed, I forget the richness of colour, characteristic of duration that is lived, to see only the simple movement by which consciousness passes from one shade to another. The inner life is all this at once: variety of qualities, continuity of progress, and unity of direction. It cannot be represented by images.

But it is even less possible to represent it by *concepts*, that is by abstract, general or simple ideas. It is true that no image can reproduce exactly the original feeling I have of the flow of my own conscious life. But it is not even necessary that I should attempt to render it. If a man is incapable of getting for himself the intuition of the constitutive duration of his own being, nothing will ever give it to him, concepts no more than images. Here the single aim of the philosopher should be to promote a certain effort, which in most men is usually fettered by habits of mind more useful to life. Now the image has at least this advantage, that it keeps us in the concrete. No image can replace the intuition of duration, but many diverse images, borrowed from very different orders of things, may, by the convergence of their action, direct consciousness to the precise point where there is a certain intuition to be seized. By choosing images as dissimilar as possible, we shall prevent any one of them from usurping the place of the intuition it is intended to call up, since it would then be driven away at once by its rivals. . . .

If I seek to *analyse* duration—that is, to resolve it into ready-made concepts—I am compelled, by the very nature of the concepts and of analysis, to take two opposing views of *duration in general*, with which I then attempt to reconstruct it. This combination, which will have, moreover, something miraculous about it—since one does not understand how two contraries would ever meet each other—can present neither a diversity of degrees nor a variety of forms; like all miracles, it is or it is not. I shall have to say, for example, that there is on the one hand a *multiplicity* of successive states of consciousness, and on the other a *unity* which binds them together. Duration will be the "synthesis" of this unity and this multiplicity, a mysterious operation which takes place in darkness, and in regard to which, I repeat, one does not see how it would admit of shades or of degrees. In this hypothesis there is, and can only be, one single duration, that in which our own consciousness habitually works. To express it more clearly—if we consider duration under the simple aspect of a movement accomplishing itself in space, and we seek to reduce to concepts, movement considered as representative of time, we shall have, on the one hand, as great a number of points on the trajectory as we may desire, and, on the other hand, an abstract unity which holds them together as a thread holds together the pearls of a necklace. Between this abstract multiplicity and this abstract unity, the combination, when once it has been posited as possible, is something unique, which will no more admit of shades than does the addition of given numbers in arithmetic. But if, instead of professing to analyse duration (*i.e.*, at bottom, to make a synthesis of it with concepts), we at once place ourselves in it by an effort of intuition, we have the feeling of a certain very determinate tension, in which the determination itself appears as a choice between an infinity of possible durations. Henceforward we can picture to ourselves as many durations as we wish, all very different from each other, although each of them, on being reduced to concepts—that is, observed externally from two opposing points of view—always comes in the end to the same indefinable combination of the many and the one.

Let us express the same idea with more precision. If I consider duration as a multiplicity of moments bound to each other by a unity which goes through them like a thread, then, however short the chosen duration may be, these moments are unlimited in number. I can suppose them as close together as I please; there will always be between these mathematical points other mathematical points, and so on to infinity. Looked at from the point of view of multiplicity then, duration disintegrates into a powder of moments, none of which endures, each being an instantaneity. If, on the other hand, I consider the unity which binds the moments together, this cannot endure either, since by hypothesis everything that is changing, and everything that is really durable in the duration, has been put to the account of the multiplicity of moments. As I probe more deeply into its essence, this unity will appear to me as some immobile substratum of that which is moving, as some intemporal essence of time; it is this that I shall call eternity;— an eternity of death, since it is nothing else than the movement emptied of the mobility which made its life. Closely examined, the opinions of the opposing schools on the subject of duration would be seen to differ solely in this, that they attribute a capital importance to one or the other of these two concepts. Some adhere to the point of view of the multiple; they set up as concrete reality the distinct moments of a time which they have reduced to powder; the unity which enables us to call the grains a powder, they hold to be much more artificial. Others, on the contrary, set up the unity of duration as concrete reality. They place themselves in the eternal. But as their eternity remains, notwithstanding, abstract, since it is empty, being the eternity of a concept which, by hypothesis, excludes from itself the opposing concept, one does not see how this eternity would permit of an indefinite number of moments coexisting in it. In the first hypothesis we have a world resting on nothing, which must end and begin again of its own accord at each instant. In the second we have an infinity of abstract eternity, about which also it is just as difficult to understand why it does not remain enveloped in itself and how it allows things to coexist with it. But in both cases, and whichever of the two metaphysics it be that one is switched into, time appears, from the psychological point of view, as a mixture of two abstractions, which admit of neither degrees nor shades. In one system as in the other, there is only one unique duration, which carries everything with it—a bottomless, bankless river, which flows without assignable force in a direction which could not be defined. Even then we can only call it a river, and the river only flows, because reality obtains from the two doctrines this concession, profiting by a moment of perplexity in their logic. As soon as they recover from this perplexity, they freeze this flux either into an immense solid sheet, or into an infinity of crystallized needles, always into a *thing* which necessarily partakes of the immobility of a *point of view*.

It is quite otherwise if we place ourselves from the first, by an effort of intuition, in the concrete flow of duration. Certainly, we shall then find no logical reason for positing multiple and diverse durations. Strictly, there might well be no other duration than our own, as, for example, there might be no other colour in the world but orange. But, just as a consciousness based on colour, which sympathised internally with orange instead of perceiving it externally, would feel itself held between red and yellow, would even perhaps suspect beyond this last colour a complete spectrum into which the continuity from red to yellow might expand naturally, so the intuition of our duration, far from leaving us suspended in the void as pure analysis would do, brings us into contact with a whole continuity of durations which we must try to follow, whether downwards or upwards; in both cases we can extend ourselves indefinitely by an increasingly violent effort, in both cases we transcend ourselves. In the first, we advance towards a more and more attenuated duration, the pulsations of which, being rapider than ours and dividing our simple sensation, dilute its quality into quantity; at the limit would be pure homogeneity, that pure *repetition* by which we define materiality. Advancing in the other direction, we approach a duration which strains, contracts and intensifies itself more and more; at the limit would be eternity. No longer conceptual eternity, which is an eternity of death, but an eternity of life. A living, and therefore still moving eternity, in which our own particular duration would be included as the vibrations are in light; an eternity which would be the concentration of all duration, as materiality is its dispersion. Between these two extreme limits intuition moves, and this movement is the very essence of metaphysics.

THE PROBLEM OF INFINITY
CONSIDERED HISTORICALLY

Bertrand Russell

Reprinted from Lecture VI of *Our Knowledge of the External World*, Allen & Unwin, Ltd., London, 1922.

The kind of way in which infinity has been used to discredit the world of sense may be illustrated by Kant's first two antinomies. In the first, the thesis states: "The world has a beginning in time, and as regards space is enclosed within limits"; the antithesis states: "The world has no beginning and no limits in space, but is infinite in respect of both time and space." Kant professes to prove both these propositions, whereas, if what we have said on modern logic has any truth, it must be impossible to prove either. In order, however, to rescue the world of sense, it is enough to destroy the proof of *one* of the two. For our present purpose, it is the proof that the world is *finite* that interests us. Kant's argument as regards space here rests upon his argument as regards time. We need therefore only examine the argument as regards time. What he says is as follows:

> For let us assume that the world has no beginning as regards time, so that up to every given instant an eternity has elapsed, and therefore an infinite series of successive states of the things in the world has passed by. But the infinity of a series consists just in this, that it can never be completed by successive synthesis. Therefore an infinite past world-series is impossible, and accordingly a beginning of the world is a necessary condition of its existence; which was the first thing to be proved.

Many different criticisms might be passed on this argument, but we will content ourselves with a bare minimum. To begin with, it is a mistake to define the infinity of a series as "impossibility of completion by successive synthesis." The notion of infinity . . . is primarily a property of *classes*, and only derivatively applicable to series; classes which are infinite are given all at once by the defining property of their members, so that there is no question of "completion" or of "successive synthesis." And the word "synthesis," by suggesting the mental activity of synthesizing, introduces, more or less surreptitiously, that reference to mind by which all Kant's philosophy was infected. In the second place, when Kant says that an infinite series can "never" be completed by successive synthesis, all that he has even conceivably a right to say is that it cannot be completed *in a finite time*. Thus what he really proves is, at most, that if the world had no beginning, it must have already existed for an infinite time. This, however, is a very poor conclusion, by no means suitable for his purposes. And with this result we might, if we chose, take leave of the first antinomy.

It is worth while, however, to consider how Kant came to make such an elementary blunder. What happened in his imagination was obviously something like this: Starting from the present and going backwards in time, we have, if the world had no beginning, an infinite series of events. As we see from the word "synthesis," he imagined a mind trying to grasp these successively, *in the reverse order* to that in which they had occurred, i.e. going from the present backwards. *This* series is obviously one which has no end. But the series of events up to the present has an end, since it ends with the present. Owing to the inveterate subjectivism of his mental habits, he failed to notice that he had reversed the sense of the series by substituting backward synthesis for forward happening, and thus he supposed that it was necessary to identify the mental series, which had no end, with the physical series, which had an end but no beginning. It was this mistake, I think, which, operating unconsciously, led him to attribute validity to a singularly flimsy piece of fallacious reasoning.

The second antinomy illustrates the dependence of the problem of continuity upon that of infinity. The thesis states: "Every complex substance in the world consists of simple parts, and there exists everywhere nothing but the simple or what is composed of it." The antithesis states: "No complex thing in the world consists of simple parts, and everywhere in it there exists nothing simple." Here, as before, the proofs of both thesis and antithesis are open to criticism, but for the purpose of vindicating physics and the world of sense it is enough to find a fallacy in *one* of the proofs. We will choose for this purpose the proof of the antithesis, which begins as follows: "Assume that a complex thing (as substance) consists of simple parts. Since all external relation, and therefore all composition out of substances, is only possible in space, the space occupied by a complex thing must consist of as many parts as the thing consists of. Now space does not consist of simple parts, but of spaces."

The rest of his argument need not concern us, for the nerve of the proof lies in the one statement: "Space does not consist of simple parts, but of spaces." This is like Bergson's objection to "the absurd proposition that motion is made up of immobilities." Kant does not tell us why he holds that a space must consist of spaces rather than of simple parts. Geometry regards space as made up of points, which are simple; and although, as we have seen, this view is not scientifically or logically *necessary*, it remains *prima facie* possible, and its mere possibility is enough to vitiate Kant's argument. For, if his proof of the thesis of the antinomy were valid, and if the antithesis could only be avoided by assuming points, then the antinomy itself would afford a conclusive reason in favour of points. Why, then, did Kant think it impossible that space should be composed of points?

I think two considerations probably influenced him. In the first place, the essential thing about space is spatial order, and mere points, by themselves, will not account for spatial order. It is obvious that his argument assumes absolute space; but it is spatial *relations* that are alone important, and they cannot be reduced to points. This ground for his view depends, therefore, upon his ignorance of the logical theory of order and his oscillations between absolute and relative space. But there is also another ground for his opinion, which is more relevant to our present topic. This is the ground derived from infinite divisibility. A space may be halved, and then halved again, and so on *ad infinitum*, and at every stage of the process the parts are still spaces, not points. In order to reach points by such a method, it would be necessary to come to the end of an unending process, which is impossible. But just as an infinite class can be given all at once by its defining concept, though it cannot be reached by successive enumeration, so an infinite set of points can be given all at once as making up a line or area or volume, though they can never be reached by the process of successive division. Thus the infinite divisibility of space gives no ground for denying that space is composed of points. Kant does not give his grounds for this denial, and we can therefore only conjecture what they were. But the above two grounds, which we have seen to be fallacious, seem sufficient to account for his opinion, and we may therefore conclude that the antithesis of the second antinomy is unproved.

The above illustration of Kant's antinomies has only been introduced in order to show the relevance of the problem of infinity to the problem of the reality of objects of sense. In the remainder of the present lecture, I wish to state and explain the problem of infinity, to show how it arose, and to show the irrelevance of all the solutions proposed by philosophers. In the following lecture,[1] I shall try to explain the true solution, which has been discovered by the mathematicians, but nevertheless belongs essentially to philosophy. The solution is definitive, in the sense that it entirely satisfies and convinces all who study it carefully. For over two thousand years the human intellect was baffled by the problem; its many failures and its ultimate success make this problem peculiarly apt for the illustration of method.

The problem appears to have first arisen in some such way as the following.[2] Pythagoras and his followers, who were interested, like Descartes, in the application of number to geometry, adopted in that science more arithmetical methods than those with which Euclid has made us familiar. They, or their contemporaries the atomists, believed, apparently, that space is composed of indivisible points, while time is composed of indivisible instants.[3] This belief would not, by itself, have raised the difficulties which they encountered, but it was presumably accompanied by another belief, that the number of points in any finite area or of instants in any finite period must be finite. I do not suppose that this latter belief was a conscious one, because probably no other possibility had occurred to them. But the belief nevertheless operated, and very soon brought them into conflict with facts which they themselves discovered. Before explaining how this occurred, however, it is necessary to say one word in explanation of the phrase "finite number." . . . for the present, it must suffice to say that I mean 0 and 1 and 2 and 3 and so on, for ever—in other words, any number that can be obtained by successively adding ones. This includes all the numbers that can be expressed by means of our ordinary numerals, and since such numbers can be made greater and greater, without ever reaching an unsurpassable maximum, it is easy to suppose that there are no other numbers. But this supposition, natural as it is, is mistaken.

Whether the Pythagoreans themselves believed space and time to be composed of indivisible points and instants is a debatable question.[4] . . .

The difficulty which beset the Pythagoreans in their attempts to apply numbers arose through their discovery of incommensurables. . . . Consider the case of a right-angled triangle whose two sides are equal, such a triangle as is formed by two sides of a square and a diagonal. Here, in virtue of the theorem, the square on the diagonal is double of the square on either of the sides. But Pythagoras or his early followers easily proved that the square of one whole number cannot be double of the square of another. Thus the length of the side and the length of the diagonal are incommensurable; that is to say, however small a unit of length you take, if it is contained an exact number of times in the side, it is not contained any exact number of times in the diagonal, and *vice versa*.

Now this fact might have been assimilated by some philosophies without any great difficulty, but to the philosophy of Pythagoras it was absolutely fatal. Pythagoras held that number is the constitutive essence of all things, yet no two numbers could express the ratio of the side of a square to the diagonal. It would seem probable that we may expand his difficulty, without departing from his thought, by assuming that he regarded the length of a line as determined by the number of atoms contained in it—a line two inches long would contain twice as many atoms as a line one inch long, and so on. But if this were the truth, then there must be a definite numerical ratio between any two finite lengths, because it was supposed that the number of atoms in each, however large, must be finite. Here there was an insoluble contradiction. The Pythagoreans, it is said, resolved to keep the existence of incommensurables a profound secret, revealed only to a few of the supreme heads of the sect, . . . if the science came to be doubted, the disciples might fall into sin, and perhaps even eat beans, which according to Pythagoras is as bad as eating parents' bones.

The problem first raised by the discovery of incommensurables proved, as time went on, to be one of the most severe and at the same time most far-reaching problems that have confronted the human intellect in its endeavour to understand the world. It showed at once that numerical measurement of lengths, if it was to be made accurate, must require an arithmetic more advanced and more difficult than any that the ancients possessed. They therefore set to work to reconstruct geometry on a basis which did not assume the universal possibility of numerical measurement—a reconstruction which, as may be seen in Euclid, they effected with extraordinary skill and with great logical acumen. The moderns, under the influence of Cartesian geometry, have reasserted the universal possibility of numerical measurement, extending arithmetic, partly for that purpose, so as to include what are called "irrational" numbers, which give the ratios of incommensurable lengths. But although irrational numbers have long been used without a qualm, it is only in quite recent years that logically satisfactory definitions of them have been given. With these definitions, the first and most obvious form of the difficulty which confronted the Pythagoreans has been solved; but other forms of the difficulty remain to be considered, and it is these that introduce us to the problem of infinity in its pure form.

We saw that, accepting the view that a length is composed of points, the existence of incommensurables proves that every finite length must contain an infinite number of points. In other words, if we were to take away points one by one, we should never have taken away all the points, however long we continued the process. The number of points therefore, cannot be *counted*, for counting is a process which enumerates things one by one. The property of being unable to be counted is characteristic of infinite collections, and is a source of many of their paradoxical qualities. So paradoxical are these qualities that until our own day they were thought to constitute logical contradictions. A long line of philosophers, from Zeno[5] to M. Bergson, have based much of their metaphysics upon the supposed impossibility of infinite collections. Broadly speaking, the difficulties were stated by Zeno, and nothing material was added until we reach Bolzano's *Paradoxien des Unendlichen*, a little work written in 1847–48, and published posthumously in 1851. Intervening attempts to deal with the problem are futile and negligible. The definitive solution of the difficulties is due, not to Bolzano, but to Georg Cantor, whose work on this subject first appeared in 1882. . . .

Zeno's four arguments against motion were intended to exhibit the contradictions that result from supposing that there is such a thing as change, and thus to support the Parmenidean doctrine that reality is unchanging.[6] Unfortunately, we only know his arguments through Aristotle, who stated them in order to refute them.[7] Those philosophers in the present day who have had their doctrines stated by opponents will realize that a just or adequate presentation of Zeno's position is hardly to be expected from Aristotle; but by some care in interpretation it seems possible to reconstruct the so-called "sophisms" which have been "refuted" by every tyro from that day to this.

Zeno's arguments would seem to be "ad hominem"; that is to say, they seem to assume premises granted by his opponents, and to show that, granting these premises, it is possible to deduce consequences which his opponents must deny. In order to decide whether they are valid arguments or "sophisms," it is necessary to guess at the tacit premises, and to decide who was the "homo" at whom they were aimed. Some maintain that they were aimed at the Pythagoreans,[8] while others have held that they were intended to refute the atomists.[9] M. Evellin, on the contrary, holds that they constitute a refutation of infinite divisibility,[10] while M. G. Noël, in the interests of Hegel, maintains that the first two arguments refute infinite divisibility, while the next two refute indivisibles.[11] Amid such a bewildering variety of interpretations, we can at least not complain of any restrictions on our liberty of choice.

The historical questions raised by the above-mentioned discussions are no doubt largely insoluble, owing to the very scanty material from which our evidence is derived. The points which seem fairly clear are the following: (1) That, in spite of MM. Milhaud and Paul Tannery, Zeno is anxious to prove that motion is really impossible, and that he desires to prove this because he follows Permenides in denying plurality;[12] (2) that the third and fourth arguments proceed on the hypothesis of indivisibles, a hypothesis which, whether adopted by the Pythagoreans or not, was certainly much advocated, as may be seen from the treatise *On Indivisible Lines* attributed to Aristotle. As regards the first two arguments, they would seem to be valid on the hypothesis of indivisibles, and also, without this hypothesis, to be such as would be valid if the traditional contradictions in infinite numbers were insoluble, which they are not.

We may conclude, therefore, that Zeno's polemic is directed against the view that space and time consist of points and instants; and that as against the view that a finite stretch of space of time consists of a finite number of points and instants, his arguments are not sophisms, but perfectly valid.

The conclusion which Zeno wishes us to draw is that plurality is a delusion, and spaces and times are really indivisible. The other conclusion which is possible, namely that the number of points and instants is infinite, was not tenable so long as the infinite was infected with contradictions. In a fragment which is not one of the four famous arguments against motion, Zeno says:

"If things are a many, they must be just as many as they are, and neither more nor less. Now, if they are as many as they are, they will be finite in number.

"If things are a many, they will be infinite in number; for there will always be other things between them, and others again between these. And so things are infinite in number."[13]

This argument attempts to prove that, if there are many things, the number of them must be both finite and infinite, which is impossible; hence we are to conclude that there is only one thing. But the weak point in the argument is the phrase: "If they are just as many as they are, they will be finite in number." This phrase is not very clear, but it is plain that it assumes the impossibility of definite infinite numbers. Without this assumption, which is now known to be false, the arguments of Zeno, though they suffice (on certain very reasonable assumptions) to dispel the hypothesis of finite indivisibles, do not suffice to prove that motion and change and plurality are impossible. They are not, however, on any view, mere foolish quibbles: they are serious arguments, raising difficulties which it has taken two thousand years to answer, and which even now are fatal to the teachings of most philosophers.

The first of Zeno's arguments is the argument of the racecourse, which is paraphrased by Burnet as follows:

"You cannot get to the end of a race-course. You cannot traverse an infinite number of points in a finite time. You must traverse the half of any given distance before you traverse the whole, and the half of that again before you can traverse it. This goes on *ad infinitum*, so that there are an infinite number of points in any given space, and you cannot touch an infinite number one by one in a finite time."[14]

Zeno appeals here, in the first place, to the fact that any distance, however small, can be halved. From this it follows, of course, that there must be an infinite number of points in a line. But Aristotle represents him as arguing, you cannot touch an infinite number of points *one by one* in a finite time. The words "one by one" are important. (1) If *all* points touched are concerned, then, though you pass through them continuously, you do not touch them "one by one." That is to say, after touching one, there is not another which you touch next: no two points are next each other, but between any two there are always an infinite number of others, which cannot be enumerated one by one. (2) If, on the other hand, only the successive middle points are concerned, obtained by always halving what remains of the course, then the points are reached one by one, and, though they are infinite in number, they are in fact all reached in a finite time. His argument to the contrary may be supposed to appeal to the view that a finite time must consist of a finite number of instants, in which case what he says would be perfectly true on the assumption that the possibility of continued dichotomy is undeniable. If, on the other hand, we suppose the argument directed against the partisans of infinite divisibility, we must suppose it to proceed as follows: "The points given by successive halving of the distances still to be traversed are infinite in number, and are reached in succession, each being reached a finite time later than its predecessor; but the sum of an infinite number of finite times must be infinite, and therefore the process will never be completed."[15] It is very possible that this is historically the right interpretation, but in this form the argument is invalid. If half the course takes half a minute, and the next quarter takes a quarter of a minute, and so on, the whole course will take a minute. The apparent force of the argument, on this interpretation, lies solely in the mistaken supposition that there cannot be anything between the whole of an infinite series, which can be seen to be false by observing that 1 is beyond the whole of the infinite series, $\frac{1}{2}, \frac{3}{4}, \frac{7}{8}, \frac{15}{16}, \ldots$

The second of Zeno's arguments is the one concerning Achilles and the tortoise, which has achieved more notoriety than the others. It is paraphrased by Burnet as follows:

"Achilles will never overtake the tortoise. He must first reach the place from which the tortoise started. By that time the tortoise will have got some way ahead. Achilles must then make up that, and again the tortoise will be ahead. He is always coming nearer, but he never makes up to it."[16]

This argument is essentially the same as the previous one. It shows that, if Achilles ever overtakes the tortoise, it must be after an infinite number of instants have elapsed since he started. This is in fact true; but the view that an infinite number of instants make up an infinitely long time is not true, and therefore the conclusion that Achilles will never overtake the tortoise does not follow.

The third argument, that of the arrow, is very interesting.[17] The text has been questioned. Burnet accepts the alterations of Zeller, and paraphrases thus:

"The arrow in flight is at rest. For, if everything is at rest when it occupies a space equal to itself, and what is in flight at any given moment always occupies a space equal to itself, it cannot move."

But according to Prantl, the literal translation of the unemended text of Aristotle's statement of the argument is as follows: "If everything, when it is behaving in a uniform manner, is continually either moving or at rest, but what is moving is always in the *now*, then the moving arrow is motionless." This form of the argument brings out its force more clearly than Burnet's paraphrase.

Here, if not in the first two arguments, the view that a finite part of time consists of a finite series of successive instants seems to be assumed; at any rate the plausibility of the argument seems to depend upon supposing that there are consecutive instants. Throughout an instant, it is said, a moving body is where it is: it cannot move during the instant, for that would require that the instant should have parts. Thus, suppose we consider a period consisting of a thousand instants, and suppose the arrow is in flight throughout this period. At each of the thousand instants, the arrow is where it is, though at the next instant it is somewhere else. It is never moving, but in some miraculous way the change of position has to occur *between* the instants, that is to say, not at any time whatever. This is what M. Bergson calls the cinematographic representation of reality. The more the difficulty is meditated, the more real it becomes. The solution lies in the theory of continuous series: we find it hard to avoid supposing that, when the arrow is in flight, there is a *next* position occupied at the *next* moment; but in fact there is no next position and no next moment, and when once this is imaginatively realized, the difficulty is seen to disappear.

The fourth and last of Zeno's arguments is the argument of the stadium.[18] . . .

Let us suppose three drill-sergeants, A, A', and A", standing in a row, while the two files of soldiers march past them in opposite directions. At the first moment which we consider, the

FIRST POSITION			SECOND POSITION		
B	B'	B"		B	B'	B"
A	A'	A"		A	A'	A"
C	C'	C"		C	C'	C"

three men B, B', B", in one row, and the three men C, C', C" in the other row, are respectively opposite to A, A', and A". At the very next moment, each row has moved on, and now B and C" are opposite A'. Thus B and C" are opposite each other. When, then did B pass C'? It must have been somewhere between the two moments which we supposed consecutive. It follows that there must be other moments between any two given moments, and therefore that there must be an infinite number of moments in any given interval of time.

The above difficulty, that B must have passed C' at some time between the two consecutive moments, is a genuine one, but is not precisely the difficulty raised by Zeno. What Zeno professes to prove is that "half of a given time is equal to double that time." The most intelligible explanation of the argument known to me is that of Gaye.[19] Since, however, his explanation is not easy to set forth shortly, I will re-state what seems to me to be the logical essence of Zeno's contention. If we suppose that time consists of a series of consecutive instants, and that motion consists in passing through a series of consecutive points, then the fastest possible motion is one which, at each instant, is at a point consecutive to that at which it was at the previous instant. Any slower motion must be one which has intervals of rest interspersed, and any faster motion must wholly omit some points. All this is evident from the fact that we cannot have more than one event for each instant. But now, in the case of our A's and B's and C's, B is opposite a fresh A every instant, and therefore the number of A's passed gives the number of instants since the beginning of the motion. But during the motion B has passed twice as many C's, and yet cannot have passed more than one each instant. Hence the number of instants since the motion began is twice the number of A's passed, though we previously found it was equal to this number. From this result, Zeno's conclusion follows.

Zeno's arguments, in some form, have afforded grounds for almost all the theories of space and time and infinity which have been constructed from his day to our own. We have seen that all his arguments are valid (with certain reasonable hypotheses) on the assumption that finite spaces and times consist of a finite number of points and instants, and that the third and fourth almost certainly in fact proceeded on this assumption, while the first and second, which were perhaps intended to refute the opposite assumption, were in that case fallacious. We may therefore escape from his paradoxes either by maintaining that, though space and time do consist of points and instants, the number of them in any finite interval is infinite; or by denying that space and time consist of points and instants at all; or lastly, by denying the reality of space and time altogether. It would seem that Zeno himself, as a supporter of Parmenides, drew the last of these three possible deductions, at any rate in regard to time. In this a very large number of philosophers have followed him. Many others, like M. Bergson, have preferred to deny that space and time consist of points and instants. Either of these solutions will meet the difficulties in the form in which Zeno raised them. But, as we saw, the difficulties can also be met if infinite numbers are admissible. And on grounds which are independent of space and time, infinite numbers, and series in which no two terms are consecutive, must in any case be admitted. Consider, for example, all the fractions less than 1, arranged in order of magnitude. Between any two of them, there are others, for example, the arithmetical mean of the two. Thus no two fractions are consecutive, and the total number of them is infinite. It will be found that much of what Zeno says as regards the series of points on a line can be equally well applied to the series of fractions. And we cannot deny that there are fractions, so that two of the above ways of escape are closed to us. It follows that, if we are to solve the whole class of difficulties derivable from Zeno's by analogy, we must discover some tenable theory of infinite numbers. What, then, are the difficulties which, until the last thirty years, led philosophers to the belief that infinite numbers are impossible?

The difficulties of infinity are of two kinds, of which the first may be called sham, while the others involve, for their solution, a certain amount of new and not altogether easy thinking. The sham difficulties are those suggested by the etymology, and those suggested by confusion of the mathematical infinite with what philosophers impertinently call the "true" infinite. Etymologically, "infinite" should mean "having no end." But in fact some infinite series have ends, some have not; while some collections are infinite without being serial, and can therefore not properly be regarded as either endless or having ends. The series of instants from any earlier one to any later one (both included) is infinite, but has two ends; the series of instants from the beginning of time to the present moment has one end, but is infinite. Kant, in his first antinomy, seems to hold that it is harder for the past to be infinite than for the future to be so, on the ground that the past is now completed, and that nothing infinite can be completed. It is very difficult to see how he can have imagined that there was any sense in this remark; but it seems most probable that he was thinking of the infinite as the "unended." It is odd that he did not see that the future too has one end at the present, and is precisely on a level with the past.

NOTES

[1] Not included in this volume.—ED.

[2] In what concerns the early Greek philosophers, my knowledge is largely derived from J. Burnet's valuable work, *Early Greek Philosophy* (2nd ed., London, 1908). I have also been greatly assisted by Mr. D. S. Robertson of Trinity College, who has supplied the deficiencies of my knowledge of Greek, and brought important references to my notice.

[3] Cf. Aristotle, *Metaphysics*, M. 6, 1080b, 18 sqq., and 1083b, 8 sqq.

[4] There is some reason to think that the Pythagoreans distinguished between discrete and continuous quantity. See G. J. Allman, in his *Greek Geometry from Thales to Euclid*, p. 23.

[5] In regard to Zeno and the Pythagoreans, I have derived much valuable information and criticism from Mr. P. E. B. Jourdain.

⁶ This interpretation is combated by Gaston Milhaud, *Les philo-sophes-géomètres de al Grèce*, p. 140 n., but his reasons do not seem to me convincing. All the interpretations in what follows are open to question, but all have the support of reputable authorities.

⁷ *Physics*, vi. 9. 2396 (R.P. 136-139).

⁸ Cf. Milhaud, *op. cit.*, p. 140 n.; Paul Tannery, *Pour l'histoire de la science hellène*, p. 249; Burnet, *op. cit.*, p. 362.

⁹ Cf. R. K. Gaye, "On Aristotle, *Physics*, Z ix." *Journal of Philology*, vol. xxxi., esp. p. 111. Also Moritz Cantor, *Vorlesungen über Geschichte der Mathematik*, 1st ed., vol. i., 1880, p. 168, who, however, subsequently adopted Paul Tannery's opinion (*Vorlesungen*, 3rd ed., vol. i., p. 200).

¹⁰ "Le mouvement et les partisans des indivisibles," *Revue de Métaphysique et de Morale*, vol. i., pp. 382-395.

¹¹ "Le mouvement et les arguments de Zénon d'Élée," *Revue de Métaphysique et de Morale*, vol. i., pp. 107-125.

¹² Cf. N. Brochard, "Les prétendus sophismes de Zénon d'Élée," *Revue de Métaphysique et de Morale*, vol. i., pp. 209-215.

¹³ Simplicius, *Phys.*, 140, 28 D (R.P. 133); Burnet, *op. cit.*, pp. 364-365.

¹⁴ Burnet, *op. cit.*, p. 367. Aristotle's words are: "The first is the one on the non-existence of motion on the ground that what is moved must always attain the middle point sooner than the end-point, on which we gave our opinion in the earlier part of our discourse." *Phys.*, vi. 9. 939B (R.P. 136). Aristotle seems to refer to *Phys.*, vi. 2. 223AB (R.P. 136A): "All space is continuous, for time and space are divided into the same and equal divisions. . . . Wherefore also Zeno's argument is fallacious, that it is impossible to go through an infinite collection or to touch an infinite collection one by one in a finite time. For there are two senses in which the term 'infinite' is applied both to length and to time, and in fact to all continuous things, either in regard to divisibility, or in regard to the ends. Now it is not possible to touch things infinite in regard to number in a finite time, but it is possible to touch things infinite in regard to divisibility: for time itself also is infinite in this sense. So that in fact we go through an infinite (space), in an infinite (time) and not in a finite (time), and we touch infinite things with infinite things, not with finite things." Philoponus, a sixth-century commentator (R.P. 136A, *Exc. Paris Philop. in Arist. Phys.*, 803, 2. Vit.), gives the following illustration: "For if a thing were moved the space of a cubit in one hour, since in every space there are an infinite number of points, the thing moved must needs touch all the points of the space: it will then go through an infinite collection in a finite time, which is impossible."

¹⁵ Cf. Mr. C. D. Broad, "Note on Achilles and the Tortoise," *Mind*, N.S., vol. xxii., pp. 318-9.

¹⁶ Burnet, *op. cit.*, Aristotle's words are: "The second is the so-called Achilles. It consists in this, that the slower will never be

overtaken in its course by the quickest, for the pursuer must always come first to the point from which the pursued has just departed, so that the slower must necessarily be always still more or less in advance." *Phys.*, vi. 9. 239B (R.P. 137).

[17] *Phys.*, vi. 9. 239B (R.P. 138).

[18] *Phys.*, vi. 9. 239B (R.P. 139).

[19] *Op. cit.*, p. 105.

II

Geometry
and
Physics

WHAT IS
GEOMETRY?

A. S. EDDINGTON

Reprinted from *Space, Time and Gravitation*, to which it is the Prologue; Cambridge University Press, Cambridge, England, 1920.

A conversation between an experimental PHYSICIST, a pure MATHEMATICIAN, a RELATIVIST, who advocates the newer conceptions of time and space in physics.

REL. There is a well-known proposition of Euclid which states that "any two sides of a triangle are together greater than the third side." Can either of you tell me whether nowadays there is good reason to believe that this proposition is true?

MATH. For my part, I am quite unable to say whether the proposition is true or not. I can deduce it by trustworthy reasoning from certain other propositions or axioms, which are supposed to be still more elementary. If these axioms are true, the proposition is true; if the axioms are not true, the proposition is not true universally. Whether the axioms are true or not I cannot say, and it is outside my province to consider.

PHYS. But is it not claimed that the truth of these axioms is self-evident?

MATH. They are by no means self-evident to me; and I think the claim has been generally abandoned.

PHYS. Yet since on these axioms you have been able to found a logical and self-consistent system of geometry, is not this indirect evidence that they are true?

MATH. No. Euclid's geometry is not the only self-consistent system of geometry. By choosing a different set of axioms I can, for example, arrive at Lobatchewsky's geometry, in which many of the propositions of Euclid are not in general true. From my point of view there is nothing to choose between these different geometries.

REL. How is it then that Euclid's geometry is so much the most important system?

MATH. I am scarcely prepared to admit that it is the most important. But for reasons which I do not profess to understand, my friend the Physicist is more interested in Euclidean geometry than in any other, and is continually setting us problems in it. Consequently we have tended to give an undue share of attention to the Euclidean system. There have, however, been great geometers like Riemann who have done something to restore a proper perspective.

REL. (to Physicist). Why are you specially interested in Euclidean geometry? Do you believe it to be the true geometry?

PHYS. Yes. Our experimental work proves it true.

REL. How, for example, do you prove that any two sides of a triangle are together greater than the third side?

PHYS. I can, of course, only prove it by taking a very large number of typical cases, and I am limited by the inevitable inaccuracies of experiment. My proofs are not so general or so perfect as those of the pure mathematician. But it is a recognised principle in physical science that it is permissible to generalise from a reasonably wide range of experiment; and this kind of proof satisfies me.

REL. It will satisfy me also. I need only trouble you with a special case. Here is a triangle ABC; how will you prove that $AB + BC$ is greater than AC?

PHYS. I shall take a scale and measure the three sides.

REL. But we seem to be talking about different things. I was speaking of a proposition of geometry—properties of space, not of matter. Your experimental proof only shows how a material scale behaves when you turn it into different positions.

PHYS. I might arrange to make the measures with an optical device.

REL. That is worse and worse. Now you are speaking of properties of light.

PHYS. I really cannot tell you anything about it, if you will not let me make measurements of any kind. Measurement is my only means of finding out about nature. I am not a metaphysicist.

REL. Let us then agree that by *length* and *distance* you always mean a quantity arrived at by measurements with material or optical appliances. You have studied experimentally the laws obeyed by these *measured lengths*, and have found the geometry to which they conform. We will call this geometry "Natural Geometry"; and it evidently has much greater importance for you than any other of the systems which the brain of the mathematician has invented. But we must remember that its subject matter involves the behaviour of material scales—the properties of matter. Its laws are just as much laws of physics as, for example, the laws of electromagnetism.

PHYS. Do you mean to compare space to a kind of magnetic field? I scarcely understand.

REL. You say that you cannot explore the world without some kind of apparatus. If you explore with a scale, you find out the natural geometry; if you explore with a magnetic needle, you find out the magnetic field. What we may call the field of extension, or space-field, is just as much a physical quality as the magnetic field. You can think of them both existing together in the aether, if you like. The laws of both must be determined by experiment. Of course, certain approximate laws of the space-field (Euclidean geometry) have been familiar to us from childhood; but we must get rid of the idea that there is anything inevitable about these laws, and that it would be impossible to find in other parts of the universe space-fields where these laws do not apply. As to how far space really resembles a magnetic field, I do not wish to dogmatise; my point is that they present themselves to experimental investigation in very much the same way.

Let us proceed to examine the laws of natural geometry. I have a tape-measure, and here is the triangle. $AB = 39\frac{1}{2}$ in., $BC = \frac{1}{8}$ in., $CA = 39\frac{7}{8}$ in. Why, your proposition does not hold!

PHYS. You know very well what is wrong. You gave the tape-measure a big stretch when you measured AB.

REL. Why shouldn't I?

PHYS. Of course, a length must be measured with a rigid scale.

REL. That is an important addition to our definition of length. But what is a rigid scale?

PHYS. A scale which always keeps the same length.

REL. But we have just defined length as the quantity arrived at by measures with a rigid scale; so you will want another rigid scale to test whether the first one changes length; and a third to test the second; and so *ad infinitum*. You remind me of the incident of the clock and time-gun in Egypt. The man in charge of the time-gun fired it by the clock; and the man in charge of the clock set it right by the time-gun. No, you must not define length by means of a rigid scale, and define a rigid scale by means of length.

PHYS. I admit I am hazy about strict definitions. There is not time for everything; and there are so many interesting things to find out in physics, which take up my attention. Are you so sure that you are prepared with a logical definition of all the terms you use?

REL. Heaven forbid! I am not naturally inclined to be rigorous about these things. Although I appreciate the value of the work of those who are digging at the foundations of science, my own interests are mainly in the upper structure. But sometimes, if we wish to add another storey, it is necessary to deepen the foundations. I have a definite object in trying to arrive at the exact meaning of length. A strange theory is floating round, to which you may feel initial objections; and you probably would not wish to let your views go by default. And after all, when you claim to determine lengths to eight significant figures, you must have a pretty definite standard of right and wrong measurements.

PHYS. It is difficult to define what we mean by rigid; but in practice we can tell if a scale is likely to change length appreciably in different circumstances.

REL. No. Do not bring in the idea of change of length in describing the apparatus for defining length. Obviously the adopted standard of length cannot change length, whatever it is made of. If a metre is defined as the length of a certain bar, that bar can never be anything but a metre long; and if we assert that this bar changes length, it is clear that we must have changed our minds as to the definition of length. You recognised that my tape-measure was a defective standard—that it was not rigid. That was not because it changed length, because, if it was the standard of length, it could not change length. It was lacking in some other quality.

You know an approximately rigid scale when you see one. What you are comparing it with is not some non-measurable ideal of length, but some attainable, or at least approachable, ideal of material constitution. Ordinary scales have defects— flexure, expansion with temperature, etc.—which can be reduced by suitable precautions; and the limit, to which you approach as you reduce them, is your rigid scale. You can define these defects without appealing to any extraneous definition of length; for example, if you have two rods of the same material whose extremities are just in contact with one another, and when one of them is heated the extremities no longer can be adjusted to coincide, then the material has a temperature-coefficient of expansion. Thus you can compare experimentally the temperature-coefficients of different metals and arrange them in diminishing sequence. In this sort of way you can specify the nature of your ideal rigid rod, before you introduce the term length.

PHYS. No doubt that is the way it should be defined.

REL. We must recognise then that all our knowledge of space rests on the behaviour of material measuring-scales free from certain definable defects of constitution.

PHYS. I am not sure that I agree. Surely there is a sense in which the statement $AB = 2CD$ is true or false, even if we had no conception of a material measuring-rod. For instance, there is, so to speak, twice as much paper between A and B, as between C and D.

REL. Provided the paper is uniform. But then, what does uniformity of the paper mean? That the amount in given length is constant. We come back at once to the need of defining length.

If you say instead that the amount of "space" between A and B is twice that between C and D, the same thing applies. You imagine the intervals filled with uniform space; but the uniformity simply means that the same amount of space corresponds to each inch of your rigid measuring-rod. You have arbitrarily used your rod to divide space into so-called equal lumps. It all comes back to the rigid rod.

I think you were right at first when you said that you could not find out anything without measurement; and measurement involves some specified material appliance.

Now you admit that your measures cannot go beyond a certain close approximation, and that you have not tried all possible conditions. Supposing that one corner of your triangle was in a very intense gravitational field—far stronger than any we have had experience of—I have good ground for believing that under those conditions you might find the sum of two sides of a triangle, as measured with a rigid rod, appreciably less than the third side. In that case would you be prepared to give up Euclidean geometry?

PHYS. I think it would be risky to assume that the strong force of gravitation made no difference to the experiment.

REL. On my supposition it makes an important difference.

PHYS. I mean that we might have to make corrections to the measures, because the action of the strong force might possibly distort the measuring-rod.

REL. In a rigid rod we have eliminated any special response to strain.

PHYS. But this is rather different. The extension of the rod is determined by the positions taken up by the molecules under the forces to which they are subjected; and there might be a response to the gravitational force which all kinds of matter would share. This could scarcely be regarded as a defect; and our so-called rigid rod would not be free from it any more than any other kind of matter.

REL. True; but what do you expect to obtain by correcting the measures? You correct measures, when they are untrue to standard. Thus you correct the readings of a hydrogen-thermometer to obtain the readings of a perfect gas-thermometer, because the hydrogen molecules have finite size, and exert special attractions on one another, and you prefer to take as standard an ideal gas with infinitely small molecules. But in the present case, what is the standard you are aiming at when you propose to correct measures made with the rigid rod?

PHYS. I see the difficulty. I have no knowledge of space apart from my measures, and I have no better standard than the rigid rod. So it is difficult to see what the corrected measures would mean. And yet it would seem to me more natural to suppose that the failure of the proposition was due to the measures going wrong rather than to an alteration in the character of space.

REL. Is not that because you are still a bit of a metaphysicist? You keep some notion of a space which is superior to measurement, and are ready to throw over the measures rather than let this space be distorted. Even if there were reason for believing in such a space, what possible reason could there be for assuming it to be Euclidean? Your sole reason for believing space to be Euclidean is that hitherto your measures have made it appear so; if now measures of certain parts of space prefer non-Euclidean geometry, all reason for assuming Euclidean space disappears. Mathematically and conceptually Euclidean and non-Euclidean space are on the same footing; our preference for Euclidean space was based on measures, and must stand or fall by measures.

PHYS. Let me put it this way. I believe that I am trying to measure something called length, which has an absolute meaning in nature, and is of importance in connection with the laws of nature. This length obeys Euclidean geometry. I believe my measures with a rigid rod determine it accurately when no disturbance like gravitation is present; but in a gravitational field it is not unreasonable to expect that the uncorrected measures may not give it exactly.

REL. You have three hypotheses there:—(1) there is an absolute thing in nature corresponding to length, (2) the geometry of these absolute lengths is Euclidean, and (3) practical measures determine this length accurately when there is no gravitational force. I see no necessity for these hypotheses, and propose to do without them. *Hypotheses non fingo*. The second hypothesis seems to be particularly objectionable. You assume that this absolute thing in nature obeys the laws of Euclidean geometry. Surely it is contrary to scientific principles to lay down arbitrary laws for nature to obey; we must find out her laws by experiment. In this case the only experimental evidence is that measured lengths (which by your own admission are not necessarily the same as this absolute thing) sometimes obey Euclidean geometry and sometimes do not. Again it would seem reasonable to doubt your third hypothesis beyond, say, the sixth decimal place; and that would play havoc with your more delicate measures. But where I fundamentally differ from you is the first hypothesis. Is there some absolute quantity in nature that we try to determine when we measure length? When we try to determine the number of molecules in a given piece of matter, we have to use indirect methods, and different methods may give systematically different results; but no one doubts that there is a definite number of molecules, so that there is some meaning in saying that certain methods are theoretically good and others inaccurate. Counting appears to be an absolute operation. But it seems to me that other physical measures are on a different footing. Any physical quantity, such as length, mass, force, etc., which is not a pure number, can only be defined as the result arrived at by conducting a physical experiment according to specified rules.

So I cannot conceive of any "length" in nature independent of a definition of the way of measuring length. And, if there is, we may disregard it in physics, because it is beyond the range of experiment. Of course, it is always possible that we may come across some quantity, not given directly by experiment, which plays a fundamental part in theory. If so, it will turn up in due course in our theoretical formulae. But it is no good assuming such a quantity, and laying down *a priori* laws for it to obey, on the off-chance of its proving useful.

PHYS. Then you will not let me blame the measuring-rod when the proposition fails?

REL. By all means put the responsibility on the measuring-rod. Natural geometry is the theory of the behaviour of material scales. Any proposition in natural geometry is an assertion as to the behaviour of rigid scales, which must accordingly take the blame or credit. But do not say that the rigid scale is wrong, because that implies a standard of right which does not exist.

PHYS. The space which you are speaking of must be a sort of abstraction of the extensional relations of matter.

REL. Exactly so. And when I ask you to believe that space can be non-Euclidean, or, in popular phrase, warped, I am not asking you for any violent effort of the imagination; I only mean that the extensional relations of matter obey somewhat modified laws. Whenever we investigate the properties of space experimentally, it is these extensional relations that we are finding. Therefore it seems logical to conclude that space as known to us must be the abstraction of these material relations, and not something more transcendental. The reformed methods of teaching geometry in schools would be utterly condemned, and it would be misleading to set schoolboys to verify propositions of geometry by measurement, if the space they are supposed to be studying had not this meaning.

I suspect that you are doubtful whether this abstraction of extensional relations quite fulfils your general idea of space; and, as a necessity of thought, you require something beyond. I do not think I need disturb that impression, provided you realise that it is not the properties of this more transcendental thing we are speaking of when we describe geometry as Euclidean or non-Euclidean.

MATH. The view has been widely held that space is neither physical nor metaphysical, but conventional. Here is a passage from Poincaré's *Science and Hypothesis*, which describes this alternative idea of space:

"If Lobatschewsky's geometry is true, the parallax of a very distant star will be finite. If Riemann's is true, it will be negative. These are the results which seem within the reach of experiment, and it is hoped that astronomical observations may enable us to decide between the two geometries. But what we call a straight line in astronomy is simply the path of a ray of light. If, therefore, we were to discover negative parallaxes, or to prove that all parallaxes are higher than a certain limit, we should have a choice between two conclusions: we could give up Euclidean geometry, or modify the laws of optics, and suppose that light is not rigorously propagated in a straight line. It is needless to add that everyone would look upon this solution as the more advantageous. Euclidean geometry, therefore, has nothing to fear from fresh experiments."

REL. Poincaré's brilliant exposition is a great help in understanding the problem now confronting us. He brings out the interdependence between geometrical laws and physical laws, which we have to bear in mind continually. We can add on to one set of laws that which we subtract from the other set. I admit that space is conventional—for that matter, the meaning of every word in the language is conventional. Moreover, we have actually arrived at the parting of the ways imagined by Poincaré, though the crucial experiment is not precisely the one he mentions. But I deliberately adopt the alternative, which, he takes for granted, everyone would consider less advantageous. I call the space thus chosen *physical space*, and its geometry *natural geometry*, thus admitting that other conventional meanings of space and geometry are possible. If it were only a question of the meaning of space—a rather vague term—these other possibilities might have some advantages. But the meaning assigned to length and distance has to go along with the meaning assigned to space. Now these are quantities which the physicist has been accustomed to measure with great accuracy; and they enter fundamentally into the whole of our experimental knowledge of the world. We have a knowledge of the so-called extent of the stellar universe, which, whatever it may amount to in terms of ultimate reality, is not a mere description of location in a conventional and arbitrary mathematical space. Are we to be robbed of the terms in which we are accustomed to describe that knowledge?

The law of Boyle states that the pressure of a gas is proportional to its density. It is found by experiment that this law is only approximately true. A certain mathematical simplicity would be gained by conventionally redefining *pressure* in such a way that Boyle's law would be rigorously obeyed. But it would be high-handed to appropriate the word pressure in this way, unless it had been ascertained that the physicist had no further use for it in its original meaning.

PHYS. I have one other objection. Apart from measures, we have a general perception of space, and the space we perceive is at least approximately Euclidean.

REL. Our perceptions are crude measures. It is true that our perception of space is very largely a matter of optical measures with the eyes. If in a strong gravitational field optical and mechanical measures diverged, we should have to make up our minds which was the preferable standard, and afterwards abide by it. So far as we can ascertain, however, they agree in all circumstances, and no such difficulty arises. So, if physical measures give us a non-Euclidean space, the space of perception will be non-Euclidean. If you were transplanted into an extremely intense gravitational field, you would directly perceive the non-Euclidean properties of space.

PHYS. Non-Euclidean space seems contrary to reason.

MATH. It is not contrary to reason, but contrary to common experience, which is a very different thing, since experience is very limited.

PHYS. I cannot imagine myself perceiving non-Euclidean space!

MATH. Look at the reflection of the room in a polished door-knob, and imagine yourself one of the actors in what you see going on there.

REL. I have another point to raise. The distance between two points is to be the length measured with a rigid scale. Let us mark the two points by particles of matter, because we must somehow identify them by reference to material objects. For simplicity we shall suppose that the two particles have no relative motion, so that the distance—whatever it is—remains constant. Now you will probably agree that there is no such thing as absolute motion; consequently there is no standard condition of the scale which we can call "at rest." We may measure with the scale moving in any way we choose, and if results for different motions disagree, there is no criterion for selecting the true one. Further, if the particles are sliding past the scale, it makes all the difference what instants we choose for making the two readings.

PHYS. You can avoid that by defining distance as the measurement made with a scale which has the same velocity as the two points. Then they will always be in contact with two particular divisions of the scale.

REL. A very sound definition; but unfortunately it does not agree with the meaning of distance in general use. When the relativist wishes to refer to this length, he calls it the *proper-length*; in non-relativity physics it does not seem to have been used at all. You see it is not convenient to send your apparatus hurling through the laboratory—after a pair of α particles, for example. And you could scarcely measure the length of a wave of light by this convention.[1] So the physicist refers his lengths to apparatus at rest on the earth; and the mathematician starts with the words "Choose unaccelerated rectangular axes Ox, Oy, Oz, . . ." and assumes that the measuring-scales are at rest relatively to these axes. So when the term length is used some arbitrary standard motion of the measuring apparatus must always be implied.

PHYS. Then if you have fixed your standard motion of the measuring-rod, there will be no ambiguity if you take the readings of both particles at the same moment.

REL. What is the same moment at different places? The conception of simultaneity in different places is a difficult one. Is there a particular instant in the progress of time on another world, Arcturus, which is the same as the present instant on the Earth?

PHYS. I think so, if there is any connecting link. We can observe an event, say a change of brightness, on Arcturus, and, allowing for the time taken by light to travel the distance, determine the corresponding instant on the earth.

REL. But then you must know the speed of the earth through the aether. It may have shortened the light-time by going some way to meet the light coming from Arcturus.

PHYS. Is not that a small matter?

REL. At a very modest reckoning the motion of the earth in the interval might alter the light-time by several days. Actually, however, any speed of the earth through the aether up to the velocity of light is admissible, without affecting anything observable. At least, nothing has been discovered which contradicts this. So the error may be months or years.

PHYS. What you have shown is that we have not sufficient knowledge to determine in practice which are simultaneous events on the Earth and Arcturus. It does not follow that there is no definite simultaneity.

REL. That is true, but it is at least possible that the reason why we are unable to determine simultaneity in practice (or, what comes to pretty much the same thing, our motion through the aether) in spite of many brilliant attempts, is that there is no such thing as absolute simultaneity of distant events. It is better therefore not to base our physics on this notion of absolute simultaneity, which may turn out not to exist, and is in any case out of reach at present.

But what all this comes to is that time as well as space is implied in all our measures. The fundamental measurement is not the interval between two points of space, but between two points of space associated with instants of time.

Our natural geometry is incomplete at present. We must supplement it by bringing in time as well as space. We shall need a perfect clock as well as a rigid scale for our measures. It may be difficult to choose an ideal standard clock; but whatever definition we decide on must be a physical definition. We must not dodge it by saying that a perfect clock is one which keeps perfect time. Perhaps the best theoretical clock would be a pulse of light travelling in vacuum to and fro between mirrors at the ends of a rigid scale. The instants of arrival at one end would define equal intervals of time.

PHYS. I think your unit of time would change according to the motion of your "clock" through the aether.

REL. Then you are comparing it with some notion of absolute time. I have no notion of time except as the result of measurement with some kind of clock. (Our immediate perception of the flight of time is presumably associated with molecular processes in the brain which play the part of a material clock.) If you know a better clock, let us adopt it; but, having once fixed on our ideal clock there can be no appeal from its judgments. You must remember too that if you wish to measure a second *at one place*, you must keep your clock fixed at what you consider to be one place; so its motion is defined. The necessity of defining the motion of the clock emphasises that one cannot consider time apart from space; there is one geometry comprising both.

PHYS. Is it right to call this study *geometry*? Geometry deals with space alone.

MATH. I have no objection. It is only necessary to consider time as a fourth dimension. Your complete natural geometry will be a geometry of four dimensions.

PHYS. Have we then found the long-sought fourth dimension?

MATH. It depends what kind of a fourth dimension you were seeking. Probably not in the sense you intend. For me it only means adding a fourth variable, t, to my three space-variables x, y, z. It is no concern of mine what these variables really represent. You give me a few fundamental laws that they satisfy, and I proceed to deduce other consequences that may be of interest to you. The four variables may for all I know be the pressure, density, temperature and entropy of a gas; that is of no importance to me. But you would not say that a gas had four dimensions because four mathematical variables were used to describe it. Your use of the term "dimensions" is probably more restricted than mine.

PHYS. I know that it is often a help to represent pressure and volume as height and width on paper; and so geometry may have applications to the theory of gases. But is it not going rather far to say that geometry can deal directly with these things and is not necessarily concerned with lengths in space?

MATH. No. Geometry is nowadays largely analytical, so that in form as well as in effect, it deals with variables of an unknown nature. It is true that I can often see results more easily by taking my x and y as lengths on a sheet of paper. Perhaps it would be helpful in seeing other results if I took them as pressure and density in a steam-engine; but a steam-engine is not so handy as a pencil. It is literally true that I do not want to know the significance of the variables x, y, z, t that I am discussing. That is lucky for the Relativist, because although he has defined carefully how they are to be measured, he has certainly not conveyed to me any notion of how I am to picture them, if my picture of absolute space is an illusion.

PHYS. Yours is a strange subject. You told us at the beginning that you are not concerned as to whether your propositions are true, and now you tell us you do not even care to know what you are talking about.

MATH. That is an excellent description of Pure Mathematics, which has already been given by an eminent mathematician.[2]

REL. I think there is a real sense in which time is a fourth dimension—as distinct from a fourth variable. The term dimension seems to be associated with relations of *order*. I believe that the order of events in nature is one indissoluble four-dimensional order. We may split it arbitrarily into space and time, just as we can split the order of space into length, breadth and thickness. But space without time is as incomplete as a surface without thickness.

MATH. Do you argue that the real world behind the phenomena is four-dimensional?

REL. I think that in the real world there must be a set of entities related to one another in a four-dimensional order, and that these are the basis of the perceptual world so far as it is yet explored by physics. But it is possible to pick out a four-dimensional set of entities from a basal world of five dimensions, or even of three dimensions. The straight lines in three-dimensional space form a four-dimensional set of entities, i.e. they have a fourfold order. So one cannot predict the ultimate number of dimensions in the world—if indeed the expression *dimensions* is applicable.

PHYS. What would a philosopher think of these conceptions? Or is he solely concerned with a metaphysical space and time which is not within reach of measurement.

REL. In so far as he is a psychologist our results must concern him. Perception is a kind of crude physical measurement; and perceptual space and time is the same as the measured space and time, which is the subject-matter of natural geometry. In other respects he may not be so immediately concerned. Physicists and philosophers have long agreed that motion through absolute space can have no meaning; but in physics the question is whether motion through aether has any meaning. I consider that it has no meaning; but that answer, though it brings philosophy and physics into closer relation, has no bearing on the philosophic question of absolute motion. I think, however, we are entitled to expect a benevolent interest from philosophers, in that we are giving to their ideas a perhaps unexpected practical application.

Let me now try to sum up my conclusions from this conversation. We have been trying to give a precise meaning to the term *space*, so that we may be able to determine exactly the properties of the space we live in. There is no means of determining the properties of our space by *a priori* reasoning, because there are many possible kinds of space to choose from, no one of which can be considered more likely than any other. For more than 2000 years we have believed in a Euclidean space, because certain experiments favoured it; but there is now reason to believe that these same experiments when pushed to greater accuracy decide in favour of a slightly different space (in the neighbourhood of massive bodies). The relativist sees no reason to change the rules of the game because the result does not agree with previous anticipations. Accordingly when he speaks of space, he means the space revealed by measurement, whatever its geometry. He points out that this is the space with which physics is concerned; and, moreover, it is the space of everyday perception. If his right to appropriate the term space in this way is challenged, he would urge that this is the sense in which the term has always been used in physics hitherto; it is only recently that conservative physicists, frightened by the revolutionary consequences of modern experiments, have begun to play with the idea of a pre-existing space whose properties cannot be ascertained by experiment—a metaphysical space, to which they arbitrarily assign Euclidean properties, although it is obvious that its geometry can never be ascertained by experiment. But the relativist, in defining space as *measured space*, clearly recognizes that all measurement involves the use of material apparatus; the resulting geometry is specifically a study of the extensional relations of matter. He declines to consider anything more transcendental.

My second point is that since natural geometry is the study of extensional relations of natural objects, and since it is found that their space-order cannot be discussed without reference to their time-order as well, it has become necessary to extend our geometry to four dimensions in order to include time.

NOTES

[1] The proper-length of a light-wave is actually infinite.

[2] Bertrand Russell: "Pure mathematics consists entirely of such asseverations as that, if such and such a proposition is true of *anything,* then such and such a proposition is true of that thing. It is essential not to discuss whether the first proposition is really true, and not to mention what the anything is of which it is supposed to be true. . . . Thus mathematics may be defined as the subject in which we never know what we are talking about, nor whether what we are saying is true."

SPACE

AND GEOMETRY

ERNEST NAGEL

This selection is Chapter 8 of *The Structure of Science, Problems in the Logic of Scientific Explanation*, Harcourt, Brace and World, Inc., New York and Burlingame, 1961.

Even a casual examination of the Newtonian axioms of motion makes clear that some frame of spatial reference must first be stipulated before the axioms can be employed to analyze the motions of bodies. The first axiom asserts that a body continues to move with constant speed along a straight line unless some force is impressed on the body. The second axiom declares that the acceleration of a body (that is, its change of speed along a straight line or its departure from rectilinear motion) is proportional to the impressed force. What is to be understood, however, by "straight line" in these statements, and with respect to what frame of reference is a motion to be judged as rectilinear? These questions . . . must now be discussed. They have been under critical consideration since the time of Newton, and the difficulties in the Newtonian answers to them have finally led in the present century to the development of a non-Newtonian mechanics. But the logical issues they involve are relevant to the study of the structure of explanation in general, and not only in mechanics. Although we shall take the axioms of mechanics as the starting point of our discussion, we shall eventually be concerned with these more general considerations.

I. THE NEWTONIAN SOLUTION

Neither Newton nor his contemporaries had any reason for supposing that a doubt could arise as to what is to be understood by "straight line" in his formulations of the axioms of motion, for the only theory of geometry known at that time

was the system of Euclid. It was therefore taken for granted
that a line is straight if it conforms to the conditions specified
in Euclidean geometry. Let us assume for the present that
Euclidean geometry offers no difficulties. . . .

However, no such unanimity existed concerning the spatial
frame to which the motions of bodies are to be referred. Even
in Newton's day vigorous debates took place over this question. It might seem at first glance that *any* frame of reference
could be chosen, and that only convenience in handling special problems ought to dictate the selection. But a more careful
examination of Newtonian theory reveals such a conception
to be mistaken. It is of course true that in actual practice a
variety of different frames of reference are used and that considerations of convenience do control the choice of reference
frame. Thus in some problems it is convenient to take the
earth for this purpose, in other problems the sun, and still
others the fixed stars; and in each case, within the limits of
accuracy demanded by the corresponding problem, the analysis of motions effected by using Newton's axioms may be in
good agreement with experimental findings. Nevertheless, from
the point of view of Newtonian theory, these various practical reference frames are not equally satisfactory, and none
of them is wholly suitable. We must understand clearly why
this is so.

To fix our ideas, suppose we examine the motion of a body
released from an initial position of rest relative to the earth
and falling freely in the earth's gravitational field somewhere
north of the equator. If we assume the earth to be a reference
frame permitted by Newtonian theory, then according to the
theory the body should fall with accelerated speed along a
line directed toward the earth's center of mass. On the other
hand, if the sun is taken as a theoretically allowable reference
frame for describing the body's motion, the theoretical trajectory will no longer be a straight line but a more complex
curve. For now the body must be regarded as sharing the
earth's diurnal rotation as well as annual revolution around
the sun, and, instead of falling along the line just described, the
body will move along a curve that is generally to the east of
this line. Furthermore, if one of the fixed stars is next adopted
as a permissible frame of reference, the theoretical trajectory
of the body will be still different and more complex. For not
only is the body part of a physical system (i.e., the earth)
which both rotates around an axis and revolves around the

sun but it is also part of the solar system which is accelerated with respect to some of the stars. However, the stars themselves are "fixed" only by courtesy, so that the theoretical trajectory of the body will in general vary with the star (or system of stars) employed as the frame of reference. To be sure, the differences between these various trajectories are often slight, and, since they can be neglected in many practical problems, it does not much matter in those cases which of the alternative reference frames is selected. The point nevertheless remains that in theory, and sometimes in practice, it is not a matter indifference which frame of reference is adopted for the study of motions. For the magnitude of the acceleration that a physical system undergoes, and therefore the forces that must be assumed (in accordance with the second axiom) to be acting on the system, depend essentially upon the reference frame with respect to which the acceleration is specified.

Let us be more explicit. If the earth is taken as the fixed frame of reference, the assumed force which is to account for the motion of a freely falling body must be proportional to the acceleration of that body relative to the earth. If the force is assumed to be simply the gravitational force of the earth, the trajectory of the body should be a straight line directed to the earth's center of mass. But in point of fact the body is deflected from that path; and as long as the earth is regarded as "fixed," there appears to be no ready way of explaining this circumstance, short of introducing *ad hoc* "deflecting forces" to account for it. The situation is altered, however, if the sun is taken as the reference frame. For now the indicated deflection is immediately explained in terms of the rotational acceleration of the earth. The general conclusion to be drawn from this example is therefore as follows: When a certain spatial reference frame is adopted, the Newtonian axioms suffice to analyze many types of motions of bodies, if forces of a relatively simple form are assumed as determinants of accelerations. On the other hand, if an arbitrary reference frame is adopted the forces that must be assumed are in general enormously complex, vary in no easily specifiable way from case to case, and have the earmarks of *ad hoc* hypotheses. Accordingly, if forces are not to be introduced in an arbitrary manner, if the determinants of accelerations are to be specified in a uniform way for extensive classes of motions rather than postulated in different ways for different special problems, there must be a privileged or "absolute" frame of reference to which the motions of bodies must be referred. At any rate, so Newton believed, and the remarkable success of his system of mechanics persuaded several generations of physicists that he was right.

The point just made can be formulated in a more technical manner. Since this technical formulation uses a notion that plays a fundamental role in the construction of physical theories, it is desirable to present that formulation in outline. Suppose that the motion of bodies is referred to a spatial reference frame S, so that the distances of an arbitrary point-mass from three mutually perpendicular axes determined by S are x, y, and z. Then the differential equations of motion of a point-mass with mass m are $m\dfrac{d^2x}{dt^2} = F_x$, with similar equations for the other coordinates, where F_x is one component of a definite force-function. For example, if the point-mass m is in the gravitational field of a body with mass M and spatial coordinates x_1, y_1, z_1, then $F_x = \dfrac{GmM\,(x - x_1)}{r^3}$, where r^2 (the square of the distance between the two bodies) $=$ $(x - x_1)^2 + (y - y_1)^2 + (z - z_1)^2$. Now let S' be any other reference frame that is moving with respect to S in any arbitrary manner; for example, it may be rotating with respect to S or it may be moving with accelerated speed. Let x', y', z', etc., be the coordinates of the bodies referred to S'. The coordinates of S will then be related to those of S' by equations of transformation which will in general involve the time. To fix our ideas, suppose that S' is moving with respect to S with a constantly accelerated speed, so that the coordinates of the two systems are related by the equation:

$$x' = x + v_x t + \frac{a_x t^2}{2}$$

(with similar equations for the other two coordinates), where v_x is the x-component of the velocity of S' with respect to S at time $t = 0$, and a_x is the x-component of the constant acceleration of S'. A simple calculation shows that the differential equations of motion of the body referred to S' have the form:

$$m\,\frac{d^2x'}{dt^2} = \frac{GmM\,(x' - x_1')}{r'^3} = m\,\frac{d^2x}{dt^2} + a_x m$$

It is thus clear that in S' the force acting on the point-mass m differs from the force in S by a quantity proportional to the constant acceleration of S' relative to S. In brief, the equations of motion are in general *not invariant* under a transformation of coordinates from one frame of reference to another; and in particular, they are not invariant for two reference systems which are relatively accelerated. Accordingly, if S is a reference system in which, for example, the first axiom is satisfied by a certain body, that body will not satisfy the axiom if its motion is referred to S'. Thus suppose that a body, say the star Arcturus, is far removed from the influence of other bodies, so that when its motion is referred to a certain frame of reference, say the frame defined by the constellation Orion, its motion is along a straight line with constant velocity. But if Arcturus is referred to coordinate axes fixed in the earth, its motion is no longer rectilinear and uniform, but is accelerated; and by hypothesis, there is no identifiable force which accounts for its motion when it is so referred.

It was considerations of this kind, including the general *noninvariance* of the equations of motion under transformations to arbitrary reference frames, which persuaded Newton that motions must be referred to a privileged frame of reference that he called "absolute space." "Absolute space," according to him, "in its own nature and without regard to anything external, always remains similar and immovable." Absolute space is thus nonsensible and is not a material object or relation between such objects. It is an amorphous receptacle within which all physical processes occur and to which physical motions must be referred if they are to be understood in terms of the axioms of mechanics. On the other hand, Newton declared that

Relative space is some movable dimension or measure of the absolute spaces; which our senses determine by its position to bodies; and which is vulgarly taken for immovable space. . . . Absolute motion, is the translation of a body from one absolute space to another; and relative motion, the translation from one relative place into another. . . . But because the parts of space cannot be seen, or distinguished from one another by our senses, . . . instead of absolute places and motions we use relative ones; and that without any inconvenience in common affairs; but in philosophical disquisitions, we ought to abstract from our senses, and consider things themselves, distinct from what are only sensible measures of them. For it may be that there is no body really at rest, to which the places and motions of others may be referred.[1]

In effect, Newton was prepared to admit that *kinematically* all motion is relative, but he maintained that, when considered *dynamically* and in terms of the forces determining them, motions must be referred to absolute space as the frame of reference.

Newton supported his assumption of an absolute space by theological and general philosophical arguments, but he also adduced what he believed was incontrovertible experimental evidence in its favor. He recognized quite explicitly that it is impossible to ascertain by any mechanical experiments whether a body is really at rest or moving with uniform *velocity* with respect to absolute space. For the differential equations of motion *are invariant* (i.e., their form is preserved) in all reference frames that have a uniform velocity (with rest as a limiting case) relative to absolute space. In consequence, it is not possible to distinguish experimentally between absolute and relative uniform *velocity*.[2] On the other hand, Newton maintained that it *is* possible to distinguish by means of mechanical experiments between absolute and relative *acceleration*, and therefore to decide experimentally whether or not a body has an accelerated motion with respect to absolute space. The evidence he offered for this conclusion included the now-famous bucket experiment. Since Newton's interpretation of this experiment has been the focus of much subsequent criticism, we shall describe it.

A bucket filled with water is suspended from a rope, so that the rope when twisted becomes the axis of rotation for the bucket. At the outset, the water and the sides of the bucket are relatively at rest, and the surface of the water is (approximately) a plane. The bucket is then rotated. The water does not begin to rotate immediately, so that for a time the bucket has an accelerated motion with respect to the water. Nevertheless, the surface of the water during this interval remains a plane. Eventually, however, the water also acquires a rotary motion, so that finally it is at rest relative to the sides of the bucket. But now the surface of the water is concave in shape, and no longer flat. The bucket is next made to stop its rotation abruptly. However, the water does not cease to rotate immediately, and for a time has an accelerated motion relative to the sides of the bucket. Nonetheless, during this period the surface of the water continues to remain concave in shape. Finally, when the water also ceases to rotate and comes to rest relative to the bucket, its surface once more becomes a plane.

Accordingly, as Newton construed the experiment, the surface of the water can be a plane, whether it is at rest or in accelerated motion with respect to the sides of the bucket. Similarly, the surface of the water may be paraboloidal in shape, whether it is at rest or in accelerated motion relative to the bucket. He therefore concluded that the shape of the surface is independent of its state of motion relative to the bucket. On the other hand, he regarded the paraboloidal surface as a *deformation* of its normal shape, and therefore a consequence of *forces* acting on the water. According to the second axiom, however, such forces must be accompanied by *accelerated* motions. Since the state of motion of the water relative to the bucket had already been eliminated as irrelevant, Newton concluded that an acceleration relative to *absolute space* must be taken as the manifestation of the deforming forces acting on the water. In essentials Newton's argument is therefore as follows: Deformations of surfaces are evidence of impressed forces; impressed forces give rise to accelerated motions; but the deformations of surfaces are independent of the *relative* accelerations of bodies; hence the accelerations in question must be *absolute* accelerations. Since it is possible to establish by mechanical experiments whether bodies undergo deformations, it is possible to distinguish experimentally between absolute and relative accelerations, and so to identify experimentally motions that are accelerated with respect to absolute space.

Now there is something extremely puzzling about an assumption according to which it is in principle impossible to discover by mechanical means whether a body is at rest or in uniform velocity with respect to a reference frame, although it is allegedly possible to ascertain whether the body has an accelerated motion relative to that reference frame. For if a body has an acceleration with respect to a given coordinate system, it follows that the body must also have a relative velocity. If it is possible to identify the former experimentally, it seems quite mysterious why it is impossible to identify the latter. An assumption about the world, yielding a consequence that is inherently incapable of verification by experiment, appears to many minds as eminently unsatisfactory and paradoxical. Some writers have therefore concluded that the notion of absolute space is physically "meaningless." In any event, the Newtonian solution of the problem of reference frames for motions was generally regarded as something of an Achilles' heel in his system of mechanics. Although the system was accepted for more than two centuries, it was accepted primarily because a more satisfactory solution was not available.

But let us examine Newton's interpretation of the bucket experiment. Newton's argument was severely criticized by Ernst Mach, who showed that it involved a serious *non sequitur*. Newton noted quite correctly that the variations in the shape of the surface of the water are not connected with the rotation of the water relative to the sides of the *bucket*. But he concluded that the deformations of the surface must therefore be attributed to a rotation relative to *absolute space*. However, this conclusion does not follow from the experimental data and Newton's other assumptions, for there are in fact two alternative ways of interpreting those data: the change in the shape of the water's surface is a consequence either of a rotation relative to absolute space or of a rotation relative to *some system of bodies different from the bucket*. Newton adopted the first alternative, on the general assumption that inertia (i.e., the tendency of a body to continue moving uniformly along a "straight line") is an inherent property of bodies which they would continue to possess even if the remaining physical universe were annihilated.

Mach called attention to the second alternative. He argued in substance that inertial properties are contingent upon the actual distribution of bodies in the universe, so that nothing can be significantly predicated of a body's motion if the rest of the universe is assumed to vanish. He therefore maintained that it is entirely gratuitous to invoke a rotation relative to absolute space in order to account for the deformation of the water's surface, but that on the contrary it is sufficient to take a coordinate system defined by the fixed stars as the frame of reference for the rotation. Accordingly, if Mach's general approach is adopted, and if an adequate theory of mechanics can be constructed in conformity with it, it is not necessary to assume the puzzling asymmetry between absolute velocity and absolute acceleration that is so central to Newtonian theory. In terms of Mach's approach, there still may be fundamental differences between various frames of reference. Thus, the Newtonian axioms may be valid when the motions of bodies are referred to some of these reference frames, but not valid for other frames of reference. There may thus be a class of "privileged" reference frames even on Mach's view, so that motions relative to them may be called "absolute" while others are only "relative." However, absolute velocity in this sense is in principle as verifiable as is absolute acceleration.[3]

There is another way of analyzing the bucket experiment that helps make clearer just what is at stake and that throws additional light on the logical status of theories. Suppose we adopt a reference frame S, rotating relative to the earth in such a fashion that its axis of rotation is parallel to the axis of rotation of the bucket, and its constant angular velocity equal to the maximum angular velocity of the bucket. The following are then the observed data in the experiment: At the outset, the water has an accelerated rotation relative to S, with its surface a plane. Eventually, however, the water ceases to have this acceleration, its surface then being paraboloidal. Moreover, after the bucket has been abruptly stopped from rotating with respect to the earth, so that the water is finally at rest relative to the bucket, the water is once more accelerated relative to S and once more has a plane surface. Accordingly, the surface is paraboloidal only when it is at rest with respect to S, and it is a plane only when it is accelerated relative to S. The character of the water's surface is thus independent of its state of motion relative to the bucket, but not independent of its state of motion relative to S. On this analysis, therefore, a plane surface is associated with accelerated motion (relative to S), while a concave surface is connected with a state of rest (relative to S).[4]

Why not assume, in the light of this, that the "normal" surface of the water is paraboloidal and that it is the "abnormal" plane surface which is the "deformed" one? The answer is that, were this assumption adopted, we would also have to complicate in a serious way the Newtonian equations of motion. If S were generally selected as the reference frame for all motions, the angular velocity of S relative to any given system under investigation would then enter into the law about the latter. Since different systems generally possess different angular velocities relative to S, no simple formula would embrace these various special laws. The field of invariance of the differential equations of motion would in fact be extremely limited. On the Newtonian proposal of a frame of reference, or on Mach's alternative to it, the equations of motion are invariant for all so-called "Galilean frames." That is, if the equations are satisfied when motions are referred to some particular reference frame, they are satisfied in all reference frames having a constant velocity with respect to the first. On the other hand, if the equations are satisfied when motions are referred to S, they will be satisfied only in those reference frames at rest with respect to S. In short, with S as the frame of reference for all motions, the specific force-functions that would have to be supplied in order to analyze motions in terms of the Newtonian axioms would be different for nearly every special problem and would have to be invented *ad hoc* for each case.

Is not the supposition absurd, it may nevertheless be asked, that the water is in a deformed state when its surface is a plane? Do not deformations take place only when forces are acting? Is it not therefore an experimental fact that the paraboloidal surface is a consequence of such forces, and hence of the rotation of the water with respect to some frame of reference, rather than of its state of rest relative to S? Similarly, does not the rotation of the plane of Foucault's pendulum and of the axis of a gyroscope, or the flattening of the earth at its poles, or the deflection of a freely falling body from a rectilinear path to the earth's center provide *experimental* evidence that the earth must be rotating? Accordingly, is it not quite untenable to hold, as the previous paragraph suggested one might hold, that the water in the bucket and the earth itself are assumed to be "absolutely accelerated" merely because the equations of motion receive a simple, invariant form when those assumptions are made?

These queries bring us to the crux of the present discussion. The fundamental point must constantly be kept in mind that, even if the water in the bucket is declared to have an "absolute acceleration" when its surface is concave, it is not at all necessary to assume, as Newton did, that this rotation (or the rotation of the earth) takes place with respect to absolute space. Mach's critique of Newton is conclusive on this matter. The frame of reference relative to which the acceleration is said to take place can be taken to be defined by the system of the fixed stars, or by some other system of physical bodies, as it actually is in practice. The rotation of the plane of Foucault's pendulum, for example, does not establish the earth's rotation with respect to absolute space, but only with respect to the fixed stars. If the stars were concealed from us by clouds permanently surrounding the earth's surface, so that their existence would be unsuspected by us, Foucault's experiment would show only that the earth is rotating relative to the plane of the pendulum.

It is nevertheless conceivable (indeed, it happens to be the case) that, when the motions of bodies are referred to coordinate frames provided by physical bodies, the motions do not conform with complete precision to the axioms of motion. To put it differently, it is conceivable that no physical coordinate frame is a Galilean or "inertial" frame. If we decide to retain the Newtonian axioms in unmodified form, we can then introduce an "ideal frame," with respect to which the motions of bodies are in strict agreement with the axioms but to which *physical* frames of reference will at best be only good approximations. The rationale for this procedure is that, unless we adopt inertial frames for analyzing the motions of bodies in terms of the Newtonian axioms, the experimental laws of motion would undoubtedly be more complex and less manageable than if inertial frames were employed. Accordingly, the primary objective of using inertial frames, whether they are actually realized in physical systems or are only ideal constructions, is to effect a simplification in the formulation of laws. It is a happy circumstance that there are in fact physical systems which are at least approximate realizations of inertial frames. Were this not the case, the science of mechanics might perhaps have never been developed.

However, none of this can be validly construed to mean that the laws established for motions referred to inertial frames are "more real" or "more objective" than would be the less simple and noninvariant laws that might be established without introducing such frames. On the contrary, it can be shown that, if a set of relations can be affirmed to hold of a system of bodies when their motions are referred to an inertial frame, there must be definite relations between those bodies when the motions are referred to noninertial frames, even though the formulation of the latter relations may be more complex and more difficult to achieve than the formulation of the former.

For example, it is often convenient in analytical geometry to represent curves by so-called "parametric equations," in which the coordinates of points on a curve are expressed as functions of some auxiliary variable. Such parametric equations frequently make it possible to analyze the properties of a curve with much less trouble than if the curve is represented by an equation which relates the coordinates to one another directly. Nevertheless, it would be absurd to maintain that the parametric equations are "more correct" or "truer" than the equation relating the coordinates directly, or that the latter formulates the curve in a manner more "objective" (or less "objective," as the case might be) than do the parametric equations. Thus, a plane curve whose parametric equations in terms of the auxiliary variable "t" are $x = t^2 - 2t, y = t^4 + t^2 - 2t$ can also be represented by an equation which connects its coordinates directly, namely, $(y - x^2 - 9x - 8)^2 = (x + 1)(4x + 8)^2$. In many problems the former equations are much more manageable than is the latter, although the two modes of representation have the same geometric content. Analogously, the differential equations for the motion of a planet in the sun's gravitational field, when the motion is referred to the fixed stars as the coordinate frame, assume the familiar form involving the inverse square of the distance between the sun and planet. However, it is a mathematical consequence of this fact that the motion can be referred to, say, the earth as the frame of reference, so that differential equations can in principle be stated for the planet's motion when the latter is studied in this manner. These differential equations will in general be forbiddingly complex, but they will nonetheless formulate the planet's motion as objectively and completely as do the initial equations.

The introduction of inertial frames as the basis for analyzing the motions of bodies required great creative imagination, for the motions of bodies as directly observed do not exhibit patterns of changes that obviously require the use of such frames. The notion of inertia is thus not the product of "abstracting" from manifest traits of sensory experience, in the way in which the idea of a circle is commonly supposed to be such a product. On the other hand, the notion of inertia has become so completely a part of our intellectual heritage and equipment that, unless we make considerable effort to do so, it is difficult to conceive of an alternative way for interpreting the "observed facts" of motion. Moreover, the idea of inertial frames is indissolubly associated in Newtonian mechanics with the invariance of the equations of motion under transformation from one inertial frame to another. However, what is invariant is often tacitly identified with what is "objectively real," with what is permanent and not subject to spatiotemporal limitations, with what is universal.[5] Accordingly, the invariance of the equations of motion, when the motions are referred to inertial frames, gives to inertial frames a quality of importance over and above the importance they possess in making possible the analysis of mechanical phenomena in terms of a relatively simple set of force-functions. It is at least plausible that the intellectual discomfort sometimes produced by the suggestion that the water in the bucket experiment is "deformed" when the surface is flat stems in part from a disinclination to adopt frames of reference which would enormously restrict the range of invariance—and hence the "objectivity"—of the equations of motion.

It is worth recalling, finally, that the forces postulated by Newton's second axiom as determinants of accelerations cannot, in general, be measured independently of the accelerations. As was noted in the preceding chapter, the force-functions employed in Newtonian mechanics are in the main assumed hypothetically; they are explicitly characterized only by the general requirement that their magnitudes be proportional to the changes in the moment of bodies, and that they have the same direction as do these changes. Accordingly, the stimulus that usually leads to the search for forces and to the construction of force-functions is the fact that some physical system is undergoing accelerated motion. It is therefore putting the cart before the horse if we claim that we can always decide whether a body is accelerated or deformed, by ascertaining through independent experimental means what forces are impressed on it. The contrary is certainly very frequently the case. However, if we must first agree whether or not a body is accelerated or deformed before we can have grounds for believing that a force is acting on it, then at least in such cases we must first adopt a frame of reference for motions as well as a system of geometry for measuring them, before we can inquire whether a body is accelerated or deformed. Newton's procedure in assigning logical priority to the selection of a frame of reference, with respect to which motions are to be analyzed in terms of his axioms, was thus entirely cogent, however faulty may have been his arguments for absolute space.

We have now indicated at sufficient length why the question of adopting a spatial frame of reference is an important one in Newtonian mechanics, and we have also discussed the rationale for Newton's solution of the problem. We must next turn to no less important issues that arise in considering the use of geometry as a system of spatial measurement.

II. PURE AND APPLIED GEOMETRY

If we wish to determine the length of a room or the height of a moderately sized house, the usual procedure is to lay off some measuring rod (for example, a yardstick or steel tape) against the object to be measured, and so ascertain the number of times the unit length is contained in the distance under consideration. This normal method obviously assumes that the measuring rod has already been calibrated in accordance with certain rules, that the edge of the rod is straight, and that the rod is unaffected in any relevant way while it is being repeatedly moved in the process of measurement. These assumptions raise difficult questions, which we shall ignore for the present. But it is clear that this method of measuring distance is not always feasible. We usually cannot estimate the width of broad rivers in this way, nor the distances between places separated by tall mountains. We certainly cannot employ this method for measuring the distances between the stars, or the dimensions of atoms and other submicroscopic objects.

In many practical problems, and in most scientific ones, the measurement of spatial magnitudes cannot therefore be effected by such a "direct" procedure. In general, spatial measurements are made only indirectly, and require among other things the use of geometric theory. For example, if we want to determine the length of wire needed to run a line from the ridgepoles of two buildings 80 feet apart, one of which is 30 feet tall and the other 50 feet, we would most likely *calculate* the required length with the help of the Pythagorean theorem. For the length of wire needed is the hypotenuse of a right triangle whose remaining sides are 80 feet and 20 feet, respectively, so that the length in feet is equal to the square root of $80^2 + 20^2$, or $20\sqrt{17}$—approximately 83 feet.

But what justifies us in using the Pythagorean theorem in

this example? The obvious answer is that the theorem is a logical consequence of the axioms of Euclidean geometry, so that if these axioms are accepted the theorem is fully warranted. However, the question is not fully resolved by this answer. For an exactly similar question must be put concerning the axioms. The axiomatic formulation and deductive development of Euclidean geometry have the great advantage that if the question can be satisfactorily answered for the axioms it need not be faced again for any of the theorems. Nevertheless, the question must be seriously faced. What then are the grounds for accepting the axioms? In discussing such grounds we shall be compelled to examine issues that bear directly upon the logical status of theories in general, and not only upon the status of geometry.

1. Let us briefly review some of the opinions that have been held on this question. It is well known that geometry originated in the practical arts of land measurement among the ancient Egyptians. A number of useful formulas were discovered by them, which enabled their surveyors, the *harpedonaptai*, to fix definite boundaries between fields and to calculate their areas. Their formulas were simply a collection of independent rules of thumb, and the discovery that they were connected by relations of logical implication was apparently the achievement of the ancient Greeks. The Egyptian formulas were analyzed, some geometric figures were defined in terms of other figures, and additional relations between the bounding surfaces and edges of bodies were established. Moreover, after several centuries of such effort, it was shown that, if a small number of propositions about magnitudes in general, and geometric figures in particular, are accepted without proof, an indefinite number of other propositions—including those previously established—can be deduced from them. Euclid's *Elements* was thus a theoretical codification of the art of mensuration which had its roots in practices with a long prior history, and for centuries Euclid was accepted as a model of logical rigor and as the ideal form of a theoretical science.[6]

Geometry came to be employed, even before the complete emergence of modern science, as a basis not only for surveying, but also for astronomy, architecture, instrument making, and some of the engineering as well as fine arts. Newton was therefore able to regard geometry as simply a branch of a universal mechanics. As he himself put it,

> To describe right lines and circles are problems, but not geometric problems. The solution of these problems is required for mechanics; and by geometry the use of them, when so solved, is shewn; and it is the glory of geometry that from those few principles, fetched from without, it is able to produce so many things. Therefore geometry is founded in mechanical practice, and is nothing but that part of universal mechanics which accurately proposes the art of measuring. But since the manual arts are chiefly conversant in the moving of bodies, it comes to pass that geometry is commonly referred to their magnitude, and mechanics to their motion.[7]

On this view, accordingly, the axioms of geometry are true statements about certain features of physical bodies, features assumed to be specifiable in terms of definite physical procedures. Geometry is thus a hypothetico-deductive discipline which asserts that, *if* certain configurations are right lines, circles, and so on, then they must possess the properties enunciated in the various theorems.

However, two related questions are now forced on our attention, concerning which Newton had nothing explicit to say. Just what are the procedures which serve to specify, and if necessary to construct, straight lines, planes, circles, and the other figures that constitute the alleged subject matter of geometry? And in any case, on what grounds can we claim that the axioms and theorems of geometry are true of the figures that are thus identified? Newton simply referred the first question to "practical mechanics" and did not consider at all the second one. But neither question has any easy answer, and each runs into what seem like insuperable difficulties.

Straight lines can be readily constructed if we once possess a straightedge, and circles can be easily drawn if we use a pair of compasses whose points remain at a constant distance from one another. But how do we establish the "straightness" of an alleged straightedge or the constancy of the distance between compass points? On what evidence do we claim to know that the assumptions about straight lines and circles contained in the Euclidean axioms actually do hold of the figures obtained in this way? It will not do simply to say "Make measurements on these figures, and see if they conform to the Euclidean requirements." For to make measurements we must possess instruments that have straight edges and possess constant distances between their parts. We thus appear to be caught in a hopeless infinite regress. Neither does it seem satisfactory to fall back upon a direct inspection of an edge to determine whether it is straight, even if one adopts the somewhat sophisticated procedure of "sighting" along it in the manner of carpenters when they plane a piece of wood. Such direct inspection can be used only when relatively small segments of lines and surfaces are under consideration; the conclusions obtained by this procedure are not uniform for different observers or for the same observer at different times; and the procedure may even involve the same type of regress already noted. For when an edge is judged to be straight by direct inspection, what standard is employed in making the judgment? If it is some *image* of straightness, the original problem seems to present itself once more with respect to this image. On the other hand, if an edge is said to be straight on the basis of sighting along it, does not the judgment rest upon the tacit postulate that optical rays are rectilinear? An infinite regress thus appears to be unavoidable. Indeed, the regress cannot be circumvented until it is recognized, as we shall soon see, that the questions generating it are ambiguous, and that they confound issues concerning matters of *empirical fact* with issues concerning matters of *definition*.

However this may be, Newton's conception of geometry as simply a branch of an empirical science of mechanics is by no means the only view that has been taken of the subject. In classical antiquity most of the axioms were regarded as self-evident necessary truths, and the lack of "obviousness" of the parallel postulate was the chief stimulus to centuries of effort at demonstrating it from self-evident premises. Leibniz, a contemporary of Newton, explicitly maintained the Platonic doctrine that the "truths of geometry," like those of arithmetic, are certifiable as necessary without the need for an appeal to sensory experience. According to him, geometric truths are "innate, and are in us virtually, so that we can find them there if we consider attentively and set in order what we already have in the mind, without making use of any truth learned through experience or through the traditions of another."[8] Nevertheless, with some doubtful exceptions the ancients regarded geometry as dealing with the spatial properties of material bodies, even if Plato and his followers maintained that those properties are only imperfect actualizations of the eternal objects of geometrical inquiry. Just when the view was first advanced that geometry is the science of the structure of *space* (or "pure extension"), rather than of the spatial properties of material bodies, is an unsettled historical question. But by the time of Newton this view was already influential. This conception received a forthright statement from Euler, in the eighteenth century, who declared:

> Extension is the proper object of geometry, which considers bodies only insofar as they are extended, abstractedly from impenetrability and inertia; the object of geometry, therefore, is a notion much more general than that of body, as it comprehends, not only bodies, but all things simply extended, without impenetrability, if any such there be. Hence it follows that all the properties deduced in geometry from the notion of extension must likewise take place in bodies, inasmuch as they are extended.[9]

The conception of geometry as an a priori science of the structure of space was given a different turn by Kant, in his attempt to find a *via media* between the aprioristic rationalism of Leibniz and the sensationalistic empiricism of Hume. Although there is room for some doubt concerning the interpretation of many details in Kant's doctrine, its general import is that Euclidean geometry formulates the structure of the form of our external intuition. Accordingly, the axioms of Euclid and their consequences are apodictic truths concerning the spatial form of all possible experience. Kant's views on the nature of geometry have been highly influential, not only with professional philosophers but with mathematicians and physicists as well. Although important currents of philosophic thought in the nineteenth century rejected the Kantian conception and argued for an empirical interpretation of the status of geometry, Kant's influence did not diminish until subsequent developments in logic, mathematics, and physics made his views progressively more untenable. For the view of geometry as a system of a priori knowledge concerning space had the incomparable advantage over its competitors that it appeared to explain, as alternatives to it did not, why Euclid was the only known system of geometry and why mechanics (at that time still the most perfectly developed branch of theoretical physics) was so inextricably dependent on that system.

2. However, before turning to these later developments and their consequences for a philosophy of geometry, we must make explicit a distinction that has already been briefly noted and is of paramount importance in what follows. In geometry, as in every deductive argument and every deductively formulated discipline, two questions must be sharply distinguished. The first is: Do the alleged theorems of the system follow logically from the axioms? To answer it and to discover new theorems implied by the axioms are among the prime concerns of mathematicians. To resolve it, no laboratory experiments or other empirical studies need be undertaken; the only equipment required is the technique of logical demonstration. The second question is: Are any of the axioms or theorems factually or materially true? This question does not fall within the jurisdiction of the mathematician *qua* mathematician; and answers to the first question can be sought irrespective of the answers that may be given to the second. Answers to the second question can in general be supplied only by the physicist or other empirical scientist, provided that the axioms and theorems refer to identifiable empirical subject matter. This proviso is crucial, and we must therefore discuss it at some length.

It has been a matter of common knowledge since Aristotle that the validity of a syllogistic demonstration does not depend on the special meanings of the terms occurring in its premises and conclusions. Accordingly, if a syllogistic argument is valid, it remains valid when the original terms are replaced by others. In evaluating the validity of a syllogism, it is therefore permissible to ignore completely the meanings of the specific subject-matter terms and to consider only the formal structure of the constituent statements. Formal structure can be considered most simply and effectively by replacing the specific subject-matter terms by variables. The resulting expressions will then contain only such words or symbols as signify logical relations or operations. Thus, when such replacements are made in the statement "All men are mortal," the resulting expression is "All A's are B's," in which the words "all" and "are" retain their customary meanings while no specific meanings are associated with the variables "A" and "B."[10] However, the expression "All A's are B's" is clearly no longer a statement concerning which it would be significant to ask whether it is true or false. The expression has only the *form* of a statement, one which becomes a statement when words having definite meanings are substituted for the variables. We shall call such expressions "statement-forms." A statement-form may be defined for present purposes as an expression containing one or more variables, such that if subject-matter terms are substituted for the variables the resulting expression is a statement, that is, an expression concerning which it is significant to raise questions of truth or falsity. Accordingly, to evaluate the validity of a syllogism it is sufficient to consider the statement-forms of which its premises and conclusion are instances. It is thus clear that, when we are concerned with the question whether the conclusion of a syllogism follows logically from the premises, it is irrelevant to ask whether these statements are true or false.

What has just been said about the syllogism obviously applies to any deductive argument. In particular, when Euclidean geometry is discussed as a demonstrative discipline, we can ignore the meanings of the specific geometrical terms in the axioms and theorems of the system, replace those terms with variables, and pursue the task of proving theorems by attending only to the logical relations between the resulting statement-forms. However, although this point is elementary, it appears not to have occurred to any of the ancient mathematicians and philosophers, despite the fact that they were quite familiar with it in connection with syllogistic arguments. But in any event, it is of greatest importance to distinguish between geometry as a discipline whose sole aim is to discover what is logically implied by the axioms or postulates, and geometry as a discipline which seeks to make materially true assertions about a specific empirical subject matter. In the former case, mathematicians explore logical relations between statements only insofar as the latter are instances of statement-forms, so that the meanings of specific subject-matter terms are in principle irrelevant. In the latter case, the nonlogical terms occurring in the axioms and theorems must be associated with definite elements in some subject matter, so that the truth or falsity of various statements belonging to the system may be suitably investigated. Geometry when studied in the first sense as simply a deductive system is often called "pure geometry"; when studied in the second sense as a system of factual truth, it is commonly called "applied" or "physical geometry."

Let us illustrate the main burden of this discussion by considering a formulation of Euclidean geometry that meets modern standards of logical rigor, for example, Oswald Veblen's axiomatization.[11] Veblen assumes a class of objects called "points," a triadic relation between points called the relation of "lying between" and a binary relation between pairs of points called "congruence." He then imposes upon these objects and relations a number of carefully formulated conditions, in the form of sixteen assumptions or axioms. He also defines in terms of the initial (or primitive) subject-matter expressions a number of other expressions, such as "line," "plane," "angle," "circle," employing in this process ideas that belong to general logic (such as that of set or class). These defined expressions are introduced primarily for the sake of convenience and can be eliminated in favor of the primitive terms. The defined expressions may therefore be ignored in what follows. Let us now conjoin these sixteen axioms, so that they are components in a single but very complicated statement. The axioms can then be represented by the abbreviation A(point, between, congruent). On the other hand, let us represent any statement that can be formulated in terms of the primitive expressions of the system by T(point, between, congruent), although in general the primitive terms will not all occur in every such statement. The aim of demonstrative or pure geometry may then be said to be that of finding statements "T" such that "T(point, between, congruent)" is a logical consequence of "A(point, between, congruent)."

However, the deducibility of "T" from "A" cannot depend on any special meanings associated with the expressions "point," "between," and "congruent." These terms may therefore be replaced by variables with which no meanings of any sort need be associated. Accordingly, the conjoined postulates of pure geometry in the Veblen axiomatization could in principle be stipulated as the statement-form "$A(R_1, R_3, R_2)$" where "R_1" is a predicate variable (or unary relation variable), "R_3" a triadic relation variable, and "R_2" a binary relation variable. The task of the pure geometer is then to ascertain which statement-forms "$T(R_1, R_3, R_2)$" are logical consequences of the statement-form "$A(R_1, R_3, R_2)$."

On the other hand, neither the pure geometer nor the physicist can investigate the truth or falsity of the statement-forms "A" and "T" for the patent reason that, since they are not statements, it is not even significant to ask whether they are true or false. Moreover—and this is the chief burden of the present discussion—it may be equally impossible to inquire into the truth or falsity of the Veblenian axioms, even if the latter are formulated in terms of the familiar expressions "point," "between," and "congruent" rather than variables, unless these familiar expressions are associated with definite, empirically identifiable physical objects or relations between such objects. Indeed, mathematicians often employ those familiar expressions, without thereby assigning to them any specific meanings involving such reference to empirical subject matter. Thus, although Veblen uses those expressions in his formulation of the geometric axioms, he is careful to note that the reader may associate with them "any meaning" or "any image" he pleases, as long as these meanings and images are consistent with the conditions imposed upon the use of the expressions by the axioms. The point of the proviso mentioned on page 196, in connection with answers to the question whether the axioms of geometry are factually true, is therefore this: The material truth or falsity of geometric axioms and theorems can be investigated only if, for the non-logical terms occurring in the axioms and theorems, rules of correspondence or coordinating definitions are supplied which associate with those terms empirically identifiable elements in some subject matter.

3. In the light of this distinction between pure and applied geometry, let us now reconsider some of the views previously mentioned concerning the logical status of geometry.

a. The claim that the propositions of geometry are a priori, logically necessary truths is ambiguous, and can be construed in at least three senses. It can be understood to mean (1) that the statements of *pure geometry* are a priori and logically necessary, where a statement of pure geometry is of the form: If $A(R_1, R_3, R_2)$, then $T(R_1, R_3, R_2)$; or (2) that each of the postulates and theorems of *pure geometry* has this character; or finally (3) that the statements of *applied geometry*, whether axioms or theorems, are a priori and logically necessary.

On the first interpretation, the claim is obviously correct.

But it is also trivial, since whenever a conclusion demonstrably follows from a premise, the conditional statement whose antecedent clause is the premise and whose consequent clause is that conclusion will always be a logically necessary truth. On the other hand, the claim is absurd when it is understood in the second sense. For if the postulates and theorems of pure geometry are taken to be statement-forms, they cannot be regarded as either true or false, and a fortiori neither as necessarily true nor as necessarily false.

There remains for consideration only the third way of construing the claim. The issue then resolves itself into the question whether the Veblenian postulates are necessary truths for *every* interpretation of the primitive terms or only for *some*, and, if the latter, what the character of such interpretations is. The point of the question will be clearer if we first compare two different statement-forms: the statement-form "If no S is P, then no P is S" with the statement-form "No S is P." It will be evident that, no matter what subject-matter terms are substituted for the variables "S" and "P" in the first, the resulting statement will invariably be a logically necessary truth, for example, the statement "If no triangles are equilateral figures, then no equilateral figures are triangles," even though the antecedent clause of this conditional statement happens to be false. On the other hand, the second statement-form will yield a necessary truth for some substitutions of subject-matter terms for the variables, but not for others, for example, the statement "No triangles are circles" is a necessary truth, while "No triangles whose vertices are any three fixed stars are figures with an area less than two square miles" is not. Similarly, an inspection of Veblen's postulates (or of any other postulates for Euclidean geometry) shows that none of them formulates a necessary truth under every interpretation of the primitive terms. For example, Veblen's second axiom postulates that for any three points x, y, z, if y lies between x and z, then z does not lie between x and y. If we now replace the term "point" with the term "number," and the relational expression "y lies between x and z" with the relational expression "y is greater than the difference of x and z," we obtain the statement "For any three numbers x, y, and z, if y is greater than the difference of x and z, then z is not greater than the difference of x and y," which is clearly false (since, for example, although 4 is greater than the difference of 7 and 5, nevertheless 5 is greater than the difference of 7 and 4), and hence not a necessary truth. Accordingly, if the axioms are necessary truths, they are such only under certain interpretations of their primitives but not under others.

Let us, therefore, examine some proposed interpretations of the geometric axioms, and first the one contained in Euclid's *Elements*. Euclid prefaced the formal development of his system with a large number of "definitions." Some of these are definitions of terms like "triangle" and "circle" on the basis of what are obviously the primitive terms of the system, such as "point" and "line"; the other definitions are explications of these primitives. In effect, however, these explications are proposed *interpretations* of the primitives, and are presumably intended to instruct us as to the objects or relations which are designated by the primitives. For example, a point is said to be "that which has no part," a line is declared to be a "breadthless length," and a straight line is explained as "a line which lies evenly with the points on itself." These explanations undoubtedly suggest in a vague way the sorts of things to which the various terms are to be applied. Nevertheless, they are hardly explicit enough to permit us to identify without serious question what things are designated by the corresponding terms. What is it, for example, that has no parts? It can be no ordinary material object, though it may possibly be a corner of solids having sharp edges, or perhaps even an experienced pain of short duration. Moreover, even if we assume that we know just what things are to be counted as "breadthless lengths," when does such a thing lie evenly with the points on itself? It seems therefore profitless to ask whether on Euclid's own interpretation of his axioms they are true.

It might be objected, however, that this is all useless hairsplitting, since we know very well what is meant by "point" and "straight line." Points and straight lines, so it might be said, are of course not material things; they are, however, *limits* of physical objects that can nevertheless be conceived and entertained in imagination. Moreover, we can perform experiments in imagination upon points, lines, and other geometrical objects; and when we do so, we find that we cannot form our images except in conformity with the Euclidean axioms. It has been maintained, for example, that the statement "Two straight lines cannot intersect in more than one point" cannot be established by perceptual observation, but only by exercising our imagination. As one writer has put the argument,

For in the first place it is only through imagery that we can represent a line starting from a certain point and extending indefinitely in a certain direction; and in the second place, we cannot represent in perception the infinite number of different inclinations or angles that a revolving straight line may make with a given straight line. We may, however, by a rapid act of ocular movement represent a line revolving through 360° from any one direction to which it returns. In this imaginative representation the entire range of variation, covering an infinite number of values, can be exhaustively visualized because of the continuity that characterizes the movement. It is only if such a process of imagery is possible that we can say that the axiom in its universality presents to us a self-evident truth.[12]

Two comments must be made on this general position. In the first place, if geometric objects are taken to be merely conceptual or imaginary ones, the fundamental problem under discussion has not even been broached. For that problem concerns the manner in which the conceptual structure of pure geometry may come to be used in physics and the various practical arts. Nothing is contributed to the solution of this problem either by repeating that points and lines are concepts or by identifying them with images. Just what is the relevance of lines entertained in imagination for astronomy or for the construction of precision instruments, both of which must make extensive use of geometry?

In the second place, the argument from the alleged facts of mental experiments has no force whatsoever. When we perform experiments in imagination upon straight lines, in what manner are these lines envisaged? We cannot employ any arbitrary images of lines in the experiment. We must *construct* our images in a certain manner. However, if we examine the mode of their construction in those cases in which we allegedly intuit the imagined figures as Euclidean, we soon notice that the Euclidean assumptions are tacitly being used as the *rules of construction*. For example, we can certainly imagine two distinct lines with two points in common. But such lines do not count as straight lines, simply because they do not satisfy the Euclidean requirements of straightness, so that we seek to form our images so as to satisfy those requirements. Or to change the illustration, it is possible to "prove" that all triangles are isosceles—a result known to be incompatible with Euclidean postulates—with the help of suitably drawn diagrams. However, the alleged demonstration is a spurious one because (as we usually say) the diagrams have not been drawn "correctly"—where the standards of correctness are supplied by Euclidean geometry itself. Accordingly, if the Euclidean postulates serve as the rules for constructing our mental experiments, it is not at all surprising that the experiments invariably conform to the rules. In short, if Euclidean axioms are used as implicit definitions, they are indeed a priori and necessary because they then specify what sorts of things are to be counted as their own instances.

b. The conception of geometry as a branch of experimental science seems highly plausible, if only because of the origins of geometry in the practical arts of measurement. This plausibility is not diminished by the difficulties we have now canvassed in the view that geometry is a body of a priori knowledge about the structure of space. For measurements can be conducted only with material instruments and not with parts of space. No account of applied geometry is therefore adequate which creates a puzzle out of the fact that geometry functions as a theory of mensuration. On the other hand, as has already been noted, the Newtonian view of geometry as the simplest branch of mechanics appears to have difficulties of its own: and we must now try to decide whether these difficulties are quite as insuperable as they seem.

It will be useful to distinguish between two general ways of employing geometry within experimental science. (i) The first and historically earlier approach consists in specifying *independently of Euclidean geometry* certain edges, surfaces, and other configurations of material bodies, and then showing that as a matter of observed fact the things so specified conform to the Euclidean axioms within the limits of experimental error. (ii) The second approach consists in using the Euclidean postulates as implicit definitions, so that no physical configurations (whether discovered or deliberately contrived) are *called* "points," "lines," and so on, unless they satisfy the postulates within certain limits of approximation. Both approaches face similar logical and empirical problems, but each approach introduces a distinctive emphasis into the discussion, and each assigns a different status to Euclidean geometry.

(i) Let us consider the first approach more closely. Euclidean geometry and theoretical physics are certainly not more than 3000 years old. There surely was a time, therefore, when men engaged in various practical affairs did not have available the knowledge contained in these systems. Let us imagine ourselves placed into situations these men faced. Although we would have no inkling of geometry, we could nevertheless distinguish between different forms of surfaces— initially perhaps only by unaided sight or touch, but eventually perhaps by more reliable procedures. For example, some surfaces are noticeably rounded in one or more directions, others are less so, and still others appear to be quite flat. However, these discriminations are somewhat rough, and there may be no complete agreement among us as to which surfaces are the flattest. Moreover, as long as appropriate technologies are lacking, it is only by chance that we would run across such flat surfaces.

But suppose that mechanical skills become developed, and that we learn how to grind or cut bodies so that the surface of one body can be made to fit snugly upon the surface of another body. It may finally occur to us to take three bodies, and grind their surfaces until any two of the three will fit snugly on each other. This procedure provides what seems like a good objective criterion for surfaces having maximum flatness, and in any case we decide to call surfaces satisfying it "plane surfaces." It clearly would make no sense to ask whether such surfaces "really" are planes, for they are planes by *definition*, and by hypothesis there is no other standard for being a plane surface than the one stated. It is also worth noting that in judging whether two surfaces do fit snugly upon one another we may use some optical test, for example, the test that no light shines through when the surfaces are in close fit. Nevertheless, though we may employ such an optical test, we would *not* be assuming, tacitly or otherwise, that the propagation of light is "rectilinear," so that our procedure is not in fact circular. We would simply be employing a type of observable fact as a condition for *saying* that the surfaces fit snugly. It is essential to note, therefore, that thus far the *only* issue of fact at stake when a surface is declared to be a plane is whether the surface satisfies the indicated condition of fitting closely on other surfaces. In particular, it should be observed that no assumptions associated with Euclidean geometry are involved in assigning the label "plane" to such surfaces.

We may now proceed in a similar manner to construct types of edges which we decide to call "straightedges" or "straight lines," for example, by grinding two plane surfaces on a body so that they have a common edge. Furthermore, with the help of planes and straight lines we may construct other figures for which labels like "point," "triangle," "quadrilateral," and the like are introduced. Also, two straightedges may be defined as being of equal length if they can be made to coincide end to end, and a unit length may be specified by selecting some particular straightedge for this purpose.[13]

It is now possible to construct additive scales of length, angle, area, and volume. But except for one point, the details of the constructions will be omitted. In specifying a scale of length, as well as in making measurements on the basis of such a scale, it will in general be necessary to transport the unit length repeatedly. The question may therefore arise whether in the course of its motion it might not undergo a change in length. "How do we know," it might be asked, "that when a straightedge is moved from one place to another its length remains the same? How do we know that if two straightedges are equally long in one place, and one of the edges is carried to another place, the two edges continue to have the same length?"

These questions are worth considering here because they typify a frequent confusion between what are issues of fact and what are matters of definition. It *is* a question of empirical fact whether, if two straightedges are equally long at one place (i.e., if they can be made to coincide end to end) and are then transported along the same or different paths to some other place, they are equally long at the new place. Let us assume that this indeed in general is the case. On the other hand, it is not a question of empirical fact whether, if two straightedges are equally long at one place and one of them is moved to some other place, the two edges continue to be equally long. In terms of the procedure we have adopted this question can be answered only by making a decision and introducing a *definition*. In particular, it is not a question of *knowing* (that is, having observational evidence which would enable us to establish) whether or not the standard unit length alters its length when transported from place to place; this is a matter that, within the framework of assumptions we have adopted, can be settled only by a *stipulation*. It is therefore essential to distinguish the issue whether two edges that are equally long at one place continue to be equally long when both are transported along the same or different routes to another place, from the issue whether two straightedges equally long at one place continue to be equally long when only one of them is transported to another place or whether the length of the standard unit is invariant under motion. The first issue can be decided by an appeal to observation, and thus involves questions of *knowledge*; the second issue cannot be settled in this way, and involves questions of *definition*.

"But is it not the case," so our imaginary critic might reply, "that we frequently do ascribe a change in the length of a body after it has been transported, and that we often take precautions against such changes? Indeed, we even ascribe such changes when they remain at the same place, and try to prevent alterations in length (as in the case of the standard meter or yardstick) by keeping the bodies in carefully controlled environments." The answer to this query is obviously affirmative. However, this answer is predicated on rejecting the simplifying factual assumption made in the preceding paragraph, according to which two straightedges equally long at one place (as judged by the coincidence of their respective end-points) will continue to be equally long at any other place, no matter by what paths they are transported from one locality to another. Let us therefore drop this assumption, and thereby complicate the discussion.

We must now suppose that we have learned to distinguish between various kinds of bodies, for example, between different sorts of wood, metals, and stones. We shall also assume that we know how to identify various physical sources of change in the shapes and relative sizes of bodies, such sources as compressions or variations in temperature. To fix our ideas, suppose that at time t_1 and place P_1 two straightedges a and b are equally long, where a is made of maple wood and b of copper. Assume further that at a later time t_2 the straightedge b is longer than a, but that in the interim there has been a rise in the temperature of both bodies. Let us also assume that after much experience we have come to recognize that when different substances are exposed to an identical temperate change, their relative lengths are altered, and in fact by unequal amounts for different pairs of substances. Accordingly, on the hypothesis that the sole identifiable source of change has been an increase in temperature at P_1, we attribute the alteration in relative lengths of a and b to the rise in their temperature. We are not saying, it should be noted, that the length of a has remained constant and that the length of only b has increased; we are saying only that b has become longer *relative* to a.

Suppose next that, although a and b remain equally long when they are at P_1, they are unequally long when they are transported to P_2, whether by the same or different paths. This change in relative length may again be accounted for in terms of variations in temperature that one or both of the bodies have undergone. We have used temperature changes as the source of alterations in the relative lengths of straightedges, but what has been said about temperature can obviously be repeated for other sources of change that can be experimentally identified. In any event, however, the earlier assumption that two straightedges equally long at one place remain equally long when transported to another place must now be amended. In its amended form the factual assumption contains the proviso that, when straightedges are transported from one place to another, all the known sources of changes in relative lengths are kept constant, so that features of the environment, known experimentally to be relevant to alterations in relative lengths and shapes of bodies, are the same in the initial positions of the straightedges as in their final positions. Within the framework of this amended assumption, it then makes sense to say that, on being transported from P_1 to P_2 the length of a body changes (*relative* to some specially designated body) or that two bodies equally long at P_1 cease to be so when one of them but not the other is moved to P_2.

One further point in this amended account of problems of spatial measurement requires brief attention. For the objection might be made that the discussion is based on a circular and self-defeating procedure. We have been outlining a way of instituting a scale of lengths, allegedly without employing any assumptions of Euclidean geometry; and we have indicated the need for stipulating the conditions under which two straightedges are said to be of equal length. We have, however, assumed that it is possible to detect whether or not any changes occur in these conditions, for example, whether or not the temperatures of two straightedges are the same and remain constant. Must we not therefore have thermometers in our possession, and must we not in consequence have scales of length, *before* we can detect such changes or constancies? Does not the proposed construction of a scale of length assume that the end-product of the construction is available *antecedently* to the construction? And if so, is not the procedure patently circular?

Despite appearances to the contrary, no such circle need occur. For in point of fact, it is possible to determine whether there are changes in the temperature of bodies (and more generally, whether there are changes in any of the physical conditions upon which variations in relative lengths of bodies depend), without using instruments, such as the thermometer, which employ a previously established scale of lengths. For example, at a primitive level of investigation we might rely entirely on the sensitivity of our own bodies to changes in temperature within certain ranges. At a more advanced stage of knowledge we might use as a detector of temperature changes the unequal expansions or contractions of two straight rods made of different substances. It is essential to observe that in this case we would use, not a quantitative measure of linear expansion or contraction (for this would indeed involve us in a circular argument), but only the qualitative fact that two such rods initially of equal length become unequally long in variable temperature fields. At a still more sophisticated level of knowledge, we could recognize changes in temperature by using the fact that, when two different metals form a closed circuit, a magnetized needle near the circuit will be deflected when the temperature at the juncture of the metals is altered. The construction and use of such complicated detectors involves details into which we cannot enter. However, even the schematized account we have given of them suffices to indicate that an additive scale of lengths can be constructed without circularity and without employing some antecedent theory of geometry.

Once an additive scale of lengths has been instituted, and with these various difficulties out of the way, we can then construct certain figures that will be called "circles," and with the help of these figures a scale of angular measure. We have therefore outlined how, in principle, a class of figures and certain measures for them can be specified without using any assumptions of Euclidean geometry. The remaining problem is whether these figures (and others that can be constructed in an analogous manner) satisfy the axioms and theorems of

Euclidean geometry; or, conversely, whether Euclidean geometry, when its terms "point," "line," and so on are interpreted as referring to the similarly named constructed figures, is true of them. This problem, however, is a straightforward empirical one, and there is no way of knowing the answer to it before making actual empirical inquiry. Moreover, it is clear that the evidence obtainable from such inquiry will at best show only an approximate agreement between Euclidean statements and the constructed figures. For in the first place, uncontrollable disturbing factors cannot always be eliminated in conducting measurements, so that random or experimental "errors" are likely to occur. In the second place, measuring instruments are capable only of limited discriminations. For example, at a given stage of technological development we are unable to distinguish between lengths falling below a certain minimum extension. On the other hand, Euclidean geometry postulates an unlimited discriminability of lengths when it asserts that certain lengths have relative magnitudes which can be expressed only by irrational numbers. Accordingly, no overt measurement can determine whether, as geometric theory requires, the magnitude of certain lengths is indeed irrational. And finally, Euclidean statements sometimes make assertions that cannot possibly be shown to hold of actual figures by direct measurement. For example, the statement that if the alternate interior angles formed by a transversal to two lines in a plane are equal the lines *never* intersect, is such a statement. For every plane that we may construct is of finite extent, and we cannot therefore determine by observation or overt measurement whether two lines do not intersect no matter how far they are produced. Nevertheless, within regions accessible to experiment, and subject to the qualifications mentioned, the agreement between figures constructed in the manner outlined and the statements of applied Euclidean geometry is in point of fact excellent. In consequence, and until fairly recently, the theory of mechanics as well as other branches of physics has been based squarely on the assumption that Euclidean geometry is true of a class of physical configurations constructed in a way more or less analogous to the one we have sketched. Moreover, although a different system of geometry is employed in Einsteinian relativity theory, the engineering arts as well as the manufacture of laboratory instruments will undoubtedly continue to make that assumption during the foreseeable future.

(ii) We have now completed our discussion of the first approach in geometry mentioned on page 203. We must now examine the second alternative, according to which Euclidean postulates are used as implicit definitions for certain figures that constitute the domain of application of those postulates. Our examination will be relatively brief, since most of the relevant problems have already been discussed.

The essential difference between these alternative approaches is that, while on the first of them, expressions like "point," "line," and so on are applied to physical configurations that are constructed or identified in accordance with rules specifiable *independently* of the Euclidean axioms, on the second approach those expressions are applied only to such configurations that satisfy the Euclidean requirements. On the former approach, accordingly, we are in principle committed to abandoning Euclidean geometry if actual observations and measurements on independently specified lines, angles, circles, and so on reveal a significant discrepancy between the properties of these figures and what Euclidean geometry leads us to expect. On the latter approach, on the other hand, we are in principle committed to retaining Euclidean geometry at all costs, and to alter our methods for constructing figures should those methods not yield configurations in conformity with Euclid. On the first alternative, Euclidean geometry is a system of contingent, a posteriori statements concerning spatial properties of bodies antecedently classified and named. On the second alternative, Euclidean geometry is a system of a priori rules for classifying and naming such properties.

Let us indicate in outline how Euclidean geometry can be used in this latter way. If we accept the Euclidean postulates as implicit definitions, we must find or construct figures that will satisfy the conditions stated by the postulates. Suppose, then, we begin by constructing surfaces, edges, and so on in the manner proposed in connection with the above discussion of the first approach. However, we are not yet entitled to *call* those configurations "planes," "straight lines," and the like, and must first make observations and measurements on them. We may find that the outcome of such an inquiry shows those figures to possess properties in good agreement with what Euclidean geometry requires of planes, straight lines, etc. In that eventuality, we are entitled to the hypothesis that those figures *are* planes, straight lines, and the rest. On the other hand, suppose that the outcome of the inquiry shows those figures to possess traits that deviate considerably from Euclidean requirements. For example, suppose that the angle sum of certain three-sided figures differs from two right angles (as defined by a stated scale of angular magnitude) by more than 10°, a difference far greater than a possible experimental error. In this eventuality, the constructed figures would not receive the familiar geometrical names, and in particular the three-sided figure would not be called a "triangle." On the contrary, we would modify our rules for constructing figures and for measuring their spatial magnitudes until we obtained configurations that were at least approximately Euclidean.

It may turn out, however, to be extremely difficult to construct Euclidean figures, and that no matter how much we alter the rules for manufacturing the desired kinds of surfaces and straightedges we rarely if ever succeed in obtaining anything even approximating Euclidean planes and straight lines. Such a situation would still not "disprove" Euclidean geometry, though it might make the retention of Euclidean geometry as a theory of mensuration highly inconvenient. We could, of course, put up with the inconvenience, and resign ourselves to the fact that calculations about spatial dimensions on the basis of that theory are rarely if ever in agreement with the results of overt measurement. Two other alternatives would nevertheless be open to us. We might succeed in developing *physical theories* based on Euclidean geometry, so that our persistent failure to construct (or find) Euclidean configurations would be systematically explained by those theories, while at the same time the spatial magnitudes of bodies as determined by actual measurement would be in good agreement with the numerical values calculated from those theories. Or alternately, we might abandon Euclidean geometry as the a priori system of rules for classifying and naming spatial configurations, and devise some other system of pure geometry with this end in view.

The discussion thus indicates that the apparently incompatible conceptions of Euclidean geometry as an empirical science on the one hand, and of Euclidean geometry as a system of a priori rules on the other hand, can both be accepted as legitimate. Geometry is a branch of empirical science when planes, straight lines, etc., are constructed or identified as features of physical bodies in accordance with rules that can be formulated and applied without reference to Euclidean geometry. Euclidean geometry is a system of a priori rules when the construction or identification of configurations that are to bear Euclidean labels is guided and controlled by the Euclidean postulates. Nevertheless, on each approach both empirical as well as a priori assumptions play a role. On the first alternative the rules for constructing the figures designated as "planes," "straight lines," etc., are a priori, and the Euclidean statements are empirical. On the second alternative the Euclidean postulates are a priori, and the assertions that certain figures (constructed or identified in accordance with specified rules) are planes, straight lines, etc., are empirical. In brief, the difference between the two alternatives is a difference in the locus at which conventions or definitions are introduced into a body of knowledge.

NOTES

[1] Isaac Newton, *Mathematical Principles of Natural Philosophy* (ed. by Florian Cajori), Berkeley, Calif., 1947, Book 1, Scholium. [See the selection in this volume.—ED.]

[2] This follows directly from what has already been said. If in the above discussion S is the frame of reference supplied by absolute space, and S' is any reference frame moving with uniform velocity relative to S, then the equations of transformation from S to S' are: $x' = x + v_x t + x_0$, where x_0 is the component along the x-axis of the distance between the origins of the two systems at time $t = 0$, with similar equations for the other coordinates. But under these transformations the differential equations of motion *are* invariant, so that it is impossible to determine whether a body is at rest or in uniform motion with respect to S. The fact that the equations of motion are invariant in all frames of reference moving with uniform velocity relative to one another is commonly called the "Newtonian principle of relativity."

[3] Cf. Ernst Mach, *Science of Mechanics*, La Salle, Ill., 1942, Chap. 2, Sec. 4, pp. 271-298. Frames of reference belonging to the privileged class are commonly called "inertial" or "Galilean" frames of reference. As is well known, Mach's critique of Newton profoundly influenced Einstein and prepared the way for the latter's general theory of relativity. [See pp. 126-131 of this volume.—ED.]

[4] For this way of analyzing the experiment, see Peter G. Bergmann, *Introduction to the Theory of Relativity*, New York, 1942, p. xiv. For a similar analysis, but one employed as an argument for the earth's absolute motion, cf. J. C. Maxwell, *Matter and Motion*, Art. 105, pp. 84-86.

⁵ Cf. for example, Spinoza's *Ethics,* Part 2, Prop. 38: "Those things, which are common to all, and which are equally in a part and in the whole, cannot be conceived except adequately."

⁶ It is now well known that Euclid's *Elements* do not conform to modern standards of logical rigor, for many of his theorems cannot in fact be deduced from his axioms, and additional ones must be supplied.

⁷ Newton, *op. cit.* In a passage preceding this quotation, Newton asserted that "Geometry does not teach us to draw [right lines and circles, upon which mechanics is founded], but requires them to be drawn; for it requires that the learner should first be taught to describe these accurately, before he enters upon geometry."

⁸ G. W. Leibniz, *New Essays Concerning Human Understanding* (transl. by A. G. Langley), Chicago, 1916, p. 78.

⁹ L. Euler, *Letters to a German Princess* (transl. by Brewster), Vol. 2, p. 31.

¹⁰ It is indeed possible to continue the process of abstracting from meanings still further, and to replace words like "all," "are," and other logical particles by signs which are governed by stated rules of operation. But it is not germane to the present discussion to pursue this possibility, although some of the outstanding achievements of recent logical studies are a consequence of developing this suggestion.

¹¹ See his essay, "The Foundations of Geometry," in *Monographs on Topics of Modern Mathematics* (ed. by J. W. A. Young), New York, 1911.

¹² W. E. Johnson, *Logic,* Vol. 2, London, 1922, p. 202.

¹³ This method of defining planes and straight lines is developed by W. K. Clifford, *The Common Sense of the Exact Sciences,* New York, 1946, Chap. 2; and also by N. R. Campbell, *Measurement and Calculation,* London, 1928, pp. 271-278.

NON-EUCLIDEAN
SPACES

HANS REICHENBACH

Reprinted from sections 1–2 of Chapter 1 of *The Philosophy of Space and Time*, translated by Maria Reichenbach and John Freund, with introductory remarks by Rudolf Carnap, Dover Publications, Inc., New York, 1958.

1. THE AXIOM OF THE PARALLELS AND NON-EUCLIDEAN GEOMETRY

In Euclid's work, the geometrical achievements of the ancients reached their final form: geometry was established as a closed and complete system. The basis of the system was given by the geometrical axioms,[1] from which all theorems were derived. The great practical significance of this construction consisted in the fact that it endowed geometry with a certainty never previously attained by any other science. The small number of axioms forming the foundation of the system were so self-evident that their truth was accepted without reservation. The entire construction of geometry was carried through by a skillful combination of the axioms alone, without any addition of further assumptions; the reliability of the logical inferences used in the proofs was so great that the derived theorems, which were sometimes quite involved, could be regarded as certain as the axioms. Geometry thus became the prototype of a demonstrable science, the first instance of a scientific rigor which, since that time, has been the ideal of every science. In particular, the philosophers of all ages have regarded it as their highest aim to prove their conclusions "by the geometrical method."

Euclid's axiomatic construction was also important in another respect. The problem of demonstrability of a science was solved by Euclid in so far as he had reduced the science to a system of axioms. But now arose the epistemological

214

question how to justify the truth of those first assumptions. If the certainty of the axioms was transferred to the derived theorems by means of the system of logical concatenations, the problem of the truth of this involved construction was transferred, conversely, to the axioms. It is precisely the assertion of the truth of the axioms which epitomizes the problem of scientific knowledge, once the connection between axioms and theorems has been carried through. In other words: the *implicational* character of mathematical demonstability was recognized, i.e., the undeniable fact that only the implication "if *a*, then *b*" is accessible to logical proof. The problem of the categorical assertion "*a* is true *b* is true," which is no longer tied to the "if," calls for an independent solution. The truth of the axioms, in fact, represents the intrinsic problem of every science. The axiomatic method has not been able to establish knowledge with absolute certainty; it could only reduce the question of such knowledge to a precise thesis and thus present it for philosophical discussion.

This effect of the axiomatic construction, however, was not recognized until long after Euclid's time. Precise epistemological formulations could not be expected from a naïve epoch, in which philosophy was not yet based upon well-developed special sciences, and thinkers concerned themselves with cruder things than the truth of simple and apparently self-evident axioms. Unless one was a skeptic, one was content with the fact that certain assumptions had to be believed axiomatically; analytical philosophy has learned mainly through Kant's critical philosophy to discover genuine problems in questions previously utilized only by skeptics in order to deny the possibility of knowledge. These questions became the central problems of epistemology. For two thousand years the criticism of the axiomatic construction has remained within the frame of mathematical questions, the elaboration of which, however, led to peculiar discoveries, and eventually called for a return to philosophical investigations.

The mathematical question concerned the reducibility of the axiomatic system, i.e., the problem whether Euclid's axioms represented ultimate propositions or whether there was a possibility of reducing them to still simpler and more self-evident statements. Since the individual axioms were quite different in character with respect to their immediacy, the question arose whether some of the more complicated axioms might be conceived as consequences of the simpler ones, i.e., whether they could be included among the theorems. In particular, the demonstrability of the axiom of the parallels was investigated. This axiom states that through a given point there is *one and only one* parallel to a given straight line (which does not go through the given point), i.e., one straight line which lies in the same plane with the first one and does not intersect it. At first glance this axiom appears to be self-evident. There is, however, something unsatisfactory about it, because it contains a statement about infinity; the assertion that the two lines do not intersect within a finite distance transcends all possible experience. The demonstrability of this axiom would have enhanced the certainty of geometry to a great extent, and the history of mathematics tells us that excellent mathematicians from Proclus to Gauss have tried in vain to solve the problem.

A new turn was given to the question through the discovery that it was possible to do without the axiom of parallels altogether. Instead of proving its truth the opposite method was employed: it was demonstrated that this axiom could be dispensed with. Although the existence of several parallels to a given line through one point contradicts the human power of visualization, this assumption could be introduced as an axiom, and a consistent geometry could be developed in combination with Euclid's other axioms. This discovery was made almost simultaneously in the twenties of the last century by the Hungarian, Bolyai, and the Russian, Lobatschewsky; Gauss is said to have conceived the idea somewhat earlier without publishing it.

But what can we make of a geometry that assumes the opposite of the axiom of the parallels? In order to understand the possibility of a non-Euclidean geometry, it must be remembered that the axiomatic construction furnishes the proof of a statement in terms of logical derivations from the axioms alone. The drawing of a figure is only a means to assist visualization, but is never used as a factor in the proof; we know that a proof is also possible by the help of "badly-drawn" figures in which so-called congruent triangles have sides obviously different in length. It is not the immediate picture of the figure, but a concatenation of logical relations that compels us to accept the proof. This consideration holds equally well for non-Euclidean geometry; although the drawing looks like a "badly-drawn" figure, we can with its help discover whether the logical requirements have been satisfied, just as we can do in Euclidean geometry. This is why non-Euclidean geometry has been developed from its inception in an axiomatic construction; in contradistinction to Euclidean geometry where the theorems were known first and the axiomatic foundation was developed later, the axiomatic construction was the instrument of discovery in non-Euclidean geometry.

With this consideration, which was meant only to make non-Euclidean geometry plausible, we touch upon the problem of the *visualization of geometry*. Since this question will be treated at greater length in a later section,[2] the remark about "badly-drawn" figures should be taken as a passing comment. What was intended was to stress the fact that the essence of a geometrical proof is contained in the logic of its derivations, not in the proportions of the figures. Non-Euclidean geometry is a logically constructible system—this was the first and most important result established by its inventors.

It is true that a strict proof was still missing. No contradictions were encountered—yet did this mean that none would be encountered in the future? This question constitutes the fundamental problem concerning an axiomatically constructed logical system. It is to be expected that non-Euclidean statements directly contradict those of Euclidean geometry; one must not be surprised if, for instance, the sum of the angles of a triangle is found to be smaller than two right angles. This contradiction follows necessarily from the reformulation of the axiom of the parallels. What is to be required is that the new geometrical system be self-consistent. The possibility can be imagined that a statement a, proved within the non-Euclidean axiomatic system, is not tenable in a later development, i.e., that the statement *not-a* as well as the statement a is provable in the axiomatic system. It was incumbent upon the early adherents of non-Euclidean geometry, therefore, to prove that such a contradiction could never happen.

The proof was furnished to a certain extent by Klein's Euclidean model of non-Euclidean geometry. Klein succeeded in coordinating the concepts of Euclidean geometry, its points, straight lines, and planes, its concept of congruence, etc., to the corresponding concepts of non-Euclidean geometry, so that every statement of one geometry corresponds to a statement of the other. If in non-Euclidean geometry a statement a and also a statement *not-a* could be proved, the same would hold for the coordinated statements a' and *not-a'* of Euclidean geometry; a contradiction in non-Euclidean geometry would entail a corresponding contradiction in Euclidean geometry. The result was a proof of consistency, the first in the history of mathematics: it proceeds by reducing a new system of statements to an earlier one, the consistency of which is regarded as virtually certain.[3]

After these investigations by Klein the mathematical significance of non-Euclidean geometry was recognized.[4] Compared with the natural geometry of Euclid, that of Bolyai and Lobatschewsky appeared strange and artificial; but its mathematical legitimacy was beyond question. It turned out later that another kind of non-Euclidean geometry was possible. The axiom of the parallels in Euclidean geometry asserts that to a given straight line through a given point there exists exactly one parallel; apart from the device used by Bolyai and Lobatschewsky to deny this axiom by assuming the existence of several parallels, there was a third possibility, that of denying the existence of any parallel. However, in order to carry through this assumption consistently,[5] a certain change in a number of Euclid's other axioms referring to the infinity of a straight line was required. By the help of these changes it became possible to carry through this new type of non-Euclidean geometry.

As a result of these developments there exists not one geometry but a plurality of geometries. With this mathematical discovery, the epistemological problem of the axioms was given a new solution. If mathematics is not required to use certain systems of axioms, but is in a position to employ the axiom *not-a* as well as the axiom *a*, then the assertion *a* does not belong in mathematics, and mathematics is solely the science of implication, i.e., of relations of the form "if . . . then"; consequently, for geometry as a mathematical science, there is no problem concerning the truth of the axioms. This apparently unsolvable problem turns out to be a pseudo-problem. The axioms are not true or false, but arbitrary statements. It was soon discovered that the other axioms could be treated in the same way as the axiom of the parallels. "Non-Archimedian," "non-Pascalian," etc., geometries were constructed. . . .

These considerations leave us with the problem into which discipline the question of the truth of the assertion *a* should be incorporated. Nobody can deny that we regard this statement as meaningful; common sense is convinced that real space, the space in which we live and move around, corresponds to the axioms of Euclid and that with respect to this space *a* is true, while *not-a* is false. The discussion of this statement leads away from mathematics; as a question about a property of the physical world, it is a *physical* question, not a *mathematical* one. This distinction, which grew out of the discovery of non-Euclidean geometry, has a fundamental significance: it divides the problem of space into two parts; the problem of mathematical space is recognized as different from the problem of physical space.

It will be readily understood that the philosophical insight into the twofold nature of space became possible only after mathematics had made the step from Euclid's geometry to non-Euclidean geometries. Up to that time physics had assumed the axioms of geometry as the self-evident basis of its description of nature. If several kinds of geometries were regarded as mathematically equivalent, the question arose which of these geometries was applicable to physical reality; there is no necessity to single out Euclidean geometry for this purpose. Mathematics shows a variety of possible forms of relations among which physics selects the real one by means of observations and experiments. Mathematics, for instance, teaches how the planets would move if the force of attraction of the sun should decrease with the second or third or *n*th power of the distance; physics decides that the second power holds in the real world. With respect to geometry there had been a difference; only *one* kind of geometry had been developed and the problem of choice among geometries had not existed. After the discoveries of non-Euclidean geometries the duality of *physical* and *possible* space was recognized. Mathematics reveals the possible spaces; physics decides which among them corresponds to physical space. In contrast to all earlier conceptions, in particular to the philosophy of Kant, it becomes now a task of physics to determine the geometry of physical space, just as physics determines the shape of the earth or the motions of the planets, by means of observations and experiments.

But what methods should physics employ in order to come to a decision? The answer to this question will at the same time supply an answer to the question why we are justified in speaking of a specific physical space. Before this problem can be investigated more closely, another aspect of geometry will have to be discussed. For physics the analytic treatment of geometry became even more fruitful than the axiomatic one.

2. RIEMANNIAN GEOMETRY

Riemann's extension of the concept of space did not start from the axiom of the parallels, but centered around the concept of metric.

Riemann developed further a discovery by Gauss according to which the shape of a curved surface can be characterized by the geometry within the surface. Let us illustrate Gauss' idea as follows. We usually characterize the curvature of the surface of a sphere by its deviation from the plane; if we hold a plane against the sphere it touches only at one point; at all other points the distances between plane and sphere become larger and larger. This description characterizes the curvature of the surface of the sphere "from the outside"; the distances between the plane and the surface of the sphere lie outside the surface and the decision about the curvature has to make use of the third dimension, which alone establishes the difference between curved and straight. Is it possible to determine the curvature of the surface of the sphere without taking outside measurements? Is it meaningful to distinguish the curved surface from the plane within two dimensions? Gauss showed that such a distinction is indeed possible. If we were to pursue "practical geometry" on the sphere, by surveying, for instance, with small measuring rods, we should find out very soon that we were living on a curved surface. For the ratio of circumference u and diameter d of a circle

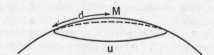

Circumference and diameter of a circle on the
surface of a sphere

we would obtain a number smaller than $\pi = 3.14 \ldots$ as is shown in the diagram. Since we stay on the surface all the time, we would not measure the "real diameter" which cuts through the inner part of the sphere, but the "curved diameter" which lies on the surface of the sphere and is longer. This diameter divided into the circumference results in a number smaller than π. Nevertheless, it is meaningful to call the point M "the center of the circle on the surface of the sphere" because it has the same distance from every point of the circle; that we find ourselves on a sphere is noticed by means of the deviation of the ratio from π. In this way we obtain a *geometry of a spherical surface* which is distinguished from the ordinary geometry by the fact that different metrical relations hold for this kind of geometry. In addition to the change in the ratio between circumference and diameter of a circle, an especially important feature is that the sum of the angles of a triangle on a sphere is greater than $180°$.

It is remarkable that this generalization of plane geometry to surface geometry is identical with that generalization of geometry which originated from the analysis of the axiom of the parallels. The leading role which has been ascribed to the axiom of the parallels in the course of the development of geometrical axiomatics cannot be justified from a purely axiomatic point of view; the construction of non-Euclidean geometries could have been based equally well upon the elimination of other axioms. It was perhaps due to an intuitive feeling for theoretical fruitfulness that the criticism always centered around the axiom of the parallels. For in this way the axiomatic basis was created for that extension of geometry in which the metric appears as an independent variable. Once the significance of the metric as the characteristic feature of the plane has been recognized from the viewpoint of Gauss' plane theory, it is easy to point out, conversely, its connection with the axiom of the parallels. The property of the straight line of being the shortest connection between two points can be transferred to curved surfaces, and leads to the concept of *straightest line*; on the surface of the sphere the great circles play the role of the shortest line of connection, and on this surface their significance is analogous to that of the straight lines on the plane. Yet while the great circles as "straight lines" share their most important property with those of the plane, they are distinct from the latter with respect to the axiom of the parallels: all great circles of the sphere intersect and therefore there are no parallels among these "straight lines." Here we encounter the second possibility of a denial (cf. § 1) of the axiom of the parallels which excludes the existence of parallels. If this idea is carried through, and all axioms are formulated on the understanding that by "straight lines" are meant the great circles of the sphere and by "plane" is meant the surface of the sphere, it turns out that this system of elements satisfies a system of axioms within two dimensions which is nearly identical in all of its statements with the axiomatic system of Euclidean geometry; *the only exception is the formulation of the axiom of the parallels.*[6] The geometry of the spherical surface can be viewed as the realization of a two-dimensional non-Euclidean geometry: *the denial of the axiom of the parallels singles out that generalization of geometry which occurs in the transition from the plane to the curved surface.*

Once this result has been recognized for two-dimensional structures, a new kind of insight is gained into the corresponding problem of several dimensions by means of a combination of the two different points of departure. The axiomatic development of non-Euclidean geometry had already been achieved for three-dimensional structures and therefore constituted an extension of three-dimensional space analogous to the relation of the plane to the curved surface. Although Euclidean space contains curved surfaces, it does not embody the degree of logical generalization that characterizes the surfaces; it can realize only the Euclidean axiom of the parallels, not the axioms contradicting the latter. This fact suggests a concept of space which contains the plane Euclidean space as a special case, but includes all non-Euclidean spaces too. Such a concept of space in three dimensions is analogous to the concept of surface in two dimensions; it has the same relation to Euclidean space as a surface has to the plane.

On the basis of these ideas Riemann could give so generalized a definition to the concept of space that it includes not only Euclidean space but also Lobatschewsky's space as special cases. According to Riemann, space is merely a three-dimensional manifold; the question is left open which axiomatic systems will hold for it. Riemann showed that it is not necessary to develop an axiomatic system in order to find the different types of space; it is more convenient to use an analytic procedure analogous to the method developed by Gauss for the theory of surfaces. The geometry of space is established in terms of six functions, the *metrical coefficients of the line element*, which must be given as a function of the coordinates; the manipulation of these functions replaces geometrical considerations, and all properties of geometry can be expressed analytically. This procedure can be likened to the method in elementary analytic geometry which establishes an equivalence between a formula with two or three variables and a curve or a surface. The imagination is thus given conceptual support that carries it to new discoveries. In analogy to the auxiliary concept of the curvature of a surface, which is measured by the reciprocal product of the main radii of curvature, Riemann introduced the auxiliary concept of *curvature of space*, which is a much more complicated mathematical structure. Euclidean space, then, has a curvature of degree zero in analogy to the plane, which is a surface of zero curvature. Euclidean space occupies the middle ground between the spaces of positive and negative curvatures: it can be shown that this classification corresponds to the three possible forms of the axiom of the parallels. In the space of positive curvature *no* parallel to a given straight line exists; in the space of zero curvature *one* parallel exists; in the space of negative curvature *more than one* parallel exists. In general, the curvature of space may vary from point to point in a manner similar to the point to point variation in the curvature of a surface; but the spaces of *constant curvature* have a special significance. The space of constant negative curvature is that of Bolyai-Lobatschewsky; the space of constant zero curvature is the Euclidean space; the space of constant positive curvature is called spherical, because it is the three-dimensional analogue to the surface of the sphere. The analytical method of Riemann has led to the discovery of more types of space than the synthetic method of Bolyai and Lobatschewsky, which led only to certain spaces of constant curvature. Modern mathematics treats all these types of space on equal terms and develops and manipulates their properties as easily as those of Euclidean geometry.

NOTES

[1] Euclid distinguished between axioms, postulates and definitions. We may be allowed for our present purpose to include all these concepts under the name of axioms.

[2] Not reprinted in this volume.—ED.

[3] Hilbert later proved the consistency of Euclidean geometry by a reduction to arithmetic. The consistency of arithmetic, which can

no longer be proved by reduction, needs a separate proof; this most important problem, which has found an elaborate treatment by Hilbert and his school, is still under discussion.

⁴ Klein did not start his investigations with the avowed purpose of establishing a proof of consistency; the proof came about inadvertently, so to speak, as a result of the construction of the model carried out with purely mathematical intentions. L. Bieberbach has shown recently that the recognition of the significance of non-Euclidean geometry was the result of long years of struggle. *Berl. Akademieber.* 1925, phys.-math. Klasse, p. 381. See Bonola-Liebmann, *Nichteuklidische Geometrie,* Leipzig, 1921, and Engel-Stäckel, *Theorie der Parallellinien von Euklid bis Gauss,* Leipzig, 1895, for the earlier history of the axiom of the parallels.

⁵ The axiom of the parallels is independent of the other axioms of Euclid only in so far as it asserts the existence of at most one parallel; that there exists at least one parallel can be demonstrated in terms of the other axioms. This fact is stated with masterful precision in Euclid's work.

⁶ It is evident, in considering the spherical surface, that two great circles will intersect in two points; hence, the denial of the axiom that two straight lines can intersect in only one point is involved. For if all of the axioms of Euclidean geometry except the parallel axiom are unchanged it is possible to prove there is at least one parallel. In the treatment of the spherical surface, however, we have seen that this theorem does not hold. This theorem depends upon the axiom that straight lines intersect in only one point; hence its denial removes the inconsistency.

GEOMETRY AND
DEFINITIONS OF CONGRUENCE

HANS REICHENBACH

This selection is taken from Reichenbach's essay "The Philosophical Significance of Relativity," as it appears in *Albert Einstein: Philosopher-Scientist*, edited by P. A. Schilpp, Library of Living Philosophers, 2nd ed., Open Court Publishing Company, La Salle, Ill., 1951.

Consider, for instance, the problem of geometry. That the unit of measurement is a matter of definition is a familiar fact; everybody knows that it does not make any difference whether we measure distances in feet or meters or light-years. However, that the comparison of distances is also a matter of definition is known only to the expert of relativity. This result can also be formulated as the definitional character of congruence. That a certain distance is congruent to another distance situated at a different place can never be proved to be true; it can only be maintained in the sense of a definition. More precisely speaking, it can be maintained as true only after a definition of congruence is given; it therefore depends on an original comparison of distances which is a matter of definition. A comparison of distances by means of the transport of solid bodies is but one definition of congruence. Another definition would result if we regarded a rod, once it had been transported to another location, as twice as long, thrice transported as three times as long, and so on. A further illustration refers to time: that the simultaneity of events occurring at distant places is a matter of definition was not known before Einstein based his special theory of relativity on this logical discovery.

The definitions employed for the construction of space and time are of a particular kind: they are co-ordinative definitions. That is, they are given by the co-ordination of a physical object, or process, to some fundamental concept. For instance, the concept "equal length" is defined by reference to a physical object, a solid rod, whose transport lays down equal distances. The concept "simultaneous" is defined by the use of light-rays which move over equal distances. The definitions of the theory of relativity are all of this type; they are coordinative definitions.

In the expositions of the theory of relativity the use of different definitions is often illustrated by a reference to different observers. This kind of presentation has led to the erroneous conception that the relativity of space-time measurements is connected with the subjectivity of the observer, that the privacy of the world of sense perception is the origin of the relativity maintained by Einstein. Such Protagorean interpretation of Einstein's relativity is utterly mistaken. The definitional character of simultaneity, for instance, has nothing to do with the perspective variations resulting for observers located in different frames of reference. That we co-ordinate different definitions of simultaneity to different observers merely serves as a simplification of the presentation of logical relationships. We could as well interchange the co-ordination and let the observer located in the "moving" system employ the time definition of the observer located in the system "at rest," and vice versa; or we could even let both employ the same time definition, for instance that of the system "at rest." Such variations would lead to different transformations; for instance, the last mentioned definition would lead, not to the Lorentz transformation, but to the classical transformation from a system at rest to a moving system. It is convenient to identify one definitional system with one observer; to speak of different observers is merely a mode of speech expressing the plurality of definitional systems. In a logical exposition of the theory of relativity the observer can be completely eliminated.

Definitions are arbitrary; and it is a consequence of the definitional character of fundamental concepts that with the change of the definitions various descriptional systems arise. But these systems are equivalent to each other, and it is possible to go from each system to another one by a suitable transformation. Thus the definitional character of fundamental concepts leads to a plurality of equivalent descriptions. A familiar illustration is given by the various descriptions of motion resulting when the system regarded as being at rest is varied. Another illustration is presented by the various geometries resulting, for the same physical space, through changes in the definition of congruence. All these descriptions represent different languages saying the same thing; equivalent descriptions, therefore, express the same physical content. The theory of equivalent descriptions is also applicable to other fields of physics; but the domain of space and time has become the model case of this theory.

The word "relativity" should be interpreted as meaning "relative to a certain definitional system." That relativity implies plurality follows because the variation of definitions leads to the plurality of equivalent descriptions. But we see that the plurality implied is not a plurality of different views, or of systems of contradictory content; it is merely a plurality of equivalent languages and thus of forms of expression which do not contradict each other but have the same content. Relativity does not mean an abandonment of truth; it only means that truth can be formulated in various ways.

I should like to make this point quite clear. The two statements "the room is 21 feet long" and "the room is 7 yards long" are quivalent descriptions; they state the same fact. That the simple truth they express can be formulated in these two ways does not eliminate the concept of truth; it merely illustrates the fact that the number characterizing a length is relative to the unit of measurement. All relatives of Einstein's theory are of this type. For instance, the Lorentz transformation connects different descriptions of space-time relations which are equivalent in the same sense as the statements about a length of 21 feet and a length of 7 yards.

Some confusion has arisen from considerations referring to the property of simplicity. One descriptional system can be simpler than another; but that fact does not make it "truer" than the other. The decimal system is simpler than the yard-foot-inch system; but an architect's plan drawn in feet and inches is as true a description of a house as a plan drawn in the decimal system. A simplicity of this kind, for which I have used the name of *descriptive simplicity*, is not a criterion of truth. Only within the frame of inductive considerations can simplicity be a criterion of truth; for instance, the simplest curve between observational data plotted in a diagram is regarded as "truer," i.e., more probable, than other connecting curves. This *inductive simplicity*, however, refers to nonequivalent descriptions and does not play a part in the theory of relativity, in which only equivalent descriptions are compared. The simplicity of descriptions used in Einstein's theory is therefore always a descriptive simplicity. For instance, the fact that non-Euclidean geometry often supplies a simpler description of physical space than does Euclidean geometry does not make the non-Euclidean description "truer."

Another confusion must be ascribed to the theory of conventionalism, which goes back to Poincaré. According to this theory, geometry is a matter of convention, and no empirical meaning can be assigned to a statement about the geometry of physical space. Now it is true that physical space can be described by both a Euclidean and a non-Euclidean geometry; but it is an erroneous interpretation of this relativity of geometry to call a statement about the geometrical structure of physical space meaningless. The choice of a geometry is arbitrary only so long as no definition of congruence is specified. Once this definition is set up, it becomes an empirical question *which* geometry holds for a physical space. For instance, it is an empirical fact that, when we use solid bodies for the definition of congruence, our physical space is practically Euclidean within terrestrial dimensions. If, in a different part of the universe, the same definition of congruence were to lead to a non-Euclidean geometry, that part of universal space would have a geometrical structure different from that of our world. It is true that a Euclidean geometry could also be introduced for that part of the universe; but then the definition of congruence would no longer be given by solid bodies.[1] The combination of a statement about a geometry with a statement of the co-ordinative definition of congruence employed is subject to empirical test and thus expresses a property of the physical world. The conventionalist overlooks the fact that only the incomplete statement of a geometry, in which a reference to the definition of congruence is omitted, is arbitrary; if the statement is made complete by the addition of a reference to the definition of congruence, it becomes empirically verifiable and thus has physical content.

Instead of speaking of conventionalism, therefore, we should speak of the relativity of geometry. Geometry is relative in precisely the same sense as other relative concepts. We might call it a convention to say that Chicago is to the left of New York; but we should not forget that this conventional statement can be made objectively true as soon as the point of reference is included in the statement. It is not a convention but a physical fact that Chicago is to the left of New York, seen, for instance, from Washington, D.C. The relativity of simple concepts, such as left and right, is well known. That the fundamental concepts of space and time are of the same type is the essence of the theory of relativity.

The relativity of geometry is a consequence of the fact that different geometries can be represented on one another by a one-to-one correspondence. For certain geometrical systems, however, the representation will not be continuous throughout, and there will result singularities in individual points or lines. For instance, a sphere cannot be projected on a plane without a singularity in at least one point; in the usual projections, the North Pole of the sphere corresponds to the infinity of the plane. This peculiarity involves certain limitations for the relativity of geometry. Assume that in one geometrical description, say, by a spherical space, we have a normal causality for all physical occurrences; then a transformation to certain other geometries, including the Euclidean geometry, leads to violations of the principle of causality, to *causal anomalies*. A light signal going from a point A by way of the North Pole to a point B in a finite time will be so represented within a Euclidean interpretation of this space, that it moves from A in one direction towards infinity and returns from the other side towards B, thus passing through an infinite distance in a finite time. Still more complicated causal anomalies result for other transformations.[2] If the principle of normal causality, i.e., a continuous spreading from cause to effect in a finite time, or *action by contact*, is set up as a necessary prerequisite of the description of nature, certain worlds cannot be interpreted by certain geometries. It may well happen that the geometry thus excluded is the Euclidean one; if Einstein's hypothesis of a closed universe is correct, a Euclidean description of the universe would be excluded for all adherents of a normal causality.

It is this fact which I regard as the strongest refutation of the Kantian conception of space. The relativity of geometry has been used by Neo-Kantians as a back door through which the apriorism of Euclidean geometry was introduced into Einstein's theory: if it is always possible to select a Euclidean geometry for the description of the universe, then the Kantian insists that it be this description which should be used, because Euclidean geometry, for a Kantian, is the only one that can be visualized. We see that this rule may lead to violations of the principle of causality; and since causality, for a Kantian, is as much an *a priori* principle as Euclidean geometry, his rule may compel the Kantian to jump from the frying pan into the fire. There is no defense of Kantianism, if the statement of the geometry of the physical world is worded in a complete form, including all its physical implications; because in this form the statement is empirically verifiable and depends for its truth on the nature of the physical world.[3]

It should be clear from this analysis that the plurality of equivalent description does not rule out the possibility of true empirical statements. The empirical content of statements about space and time is only stated in a more complicated way.

NOTES

[1] Poincaré believed that the definition of a solid body could not be given without reference to a geometry. That this conception is mistaken is shown in the present author's [Reichenbach's] *The Philosophy of Space and Time* (New York, 1958), § 5.

[2] Cf. *ibid.*, § 12. It has turned out that within the plurality of descriptions applicable to quantum mechanics the problem of causal anomalies plays an even more important part, since we have there a case where no description exists which avoids causal anomalies. (Cf. also the author's *Philosophic Foundations of Quantum Mechanics* [Berkeley, 1944], §§ 5-7, § 26.)

[3] This refutation of Kantianism was presented in the author's *Relativitätstheorie und Erkenntnis a priori* (Berlin, 1920).

GEOMETRY AS A
BRANCH OF PHYSICS

H. P. ROBERTSON

Reprinted from *Albert Einstein: Philosopher-Scientist*, edited by P. A. Schilpp, Library of Living Philosophers, 2nd ed., Open Court Publishing Company, La Salle, Ill., 1951.

Is space really curved? That is a question which, in one form or another, is raised again and again by philosophers, scientists, T. C. Mits and readers of the weekly comic supplements. A question which has been brought into the limelight above all by the genial work of Albert Einstein, and kept there by the unceasing efforts of astronomers to wrest the answer from a curiously reluctant Nature.

But what is the meaning of the question? What, indeed, is the meaning of each word in it? Properly to formulate and adequately to answer the question would require a critical excursus through philosophy and mathematics into physics and astronomy, which is beyond the scope of the present modest attempt. Here we shall be content to examine the roles of deduction and observation in the problem of physical space, to exhibit certain high points in the history of the problem and in the end to illustrate the view point adopted by presenting a relatively simple caricature of Einstein's general theory of relativity. It is hoped that this, certainly incomplete and possibly naïve, description will present the essentials of the problem from a neutral mathematico-physical viewpoint in a form suitable for incorporation into any otherwise tenable philosophical position. Here, for example, we shall not touch directly upon the important problem of form versus substance—but if one wishes to interpret the geometrical substratum here considered as a formal backdrop against which the contingent relations of nature are exhibited, one should be able to do so without distorting the scientific content.

First, then, we consider geometry as a deductive science, a branch of mathematics in which a body of theories is built up by logical processes from a postulated set of axioms (not

231

"self-evident truths"). In logical position geometry differs not in kind from any other mathematical discipline—say the theory of numbers or the calculus of variations. As mathematics, it is not the science of measurement, despite the implications of its name—even though it did, in keeping with the name, originate in the codification of rules for land surveying. The principal criterion of its validity as a mathematical discipline is whether the axioms as written down are self-consistent, and the sole criterion of the truth of a theorem involving its concepts is whether the theorem can be deduced from the axioms. This truth is clearly relative to the axioms; the theorem that the sum of the three interior angles of a triangle is equal to two right angles, true in Euclidean geometry, is false in any of the geometries obtained on replacing the parallel postulate by one of its contraries. In the present sense it suffices for us that geometry is a body of theorems, involving among others the concepts of point, angle and a unique numerical relation called distance between pairs of points, deduced from a set of self-consistent axioms.

What, then, distinguishes Euclidean geometry as a mathematical system from those logically consistent systems, involving the same category of concepts, which result from the denial of one or more of its traditional axioms? This distinction cannot consist in its "truth" in the sense of observed fact in physical science; its truth, or applicability, or still better appropriateness, in this latter sense is dependent upon observation, and not upon deduction alone. The characteristics of Euclidean geometry, as mathematics, are therefore to be sought in its internal properties, and not in its relation to the empirical.

First, Euclidean geometry is a *congruence geometry*, or equivalently the space comprising its elements is *homogeneous and isotropic*; the intrinsic relations between points and other elements of a configuration are unaffected by the position or orientation of the configuration. As an example, in Euclidean geometry all intrinsic properties of a triangle—its angles, area, etc.,—are uniquely determined by the lengths of its three sides; two triangles whose three sides are respectively equal are "congruent"; either can by a "motion" of the space into itself be brought into complete coincidence with the other, whatever its original position and orientation may be. These motions of Euclidean space are the familiar translations and rotations, use of which is made in proving many of the theorems of Euclid. That the existence of these motions (the axiom of "free mobility") is a desideratum, if not indeed a necessity, for a geometry applicable to physical space, has been forcibly argued on *a priori* grounds by Von Helmholtz, Whitehead, Russell and others; for only in a homogeneous and isotropic space can the traditional concept of a rigid body be maintained.[1]

But the Euclidean geometry is only one of several congruence geometries; there are in addition the "hyperbolic" geometry of Bolyai and Lobatchewsky, and the "spherical" and "elliptic" geometries of Riemann and Klein. Each of these geometries is characterized by a real number K, which for the Euclidean geometry is zero, for the hyperbolic negative, and for the spherical and elliptic geometries positive. In the case of 2-dimensional congruence spaces, which *may* (but need not) be conceived as surfaces embedded in a 3-dimensional Euclidean space, the constant K may be interpreted as the *curvature* of the surface into the third dimension—whence it derives its name. This name and this representation are for our purposes at least psychologically unfortunate, for we propose ultimately to deal exclusively with properties intrinsic to the space under consideration—properties which in the later physical applications can be measured within the space itself—and are not dependent upon some extrinsic construction, such as its relation to an hypothesized higher dimensional embedding space. We must accordingly seek some determination of K—which we nevertheless continue to call curvature—in terms of such inner properties.

In order to break into such an intrinsic characterization of curvature, we first relapse into a rather naïve consideration of measurements which may be made on the surface of the earth, conceived as a sphere of radius R. This surface is an example of a 2-dimensional congruence space of positive curvature $K = 1/R^2$ on agreeing that the abstract geometrical concept "distance" r between any two of its points (not the extremities of a diameter) shall correspond to the lesser of the two distances *measured on the surface* between them along the unique great circle which joins the two points.[2] Consider now a "small circle" of radius r (measured on the surface!) about a point P of the surface; its perimeter L and area A (again measured on the surface!) are clearly less than the corresponding measures $2\pi r$ and πr^2 of the perimeter and area of a circle of radius r in the Euclidean plane. An elementary calculation shows that for sufficiently small r (i.e., small compared with R) these quantities on the sphere are given approximately by:

(1)
$$L = 2\pi r \ (1 - Kr^2/6 + \ldots),$$
$$A = \pi r^2 \ (1 - Kr^2/12 + \ldots).$$

Thus, the ratio of the area of a small circle of radius 400 miles on the surface of the earth to that of a circle of radius 40 miles is found to be only 99.92, instead of 100.00 as in the plane.

Another consequence of possible interest for astronomical applications is that in spherical geometry the sum σ of the three angles of a triangle (whose sides are arcs of great circles) is *greater* than 2 right angles; it can in fact be shown that this "spherical excess" is given by

(2) $$\sigma - \pi = K\delta,$$

where δ is the area of the spherical triangle and the angles are measured in radians (in which $180° = \pi$). Further, each full line (great circle) is of finite length $2\pi R$, and any two full lines meet in two points—there are no parallels!

In the above paragraph we have, with forewarning, slipped into a non-intrinsic quasi-physical standpoint in order to present the formulae (1) and (2) in a more or less intuitive way. But the essential point is that these formulae are in fact independent of this mode of presentation; they are relations between the mathematical concepts distance, angle, perimeter and area which follow as logical consequences from the axioms of this particular kind of non-Euclidean geometry. And since they involve the space-constant K, this "curvature" may in principle at least be determined *by measurements made on the surface*, without recourse to its embedment in a higher dimensional space.

Further, these formulae may be shown to be valid for a circle or triangle in the hyperbolic plane, a 2-dimensional congruence space for which $K < 0$. Accordingly here the perimeter and area of a circle are *greater*, and the sum of the three angles of a triangle *less*, than the corresponding quantities in the Euclidean plane. It may also be shown that each full line is of infinite length, that through a given point outside a given line an infinity of full lines may be drawn which do not meet the given line (the two lines bounding the family are said to be "parallel" to the given line), and that two full lines which meet do so in but one point.

The value of the intrinsic approach is especially apparent in considering 3-dimensional congruence spaces, where our physical intuition is of little use in conceiving them as "curved" in some higher-dimensional space. The intrinsic geometry of such a space of curvature K provides formulae for the surface area S and the volume V of a "small sphere" or radius r, whose leading terms are

$$S = 4\pi r^2 (1 - Kr^2/3 + \ldots),$$

(3)

$$V = 4/3\pi r^3 (1 - Kr^2/5 + \ldots).$$

It is to be noted that in all these congruence geometries, except the Euclidean, there is at hand a natural unit of length $R = 1/|K|^{1/2}$; this length we shall, without prejudice, call the "radius of curvature" of the space.

So much for the congruence geometries. If we give up the axiom of free mobility we may still deal with the geometry of spaces which have only limited or no motions into themselves.[3] Every smooth surface in 3-dimensional Euclidean space has such a 2-dimensional geometry; a surface of revolution has a 1-parameter family of motions into itself (rotations about its axis of symmetry), but not enough to satisfy the axiom of free mobility. Each such surface has at a point $P(x, y)$ of it an intrinsic "total curvature" $K(x, y)$, which will in general vary from point to point; knowledge of the curvature at all points essentially determines all intrinsic properties of the surface.[4] The determination of $K(x, y)$ by measurements on the surface is again made possible by the fact that the perimeter L and area A of a closed curve, every point of which is at a given (sufficiently small) distance r from $P(x, y)$, are given by the formulae (1), where K is no longer necessarily constant from point to point. Any such variety for which $K = 0$ throughout is a ("developable") surface which may, on ignoring its macroscopic properties, be rolled out without tearing or stretching onto the Euclidean plane.

From this we may go on to the contemplation of 3- or higher-dimensional ("Riemannian") spaces, whose intrinsic properties vary from point to point. But these properties are no longer describable in terms of a single quantity, for the "curvature" now acquires at each point a directional character which requires in 3-space 6 components (and in 4-space 20) for its specification. We content ourselves here to call attention to a single combination of the 6, which we call the "mean curvature" of the space at the point $P(x, y, z)$, and which we again denote by K—or more fully by $K(x, y, z)$; it is in a sense the mean of the curvatures of various surfaces passing through P, and reduces to the previously contemplated space-constant K when the space in question is a congruence space.[5] This concept is useful in physical applications, for the surface area S and the volume V of a sphere of radius r about the point $P(x, y, z)$ as center are again given by formulae (3), where now K is to be interpreted as the mean curvature $K(x, y, z)$ of the space at the point P. In four and higher dimensions similar concepts may be introduced and similar formulae developed, but for them we have no need here.

We have now to turn our attention to the world of physical objects about us, and to indicate how an ordered description of it is to be obtained in accordance with accepted, preferably philosophically neutral, scientific method. These objects, which exist for us in virtue of some pre-scientific concretion of our sense-data, are positioned in an extended manifold which we call physical space. The mind of the individual, retracing at an immensely accelerated pace the path taken by the race, bestirs itself to an analysis of the interplay between object and extension. There develops a notion of the permanence of the object and of the ordering and the change in time— another form of extension, through which object and subject appear to be racing together—of its extensive relationships. The study of the ordering of actual and potential relationships, the physical problem of space and time, leads to the consideration of geometry and kinematics as a branch of physical science. To certain aspects of this problem we now turn our attention.

We consider first that proposed solution of the problem of space which is based upon the postulate that space is an *a priori* form of the understanding. Its geometry must then be a congruence geometry, independent of the physical content of space; and since for Kant, the propounder of this view, there existed but one geometry, space must be Euclidean— and the problem of physical space is solved on the epistemological, pre-physical, level.

But the discovery of other congruence geometries, characterized by a numerical parameter K, perforce modifies this view, and restores at least in some measure the objective aspect of physical space; the *a posteriori* ground for this space-constant K is then to be sought in the contingent. The means for its intrinsic determination is implicit in the formulae presented above; we have merely (!) to measure the volume V of a sphere of radius r or the sum σ of the angles of a triangle of measured area δ, and from the results to compute the value of K. On this modified Kantian view, which has been expounded at length by Russell,[6] it is inconceivable that K might vary from point to point—for according to this view the very possibility of measurement depends on the constancy of space-structure, as guaranteed by the axiom of free mobility. It is of interest to mention in passing, in view of recent cosmological findings, the possibility raised by A. Calinon (in 1889!) that the space-constant K might vary with time.[7] But this possibility is rightly ignored by Russell, for the same arguments which would on this *a priori* theory require the constancy of K in space would equally require its constancy in time.

In the foregoing sketch we have dodged the real hook in the problem of measurement. As physicists we should state clearly those aspects of the physical world which are to correspond to elements of the mathematical system which we propose to employ in the description ("realization" of the abstract system). Ideally this program should prescribe fully the operations by which numerical values are to be assigned to the physical counterparts of the abstract elements. How is one to achieve this in the case in hand of determining the numerical value of the space-constant K?

Although K. F. Gauss, one of the spiritual fathers of non-Euclidean geometry, at one time proposed a possible test of the flatness of space by measuring the interior angles of a terrestrial triangle, it remained for his Göttingen successor K. Schwarzschild to formulate the procedure and to attempt to evaluate K on the basis of astronomical data available at the turn of the century.[8] Schwarzschild's pioneer attempt is so inspiring in its conception and so beautiful in its expression that I cannot refrain from giving here a few short extracts from his work. After presenting the possibility that physical space may, in accordance with the neo-Kantian position outlined above, be non-Euclidean, Schwarzschild states (in free translation):

One finds oneself here, if one but will, in a geometrical fairyland, but the beauty of this fairy tale is that one does not know but what it may be true. We accordingly bespeak the question here of how far we must push back the frontiers of this fairyland; of how small we must choose the curvature of space, how great its radius of curvature.

In furtherance of this program Schwarzschild proposes:

> A triangle determined by three points will be defined as the paths of light-rays from one point to another, the lengths of its sides a, b, c, by the times it takes light to traverse these paths, and the angles a, β, γ will be measured with the usual astronomical instruments.

Applying Schwarzschild's prescription to observations on a given star, we consider the triangle ABC defined by the position A of the star and by two positions B, C of the earth—say six months apart—at which the angular positions of the star are measured. The base $BC = a$ is known, by measurements within the solar system consistent with the prescription, and the interior angles β, γ which the light-rays from the star make with the base-line are also known by measurement. From these the *parallax* $p = \pi - (\beta + \gamma)$ may be computed; in Euclidean space this parallax is simply the inferred angle a subtended at the star by the diameter of the earth's orbit. In the other congruence geometries the parallax is seen, with the aid of formula (2) above, to be equal to

$$(2')\qquad p = \pi - (\beta + \gamma) = a - K\delta,$$

where a is the (unknown) angle at the star A, and δ is the (unknown) area of the triangle ABC. Now in spite of our incomplete knowledge of the elements on the far right, certain valid conclusions may be drawn from this result. First, if space is hyperbolic ($K < 0$), for distant stars (for which $a \sim 0$), the parallax p will remain positive; hence if stars are observed whose parallax is zero to within the errors of observation, this estimated error will give an upper limit to the absolute value $-K$ of the curvature. Second, if space is spherical ($K > 0$), for a sufficiently distant star (more distant than one-quarter the circumference of a Euclidean sphere of radius $R = 1/K^{1/2}$, as may immediately be seen by examining a globe) the sum $\beta + \gamma$ will exceed two right angles; hence the parallax p of such a star should be negative, and if no stars are in fact observed with negative parallax, the estimated error of observation will give an upper limit to the curvature

K. Also, in this latter case the light sent out by the star must return to it after traversing the full line of length $2\pi R$ (πR in elliptic space), and hence we should, but for absorption and scattering, be able to observe the returning light as an anti-star in a direction opposite to that of the star itself!

On the basis of the evidence then available, Schwarzschild concluded that if space is hyperbolic its radius of curvature $R = 1/(-K)^{\frac{1}{2}}$ cannot be less than 64 light-years (i.e., the distance light travels in 64 years), and that if the space is elliptic its radius of curvature $R = 1/K^{\frac{1}{2}}$ is at least 1600 light-years. Hardly imposing figures for us today, who believe on other astronomical grounds that objects as distant as 500 million light-years have been sighted in the Mt. Wilson telescope, and who are expecting to find objects at twice that distance with the new Mt. Palomar mirror! But the value for us of the work of Schwarzschild lies in its sound operational approach to the problem of physical geometry—in refreshing contrast to the pontifical pronouncement of H. Poincaré, who after reviewing the subject stated:[9]

> If therefore negative parallaxes were found, or if it were demonstrated that all parallaxes are superior to a certain limit, two courses would be open to us; we might either renounce Euclidean geometry, or else modify laws of optics and suppose that light does not travel rigorously in a straight line.
>
> It is needless to add that all the world would regard the latter solution as the more advantageous.
>
> The Euclidean geometry has, therefore, nothing to fear from fresh experiments. [!]

So far we have tied ourselves into the neo-Kantian doctrine that space must be homogeneous and isotropic, in which case our proposed operational approach is limited in application to the determination of the numerical value of the space-constant K. But the possible scope of the operational method is surely broader than this; what if we do apply it to triangles and circles and spheres in various positions and at various times and find that the K so determined is in fact dependent on position in space and time? Are we, following Poincaré, to attribute these findings to the influence of an external force postulated for the purpose? Or are we to take our findings at face value, and accept the geometry to which we are led as a natural geometry for physical science?

The answer to this methodological question will depend largely on the *universality* of the geometry thus found—whether the geometry found in one situation or field of physical discourse may consistently be extended to others—and in the end partly on the predilection of the individual or of his colleagues or of his times. Thus Einstein's special theory of relativity, which offers a physical kinematics embracing measurements in space and time, has gone through several stages of acceptance and use, until at present it is a universal and indispensable tool of modern physics. Thus Einstein's general theory of relativity, which offers an extended kinematics which includes in its geometrical structure the universal force of gravitation, was long considered by some contemporaries to be a *tour de force*, at best amusing but in practice useless. And now, in extending this theory to the outer bounds of the observed universe, the kind of geometry suggested by the present marginal data seems to many so repugnant that they would follow Poincaré in postulating some *ad hoc* force, be it a double standard of time or a secular change in the velocity of light or Planck's constant, rather than accept it.

But enough of this general and historical approach to the problem of physical geometry! While we should like to complete this discussion with a detailed operational analysis of the solution given by the general theory of relativity, such an undertaking would require far more than the modest mathematical background which we have here presupposed. Further, the field of operations of the general theory is so unearthly and its *experimenta crucis* so delicate that an adequate discussion would take us far out from the familiar objects and concepts of the workaday world, and obscure the salient points we wish to make in a welter of unfamiliar and esoteric astronomical and mathematical concepts. What is needed is a homely experiment which could be carried out in the basement with parts from an old sewing machine and an Ingersoll watch, with an old file of *Popular Mechanics* standing by for reference! This I am, alas, afraid we have not achieved, but I do believe that the following example of a simple theory of measurement in a heat-conducting medium is adequate to expose the principles involved with a modicum of mathematical background. The very fact that it will lead to a rather bad and unacceptable physical theory will in itself be instructive, for its very failure will emphasize the requirement of universality of application—a requirement most satisfactorily met by the general theory of relativity.

The background of our illustration is an ordinary laboratory, equipped with Bunsen burners, clamps, rulers, micrometers and the usual miscellaneous impedimenta there met—at the turn of the century, no electronics required! In it the practical Euclidean geometry reigns (hitherto!) unquestioned, for even though measurements are there to be carried out with quite reasonable standards of accuracy, there is no need for sophisticated qualms concerning the effect of gravitational or magnetic or other general extended force-fields on its metrical structure. Now that we feel at home in these familiar, and disarming, surroundings, consider the following experiment:

Let a thin, flat metal plate be heated in any way—just so that the temperature T is not uniform over the plate. During the process clamp or otherwise constrain the plate to keep it from buckling, so that it can reasonably be said to remain flat by ordinary standards. Now proceed to make simple geometrical measurements on the plate with a short metal rule, which has a certain coefficient of expansion c, taking care that the rule is allowed to come into thermal equilibrium with the plate at each setting before making the measurement. The question now is, what is the geometry of the plate *as revealed by the results of these measurements?*

It is evident that, unless the coefficient of expansion c of the rule is zero, the geometry will not turn out to be Euclidean, for the rule will expand more in the hotter regions of the plate than in the cooler, distorting the (Euclidean) measurements which would be obtained by a rule whose length did not change according to the usual laboratory standards. Thus the perimeter L of a circle centered at a point at which a burner is applied will surely turn out to be greater than π times its measured diameter $2r$, for the rule will expand in measuring through the hotter interior of the circle and hence give a smaller reading than if the temperature were uniform. On referring to the first of formulae (1) above it is seen that the plate would seem to have a negative curvature K at the center of the circle—the kind of structure exhibited by an ordinary twisted surface in the neighborhood of a "saddle-point." In general the curvature will vary from point to point in a systematic way; a more detailed mathematical analysis of the situation shows that, on removing heat sources and neglecting radiation losses from the faces of the plate, K is everywhere negative and that the "radius of curvature" $R = 1/(-K)^{1/2}$ at any point P is inversely proportional to the rate s at which heat flows past P. (R is in fact equal to k/cs, where k is the coefficient of heat conduction *of the plate* and c is as before the coefficient of expansion *of the rule*.) The hyperbolic geometry is accordingly realized when the heat flow is constant throughout the plate, as when the long sides of an elongated rectangle are kept at different fixed temperatures.[10]

And now comes the question, what is the true geometry of the plate? The flat Euclidean geometry we had uncritically agreed upon at the beginning of the experiment, or the un-Euclidean geometry revealed by measurement? It is obvious that the question is improperly worded; the geometry is determinate only when we prescribe the method of measurement, i.e., when we set up a correspondence between the physical aspects (here readings on a definite rule obtained in a prescribed way) and the elements (here distances, in the abstract sense) of the mathematical system. Thus our original common-sense requirement that the plate not buckle, or that it be measured with an invar rule (for which $c \sim o$), leads to Euclidean geometry, while the use of a rule with a sensible coefficient of expansion leads to a locally hyperbolic type of Riemannian geometry, which is in general not a congruence geometry.

There is no doubt that anyone examining this situation will prefer Poincaré's common-sense solution of the problem of the physical geometry of the plate—i.e., to attribute to it Euclidean geometry, and to consider the measured deviations from this geometry, as due to the action of a force (thermal stresses in the rule). Most compulsive to this solution is the fact that this disturbing force lacks the requirement of universality; on employing a brass rule in place of one of steel we would find that the local curvature is trebled—and an ideal rule ($c = o$) would, as we have noted, lead to the Euclidean geometry.

In what respect, then, does the general theory of relativity differ in principle from this geometrical theory of the hot plate? The answer is: *in its universality*; the force of gravitation which it comprehends in the geometrical structure acts equally on all matter. There is here a close analogy between the gravitational mass M of the field-producing body (Sun) and the inertial mass m of the test-particle (Earth) on the one hand, and the heat conduction K of the field (plate) and the coefficient of expansion c of the test-body (rule) on the other. *The success of the general relativity theory of gravitation as a physical geometry of space-time is attributable to the fact that the gravitational and inertial masses of any body are observed to be rigorously proportional for all matter.* Whereas in our geometrical theory of the thermal field the ratio of heat conductivity to coefficient of expansion varies from substance to substance, resulting in a change of the geometry of the field on changing the test-body.

From our present point of view the great triumph of the theory of relativity lies in its absorbing the universal force of gravitation into the geometrical structure; its success in accounting for minute discrepancies in the Newtonian description of the motions of test-bodies in the solar field, although gratifying, is nevertheless of far less moment to the philosophy of physical science.[11] Einstein's achievements would be substantially as great even though it were not for these minute observational tests.

Our final illustration of physical geometry consists in a brief reference to the cosmological problem of the geometry of the observed universe as a whole—a problem considered in greater detail elsewhere in this volume. *If* matter in the universe can, taken on a sufficiently large scale (spatial gobs millions of light-years across), be considered as uniformly distributed, and if (as implied by the general theory of relativity) its geometrical structure is conditioned by matter, then to this approximation our 3-dimensional astronomical space must be homogeneous and isotropic, with a spatially-constant K which may however depend upon time. Granting this hypothesis, how do we go about measuring K, using of course only procedures which can be operationally specified, and to which congruence geometry are we thereby led? The way to the answer is suggested by the second of the formulae (3), for if the nebulae are by-and-large uniformly distributed, then the number N within a sphere of radius r must be proportional to the volume V of this sphere. We have then only to examine the dependence of this number N, as observed in a sufficiently powerful telescope, on the distance r to determine the deviation from the Euclidean value. But how is r operationally to be defined?

If all the nebulae were of the same intrinsic brightness, then their apparent brightness as observed from the Earth should be an indication of their distance from us; we must therefore examine the exact relation to be expected between apparent brightness and the abstract distance r. Now it is the practice of astronomers to assume that brightness falls off inversely with the square of the "distance" of the object—as it would do in Euclidean space, if there were no absorption, scattering, and the like. We must therefore examine the relation between this astronomer's "distance" d, as inferred from apparent brightness, and the distance r which appears as an element of the geometry. It is clear that *all* the light which is radiated at a given moment from the nebula will, after it has traveled a distance r, lie on the surface of a sphere whose area S is given by the first of the formulae (3). And since the practical procedure involved in determining d is equivalent to assuming that all this light lies on the surface of a Euclidean sphere of radius d, it follows immediately that the relationship between the "distance" d used in practice and the distance r dealt with in the geometry is given by the equation

$$4\pi d^2 = S = 4\pi r^2 (1 - Kr^2/3 + \ldots);$$

whence, to our approximation

$$d = r (1 - Kr^2/6 + \ldots), \text{ or}$$

(4)

$$r = d (1 + Kd^2/6 + \ldots).$$

But the astronomical data give the number N of nebulae counted out to a given inferred "distance" d, and in order to determine the curvature from them we must express N, or equivalently V, to which it is assumed proportional, in terms of d. One easily finds from the second of the formulae (3) and the formula (4) just derived that, again to the approximation here adopted,

$$(5) \qquad V = 4/3\,\pi\,d^3\,(1 + 3/10\,Kd^2 + \ldots).$$

And now on plotting N against inferred "distance" d and comparing this empirical plot with the formula (5), it should be possible operationally to determine the "curvature" K.[12]

The search for the curvature K indicates that, after making all known corrections, the number N seems to increase faster with d than the third power, which would be expected in Euclidean space, hence K is positive. The space implied thereby is therefore bounded, of finite total volume, and of a present "radius of curvature" $R = 1/K^{1/2}$ which is found to be of the order of 500 million light-years. Other observations, on the "red-shift" of light from these distant objects, enable us to conclude with perhaps more assurance that this radius is increasing in time at a rate which, if kept up, would double the present radius in something less than 2000 million years.

With this we have finished our brief account of Geometry as a branch of Physics, a subject to which no one has contributed more than Albert Einstein, who by his theories of relativity has brought into being physical geometries which have supplanted the tradition-steeped *a priori* geometry and kinematics of Euclid and Newton.

NOTES

[1] Technically this requirement, as expressed by the axiom of free mobility, is that there exist a motion of the 3-dimensional space into itself which takes an arbitrary configuration, consisting of a point, a direction through the point, and a plane of directions containing the given direction, into a standard such configuration. For an excellent presentation of this standpoint see B. A. W. Russell's *The Foundations of Geometry* (Cambridge, 1897), or Russell and A. N. Whitehead's article "Geometry VI: Non-Euclidean Geometry," *Encyclopaedia Brittanica,* 11th ed.

[2] The motions of the surface of the earth into itself, which enable us to transform a point and a direction through it into any other point and direction, as demanded by the axiom of free mobility, are here those generated by the 3-parameter family of rotations of the earth about its center (not merely the 1-parameter family of diurnal rotations about its "axis"!).

[3] We are here confining ourselves to metric (Riemannian) geometries, in which there exists a differential element ds of distance, whose square is a homogeneous quadratic form in the coordinate differentials.

[4] That is, the "differential," as opposed to the "macroscopic," properties. Thus the Euclidean plane and a cylinder have the same differential, but not the same macroscopic, structure.

[5] The quantities here referred to are the six independent components of the Riemann-Christoffel tensor in 3 dimensions, and the "mean curvature" here introduced (not to be confused with the mean curvature of a surface, which is an extrinsic property depending on the embedment) is $K = - R'/6$, where R' is the contracted Ricci tensor. I am indebted to Professor Herbert Busemann, of the University of Southern California, for a remark

which suggested the usefulness for my later purposes of this approach. A complete exposition of the fundamental concepts involved is to be found in L. P. Eisenhart's *Riemannian Geometry* (Princeton, 1926).

[6] In the works already referred to in fn. 1 above.

[7] "Les espaces géometriques," *Revue Philosophique,* vol. 27 (1889), pp. 588-595. The possibilities at which Calinon arrives are, to quote in free translation:

"1. Our space is and remains rigorously Euclidean;

"2. Our space realizes a geometrical space which differs very little from the Euclidean, but which always remains the same;

"3. Our space realizes successively in time different geometrical spaces; otherwise said, our spatial parameter varies with the time, whether it departs more or less away from the Euclidean parameter or whether it oscillates about a definite parameter very near to the Euclidean value."

[8] "Über das zulässige Krümmungsmaass des Raumes," *Vierteljahrsschrift der astronomischen Gesellschaft,* vol. 35 (1900), pp. 337-347. The *annual parallax,* as used in practice, is one-half that defined below.

[9] *Science and Hypothesis,* transl. by G. B. Halsted (Science Press, 1929), p. 81.

[10] This case, in which the geometry is that of the Poincaré half-plane, has been discussed in detail by E. W. Barankin "Heat Flow and Non-Euclidean Geometry," *American Mathematical Monthly,* vol. 49 (1942), pp. 4-14. For those who are numerically-minded it may be noted that for a steel plate ($k = 0.1$ cal/cm deg) 1 cm thick, with a heat flow of 1 cal/cm^2 sec, the natural unit of length R of the geometry, as measured by a steel rule ($c = 10^{-5}$/deg), is 10^4 cm~328 feet!

[11] Even here an amusing and instructive analogy exists between our caricature and the relativity theory. On extending our notions to a 3-dimensional heat-conducting medius (without worrying too much about how our measurements are actually to be carried out!), and on adopting the standard field equation for heat conduction, the "mean curvature" introduced above is found at any point to be $-(cs/k)^2$, which is of second order in the characteristic parameter c/k. (The case in which the temperature is proportional to $a^2 - r^2$, which requires a continuous distribution of heat sources, has been discussed in some detail by Poincaré, *op. cit.,* pp. 76-78, in his discussion of non-Euclidean geometry.) The field equation may now itself be given a geometrical formulation, at least to first approximation, by replacing it by the requirement that the mean curvature of the space *vanish* at any point at which no heat is being supplied to the medium—in complete analogy with the procedure in the general theory of relativity by which the classical field equations are replaced by the requirement that

the Ricci contracted curvature tensor vanish. Here, as there, will now appear certain deviations, whose magnitude here depends upon the ratio c/k, between the standard and the modified theories. One curious consequence of this treatment is that on solving the modified field equation for a spherically-symmetric source (or better, sink) of heat, one finds precisely the same spatial structure as in the Schwarzschild solution for the gravitational field of a spherically-symmetric gravitational mass—the correspondence being such that the geometrical effect of a sink which removes 1 calorie per second from the medium is equivalent to the gravitational effect of a mass of 10^{23} gm, e.g., of a chunk of rock 200 miles in diameter!

[12] This is, of course, an outrageously over-simplified account of the assumptions and procedures involved. All nebulae are *not* of the same intrinsic brightness, and the modifications required by this and other assumptions tacitly made lead one a merry astronomical chase through the telescope, the Earth's atmosphere, the Milky Way and the Magellanic Clouds to Andromeda and our other near extra-galactic neighbors, and beyond. The story of this search has been delightfully told by E. P. Hubble in his *The Realm of the Nebulae* (Yale, 1936) and in his *Observational Approach to Cosmology* (Oxford, 1937), the source of the data mentioned below.

III

Space-Time
and
Relativity

RELATIVITY

R. E. Peierls

This is Chapter 6 (with some omissions) of *The Laws of Nature* (New York: Charles Scribner's Sons, 1956).

General Remarks

The ideas of relativity at first met with rather strong opposition both amongst physicists and amongst philosophers. The physicists met the new hypothesis in the critical spirit in which it is their business to regard any new idea, until it has passed the triple test which physics requires. It must firstly leave undisturbed the successes of earlier work and not upset the explanations of observations that had been used in support of earlier ideas. Secondly it must explain in a reasonable manner the new evidence which brought the previous ideas into doubt and which suggested the new hypothesis. And thirdly it must predict new phenomena or new relationships between different phenomena, which were not known or not clearly understood at the time when it was invented. This process took some time because relativity is important only for objects moving with a speed comparable to that of light. As such objects were not readily available the opportunities for tests were few, and in many cases the tests required very difficult observations of high precision. Since then particles moving with high velocity have become commonplace in any physics laboratory. We are no longer concerned with small corrections in the behaviour of these particles which require measurements of high precision, but the relativistic features of their motion have large effects which it is quite impossible to overlook. To put the point in its crudest form: in developing machines for physical research in which particles are accelerated to very high speeds, engineers have to incorporate devices, costing many thousands of pounds, which are required only because of the relativistic features of the particle motion. They are scarcely inclined to regard such features as a result of idle or mistaken speculation. To-day no physicist who has practical knowledge of work with fast particles would question the principles of relativity.

The opposition from philosophers arose because the theory of relativity called in question statements that had been regarded as the concern of philosophers. It was doubted whether the physicist had the right to query ideas which the philosophers had regarded as evident and unquestionable truth.

By now it appears to be widely recognized that our ideas about space and time are derived ultimately from our experience of the outside world and that many statements which we regard as evident are true only within the limitations of our practical experience. They may turn out to be unjustified prejudices when extended to situations not normally familiar to us.

In the history of physics such situations are common. The discovery that the earth was revolving about its own axis and about the sun ran counter to mechanical intuition built on simple experience. When we learn at school about the opposition first encountered by this idea we tend to regard the scholars of the Middle Ages, who failed to accept it, as incredibly narrow-minded in their outlook, but the only reason why we ourselves find the thought easier is that we have become familiar with it through our early education and have accepted it before our critical power was developed far enough to question it seriously.

The idea of Galileo's law that a moving body tends to continue in its motion runs counter to everyday experience and was at first quite legitimately questioned by the scientists of the time. It was accepted only with difficulty by thinkers who generalized abstract ideas about motion.

The fact that light travels with finite speed is another fact which runs counter to our intuition. It is hard even to get used to the fact that the speed of sound is finite, and even the simple experience of watching a man chop wood at a distance, when the sound of the blow reaches us later than the sight of the falling axe, requires a conscious adjustment of our senses to an unfamiliar situation. We have learned even more strongly to rely on our eyes as telling us what goes on in the outer world and to regard the information they give us as true information about what goes on at the instant we see it. There is nothing in everyday experience on the surface of the earth which will disprove this illusion, but three hundred years of experience in astronomy and more recent methods for measuring times to enormous precision have got us to accept the finite speed of light. . . .

The acceptance of relativity was probably delayed by its name, which suggested a superficial connection with the philosophical concept of relativity, according to which all truth was regarded as relative. As we shall see, nothing is further from the contents of the new development. In relativity, the laws of physics have a precise and absolute form, only certain specific statements that our intuition leads us to regard as absolute, turn out to be prejudiced.

Motion and Rest

Long before the developments which led to Einstein's formulation of the theory of relativity it was known that many of the laws of nature, in particular the laws of mechanics, did not alter their appearance if the observer, instead of standing still, was moving with uniform velocity and direction.

It is well known that the passenger in a railway train is not aware of the motion of the train unless the vibration of the coach, due to the roughness of the track, or the acceleration caused by starting or slowing down, or by the curvature of the track, show him that the train is moving. Frequently, when we look out of the train window at another train in the station we are confused as to whether it is our train or the adjacent one which has started moving. The question is settled when we catch a glimpse of the station buildings or other objects, which, we know from experience, usually stand still. If a fellow passenger asserted that in fact the station, with the track and all the landscape, was moving and the train was standing still, and if he was not impressed by the argument that stations and landscapes don't usually behave like that, we would not find it easy to prove him wrong. Any mechanical experiment which we carry out inside the compartment would proceed precisely in the same way as if the train was standing still (apart from the vibrations and curves to which reference has already been made). And this fact is usually expressed by saying that the laws of mechanics are the same for two observers who move relatively to each other with a uniform velocity. . . .

While the laws of mechanics therefore do not allow us to distinguish uniform motion from rest, this would appear to be different when our description of nature includes electricity and light. In discussing the laws of the electromagnetic field, . . . we found that electromagnetic waves, which include light waves, always travel with a fixed velocity, which we called *c*. We should therefore expect to find some difference in the apparent speed of propagation of light, owing to the fact that we are moving ourselves. For instance, if the only motion was due to the earth's rotation, light from a lamp placed to the east of us should reach our eye more quickly because our eye is in fact moving towards the lamp and therefore coming to meet the light waves. Similarly light from a lamp placed in a westerly direction should appear to travel more slowly because it has to overtake our eye, which is moving away. If *u* is the velocity with which the surface of the earth is moving, the apparent velocity from a light source in the east should in fact be $c + u$ and that from a light source in the west $c - u$. The difference is of course very small, because *u* is about a million times smaller than *c*, but careful experiment ought to reveal the difference.

This expected result was sometimes called the ether wind, because one thought at that time that all space was filled with a hypothetical substance called ether which acted as a carrier for electromagnetic waves. The effect of our motion through the ether on the propagation of light was similar to the effect of a strong wind on the propagation of sound, when it is well known that sound will appear to travel more rapidly down-wind than up-wind.

In their famous experiment Michelson and Morley set out to detect this "ether wind." To avoid the difficulty of measuring distances and times to the required high precision they used in fact a light beam which was split in two. One part was travelling, say, in an east-west direction and back after reflection from a mirror, and another north-south to another mirror and back. On being combined again these two light rays would form interference fringes of the kind we have discussed [earlier]. The position of these fringes depends on how many oscillations each light wave has undergone during its travel. It is not difficult to work out the times taken for the return journey by either light beam. This calculation, which we shall omit here, shows that the effect of the earth's motion is the same as if the east-west arm of the apparatus was

lengthened in the ratio of 1 to $\sqrt{1 - \dfrac{u^2}{c^2}}$. It is difficult to measure the length of the light paths to the required accuracy, but Michelson's reasoning was that on turning his apparatus through a right angle, so that now the other arm pointed in the east-west direction, this was bound to lengthen the time of travel of the one light ray and shorten the other, with a resultant shift of the interference fringes. Since the earth not merely rotates, but also revolves in its orbit about the sun, the effective speed with which we are moving should also alter its direction between day and night. Therefore even if the apparatus was left still, the fringes should alter their position in the course of a day.

The result of this experiment was completely negative. Whichever way the apparatus was turned and however long one looked at it no shift in the fringes was observed. All other attempts to observe the "ether wind" failed similarly. And hence the one kind of experiment which appeared to make it possible to distinguish motion from rest had failed.

The Lorentz Contraction

It took a long time to understand completely the implications of this result. Various explanations were tried. It was for example suggested that the velocity of light might be dependent on the motion of the light source. Since the Michelson experiment employed a lamp which was fixed on the earth's surface one would then expect the light to travel with a given speed relatively to this source, i.e. relatively to the surface of the earth. But this explanation had to be given up on reflection, since we know that the light from the sun, for example, really comes from atoms in the atmosphere of the sun, which . . . are moving about irregularly and with great speed. The light from different atoms should therefore, if this explanation were right, have different speeds. If we looked not at the sun but at a distant star, which is similar to the sun, but further away, we should see this star not as a point, but as a streak, since light from different atoms on the star should have come to us with different speeds and should therefore have started when the star was in quite different positions in relation to our telescope.

It was also suggested that in the neighbourhood of the earth the ether should take some part in the motion of the earth, so that there was an "atmosphere" of ether around us, which moved in the same way as the earth itself. This hypothesis, too, was untenable, because if it were true there had to be somewhere outside the earth a transition region where the ether's speed changed from that of the earth to that of free space. This would lead to a refraction of light rays and an apparent displacement in the position of stars, which is not observed.

More such alternatives were considered and found equally unsuccessful. Nearer to the truth came a suggestion by H. A. Lorentz, which at first sight seems very far fetched. Lorentz suggested that every moving object contracts in the direction of motion in the ratio of $\sqrt{1 - \dfrac{u^2}{c^2}}$ to 1, if u is its speed.

If the arm of Michelson's apparatus which points in the direction of motion had in fact contracted by that amount this would just make up for the effect of the "ether wind." When one first hears of this "Lorentz contraction" it sounds most artificial and unreasonable. It sounds, in fact, as if nature had conspired to adjust its laws in such a way as to make it hard for us to observe the ether wind. There seemed no reason in mechanics why bodies should undergo this peculiar contraction.

However, there was a much better basis for the suggestion than appeared at first sight, because Lorentz knew already that all matter consists of atoms . . . and that atoms contain positively and negatively charged particles, the forces holding them together being electric forces. Now in the laws of electricity and magnetism the velocity of light occurs as a constant. If charged particles move with a speed not very small compared with that of light, then their electric fields become more complicated. A point charge, for example, such as that of an atomic nucleus, has an electric field in which the potential depends only on the distance from the charge, so that all points on a sphere centred on the charge are at the same potential. If the charge is moving it produces not merely an electric, but also a magnetic field, because the transport of electric charges means the same thing as an electric current. In addition the electric and magnetic fields at any point in space near which the charge passes vary with time. We saw [earlier] that complications arise in the laws of the electric and magnetic field if the fields vary in time. This more complicated mathematical problem can be solved, and the result is that the surfaces of constant potential are no longer spheres, but are flattened in the direction of motion, by just the same amount that would be required by the Lorentz contraction.

This change in the electric forces no doubt would cause a change in the size and shape of an atom and in the distances of different atoms from each other. Not enough was known at the time about atomic structure to be sure what the overall effect would be. At least the possibility emerged that this Lorentz contraction could just come out as a consequence of the effect of the motion on the electric forces which hold atoms together.

This idea of the Lorentz contraction would, if confirmed, account for the negative result of the Michelson experiment, but at first sight it would open up a new possibility of finding out whether a body was at rest or in motion. Consider again for simplicity only the motion due to the rotation of the earth, and take a very accurately measured cube, which has one of its edges in the north-south direction. If we turn it through a right angle, the edge which was in a north-south direction before, and now is in an east-west direction, will have undergone the Lorentz contraction, whereas the other horizontal edge which previously was east-west and therefore was contracted, will have expanded. In the new position measurement should show us therefore that the body was no longer a cube. Now first of all, the amount of the Lorentz contraction is so small that it would be extremely hard to detect the change, but we can imagine that we had measuring instruments of sufficient accuracy to do so.

We should, however, then find a much more fundamental difficulty. The only way to measure a length is by comparing it directly or indirectly with a measuring rod or tape. This measuring rod itself, like any other rigid body, would be subject to the Lorentz contraction and therefore, as we turn it to place it first against the one edge and then against the other edge it would follow the Lorentz contraction. The two edges would still appear to be of the same length in terms of the marks on the measuring rod.

If we therefore apply consistently the postulate of Lorentz to all solid bodies without exception then it would be impossible in principle to detect the contraction by any direct measurement.

How to Compare Lengths and Times

But one might think that the Lorentz contraction would still be a suitable way to find out which of the two passing trains with which we started this discussion, is standing still, and which is moving. Suppose that the coaches of both trains are of identical construction, then that of the moving train should have contracted and therefore be shorter than that of the stationary train. To make matters easier imagine for the moment that the speed of the moving train was not one-millionth, but one-third of the velocity of light; then the Lorentz contraction would shorten it by 5 per cent, which should be easy to see. Thus all we would have to do is to watch carefully as two coaches pass each other and to see which is the longer.

However, this also is much more involved than it seems, if we remember that we are dealing with very high velocities and very short times. To make this point clear consider Figure 1. A is meant to indicate the stationary coach and B the moving one.

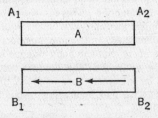

Fig. 1. Passing trains.

If B is really shorter, then at the instant when its front end B_1 is in line with the end A_1 of the stationary coach, its rear end B_2 will already have passed the end A_2 of the stationary coach. But to be sure of this we must be very sure of our time. If we look at B_2 a very little too early it will not yet have passed A_2 and we would then think the coach too long. In other words, comparing the lengths of the two coaches boils down to being able to tell with great accuracy the time of passage of the ends past each other. The moving coach is longer, or shorter, than the other if its end B_2 passes A_2 after, or before, the end B_1 passes A_1.

We may, of course, look at the two ends simultaneously,

but we then get involved in complications due to the finite speed of propagation of light. Supposing we stand at A_1. At the time when we see B_1 passing, we shall also see the end B_2, not as it is at that instant, but as it was a very short time earlier, namely earlier by the amount of time which the light has taken to travel from B_2 to A_1. It is true that we can correct for this, since we know the speed of propagation of light, but in working out this correction we must firstly know the length of our own coach, i.e. we must assume that our own coach is standing still, and thus no Lorentz contraction has to be applied, and also that the light travels with its normal speed and no ether wind has to be allowed for.

We might try to get some time signal across from A_2 to A_1 in some other way. We could imagine that an assistant stationed at A_2 presses a button when B_2 passes and that this gives some kind of signal at A_1, but this transmission must either be by mechanical means or by an electric current or some other device, and all these means of communication are in fact slower than the speed of light. To work out their speed of propagation with the required accuracy gets us into the same difficulty.

Let us then limit ourselves to light signals and consider how the whole process would appear from the point of view of an observer on the other train who is trying to apply the same test. He would also watch the passage of A_2 past B_2 and A_1 past B_1 and he would also transmit a light signal from one end of the coach to the other and apply a correction for the time this has taken. He would naturally assume that his train was at rest and would apply no correction for ether wind or for the Lorentz contraction. The interesting result of working this out in detail is that on the same basis he would come to the conclusion that our coach has been shortened by the Lorentz constraction, i.e. that A_1 and B_1 are in line *before* A_2 is in line with B_2.

The important point to realize is that the observations of both observers are completely interchangeable. We are coming to the conclusion that the coach B is moving and therefore shortened by Lorentz contraction, whereas the other observer, starting from the assumption that his coach is at rest, would conclude that our coach A is moving and therefore shortened. No doubt we will criticize his measurement because he has made assumptions about the length of his coach and about the absence of an ether drift which to us appear wrong, but he will answer by pointing out that from his point of view the same criticism applies to our method.

The essence of the argument, which is a simplified version of that used by Einstein, is that the length of an object has no absolute meaning, but depends on the state of motion of the observer by whom it is determined. Similarly the question of time, i.e. which of two events occurring at two distant points is the earlier, has no absolute meaning, but depends on the point of view of the observer.

These ideas certainly are in conflict with our intuition, which regards the length of an object or the time at which an event takes place as absolute concepts, not necessarily related to any actual process of measurement. One has to get accustomed to the realization that the certainty with which we take these things for granted comes only from the experience of everyday life. In practice any doubt about the meaning of the exact length of a moving object, such as a railway coach, is completely negligible within the limits of error of any practical determination. In practice also we never get into trouble by assuming that any event which we can see takes place when we see it, since for practical purposes the speed of light may be regarded as infinitely large.

It is one of the merits of Einstein's theory to have brought out clearly the fact that our ideas of space and time are abstractions from experience, and that in applying them to situations involving greater speeds or shorter times than those to which we are normally accustomed we cannot take their meaning for granted without specifying them precisely. This is done only by formulating questions which can, in principle, be answered by an actual observation.

So far we find, then, that the postulate of the Lorentz contraction is consistent with the principle that the laws of physics are precisely the same for two observers moving uniformly with respect to each other and that there is no observation by which the point of view of one observer can be preferred over that of another.

The connection between the observations of different observers may be expressed by means of diagrams. Figure 2 shows a kind of graph in which distances are plotted horizontally and time vertically. The horizontal line OX shows some line, for example the railway track under discussion, at some particular instant which we take as the zero of time. Along this we have marked out equal distances with "mile-

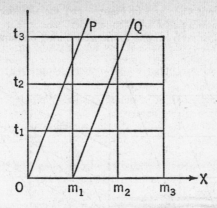

Fig. 2. Distance-time graph.

stones" m_1, m_2, m_3, etc. The vertical line Ot indicates the
passage of time at the starting-point O and equal time intervals have been marked off at t_1, t_2, t_3. Subsequent horizontal
lines show the track again at the later times and subsequent
vertical lines the passage of time at the various milestones.
Such diagrams are in fact used in constructing timetables,
since one can indicate the passage of a train in such a diagram
by a sloping line, such as OP or m_1Q and one can then tell at
a glance at what time the train will pass a particular point,
or where two trains will meet. In our case OP and m_1Q are
meant to show the two ends of one coach of the moving train.
    If the observer on the moving train also sets up distance
marks on the train and records times, how would his finding
compare with ours? For the case of a real train, when all
velocities are small compared to that of light, and relativity is
therefore unimportant, this is represented by Figure 3(a). The
sloping lines now indicate points which to the observer on the
train appear to stand still and which he regards as marking
out distances; indeed the end of each coach is to him a fixed
landmark. Our own milestones are shown again as broken
lines for comparison; the time marks are the same for both
of us, since apart from relativity the time is the same for all
observers.
    Compare with this the relativistic picture which arises when
we are dealing with really high velocities. For this we must go

to a different scale and in Figure 3(b) the situation is drawn
again, assuming that our "milestones" are so far apart that
light will just go from one to the next during one time interval.

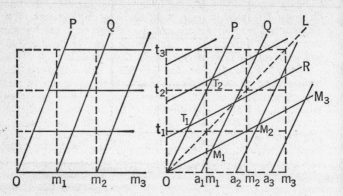

Fig. 3(a).   Moving observer; low speed.
     (b).   Moving observer; one-half light velocity.

(For instance, if the time interval is one second, the distance
between the milestones must be 300,000 kilometres.) On this
scale the motion of any real train would appear as a practically
vertical line, but we have shown in the drawing the case of a
"train" moving with one-half the speed of light. Again OP
and a_1Q show the movement of rear and front of one coach.
a_1 now does not coincide with m_1 because, owing to the
Lorentz contraction, the coach appears to have shortened, and
at one instant of time it occupies less than the distance between O and m_1. However, the time marks of the moving
observer are now given by the sloping lines like $O M_1 M_2 M_3$,
etc., and therefore to find the length of our coach the moving
observer would see how much of this line is cut by the lines
indicating the ends of our coach, i.e. the two vertical broken
lines through O and m_1. The second vertical line passes to
the left of M_1 and the moving observer will therefore find our
coach shorter than this.

So far the diagram only repeats in graphical form what we
have discussed in words already, but it also permits a further
conclusion. Consider the dotted line OL. This indicates the
passage of a light signal, since a movement along this line
describes a motion which takes just one of our time intervals

to travel to the next milestone, and this by convention just represents the velocity of light. Now we know from the Michelson experiment that the velocity of light must appear to be the same for the moving observer and therefore the line *OL* must also pass through the corners of his network. The diagram has been drawn in accordance with this, and we see that the later time marks of the moving observer such as T_1R will intersect our time mark between Ot_1 and OP.

From this we can deduce the effect of motion on the measurement of time, i.e. on a clock. If the observer carries a clock which passed our stationary clock at O and was then synchronized with it, it will "strike," i.e. indicate the passage of exactly one time interval at T_1, but to us this is later than t_1. We therefore conclude that the clock carried by the moving observer is slow. From the symmetry of the figure it is slow compared to ours in the same ratio in which the distances have been contracted by the Lorentz contraction. This effect on moving clocks is known as the time dilation. It is again true that with exactly equal right the observer on the train will conclude that our clock is slow, since its "stroke," t_1, lies above the line T_1R, i.e. occurs from his point of view after the end of the first time interval.

This predicted time dilation can be linked very directly with the known behaviour of fast particles. . . .

Composition of Velocities

With this new relationship between the records of time and distance kept by different observers there must also go a new rule for the composition of velocities. If a man on a train moving with velocity u fires a gun in the forward direction, which gives the bullet a speed v with respect to the train, then the total velocity of the bullet as seen from the track would be $u + v$. But in relativity we must translate the path of the bullet, which to the observer on the train is just motion with velocity v, by correcting for the difference in viewpoint about times and distances and if this is carried out we find the result $\dfrac{u + v}{1 + \dfrac{uv}{c^2}}$. If u and v are both small compared to c, the velocity of light, then the second part of the denominator is very small compared to 1, and the result is again that the velocities just add. But if u and v are not so small then the answer is different. For example if u and v are each one-half of the velocity of light the combined effect of the two velocities is $\frac{4}{5}c$. One sees easily that if u and v are both less than the velocity of light, however close they may be to it, the resultant velocity will still be less than c.

This rule is correct only when the two velocities are in the same direction. For the more general case the rule is more complicated and I shall not give it here.

One particular consequence of this law of composition of velocities is evident. If $v = c$, i.e. if instead of firing a gun the man on the train sends out a light signal, then the combined velocity is again equal to c, and this brings us back to the starting point, namely that light appears to be travelling to both observers with the same velocity.

This principle of the constant velocity of light is the most basic law of relativity. To avoid misunderstanding it should be stressed that in all this when we have spoken of the velocity of light we ought really to have said "the velocity of light in empty space." In any material medium, even in air, and to a greater extent in such dense substances as water or glass, we know from the refraction of light . . . that the velocity of light is different. This is caused by the effect of the light wave on the electric charges contained in the atoms. A light ray passing through glass will therefore not appear to different observers as propagating with the same speed. This is not surprising since the situations seen by the two observers are then different. For the one, glass or water is at rest, whereas for the other one it is moving. It was in fact in a study of the propagation of electromagnetic waves through moving bodies that Einstein was first led to a clear understanding of the arguments I have sketched above.

Mechanics of Fast-moving Objects

This new application of the properties of distances and times also requires us to change the laws of mechanics as applied to fast-moving bodies; to see what modifications are required one can reason in the following way: look at the second law of Newton in the form that the force equals the rate of change of momentum. Before we can generalize this law we must be clear that both the ideas of momentum and force may need revision; the form which the mechanical law will take depends to some extent on which of the two, if either, we want to keep unchanged.

It turns out that the clearest statement of the laws is obtained if we retain the law of action and reaction, which says that the forces which two objects exert on each other must be opposite and equal. In consequence, the interaction between two objects does not change the total momentum. We also want to retain the law of conservation of energy, and in particular of mechanical energy, if we are dealing with an elastic collision. Consider then the collision between two elastic identical objects, for example two billiard balls. If these collide with equal and opposite speeds their combined momentum must be zero, and if momentum is to be conserved it must be zero even after the collision. This means after the collision the two billard balls must still travel with opposite and equal velocity.

In Figure 4 we have shown two such balls colliding, with

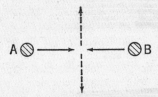

Fig. 4.   Collision of two objects.

the arrows indicating their initial velocity of approach; the equal length of the arrows is meant to indicate that their speeds are equal. Now suppose that they have collided in such a way that they separate again in a direction at right angles to the original. This is indicated in the figure by the broken arrows. The speeds must again be equal. Moreover, energy conservation requires the speed with which the balls separate to be equal to that with which they collided: If the velocity of approach is u the total kinetic energy before the collision is twice the kinetic energy of a ball moving with velocity u, and if they separate again with velocity u this will also be the final energy. Note that this argument is valid regardless of how the kinetic energy depends on velocity.

We conclude that the situation sketched in Figure 4 is

mechanically possible, i.e. preserves momentum and energy. If the mechanical laws are to be the same for all observers, the laws of conservation of energy and momentum must also appear satisfied if this same process is seen by a moving observer. Supposing that the moving observer is travelling with speed v from right to left, so that to him the ball A will appear to travel faster and B more slowly than before; after the collision both will apparently not be travelling at right angles to the original line, but more towards the right. This is shown in Figure 5. The important thing is that from the law

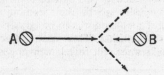

Fig. 5.   The collision of Fig. 4, seen by a moving observer.

of composition of velocity we can work out the directions and magnitudes of all the velocities as seen by this observer, i.e. the directions and lengths of all the arrows of Figure 5. Now the momentum and energy must still be conserved, and this tells us something about the momentum and energy of objects moving with various velocities. I shall not give the details of this calculation here, which requires a little more mathematics than we are using, but from it the dependence of energy and momentum on the speed of a particle can be determined quite unambiguously.

The result is that the momentum p of an object of mass m moving with velocity u is $p = \dfrac{mu}{\sqrt{1 - \dfrac{u^2}{c^2}}}$. This relation be-

tween p and u is shown in graphical form in Figure 6. The full line shows the relativistic relation, whereas the broken line represents the law of Newtonian mechanics, where $p = mu$. It is seen that the relativistic momentum for small u is practically the same as before, but for velocities near that of light, very much larger than the momentum of Newtonian mechanics. This same result can be expressed differently; we may still retain the definition of momentum as mass times velocity, provided we are prepared to regard the mass of the object as variable.

If we denote by M the quantity $M = \dfrac{m}{\sqrt{1 - \dfrac{u^2}{c^2}}}$ then indeed

the momentum is $p = Mu$. To distinguish the two definitions of mass one sometimes calls m the "rest mass" of the object, i.e. the mass which we determine by dynamical measurements when the object is at rest or moving very slowly. Even if we work with this idea of a variable mass we must, however, remember that we should not express the dynamical law in

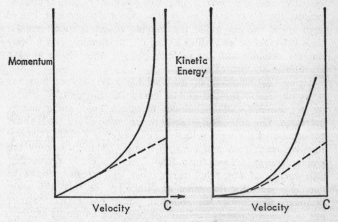

Fig. 6. Momentum and velocity.   Fig. 7. Kinetic energy and velocity.

the form that force equals mass times acceleration. The force must be taken as the rate of change of the momentum, i.e. the rate of change of mass times velocity. These two are not the same when the mass also changes in the course of the motion.

This result says that as the speed of an object increases towards that of light, it becomes harder and harder to accelerate it. In fact, to make it reach the velocity of light in any finite time an infinitely strong force would be required. It is therefore evident that it is impossible for any material object ever to reach the velocity of light, let alone exceed it. This result is satisfactory for the consistency of our views. If it were possible to apply our considerations about the relations between different observers to a man on a train travelling with more than the velocity of light we would immediately become involved in paradoxes.

As regards the kinetic energy, the argument which I have sketched indicates that $E_{kin} = \dfrac{mc^2}{\sqrt{1 - \dfrac{u^2}{c^2}}} - mc^2$. This rela-

tionship is shown graphically in Figure 7, where again the full line gives the relativistic relation between kinetic energy and velocity, and the broken line the relation $E_{kin} = \frac{1}{2}mu^2$. As before, the Newtonian law is adequate at low speeds, but at high speed the energy of the moving object is much greater than it would be according to the old mechanics, and a body moving actually with the velocity of light would have to have an infinite amount of energy.

Using the definition of the variable mass, M, we may express our result for the kinetic energy as $E_{kin} = (M - m)c^2$ or, as the mass of the object increases the kinetic energy increases. The amount is exactly the increase of the mass times the square of the light velocity. This suggests a close relationship between mass and energy, though up to now only the kinetic energy entered into this. One can easily prove that also energy of any other form must lead to an increase in mass by precisely the same amount. To see this it is only necessary to consider the collision, not of two elastic objects, but of two completely inelastic objects, say to lumps of clay, of the same mass. If these collide with equal and opposite velocities they will just stick together. As the total momentum is zero the fused lump has no momentum and therefore is standing still. We may look at this process from the point of view of a moving observer and again require that momentum be conserved. If we work out the apparent velocities for the new observer we find that momentum would not be conserved if the fused lump had simply the combined rest mass of the separate lumps, but only if its rest mass had increased precisely by the amount of energy which had been converted into heat, divided by c^2. Hence the relationship between mass and energy includes also heat energy, and since we may arrange for the heat to cause some chemical reaction or to generate electricity by some internal mechanism, the same must apply to energy in any other form. . . .

Applications and Confirmation

From reading so far the reader will have got the impression that we have built an elaborate structure of reasoning on one experimental fact, namely the negative result of the Michelson experiment. However reliable the experiment, and however attractive the general principle of the independence of all laws of physics of the state of motion of the observer, one would not have accepted such far-reaching conclusions without a great deal of further support. In fact other physicists have claimed from time to time that on repeating the Michelson experiment they did find a positive answer, but in the meantime other evidence for the theory of relativity had become so strong that we would have no cause to change our views if some fundamental flaw was discovered in Michelson's reasoning.

Such support comes from practically any observation of the behaviour of particles moving at high velocity. Historically the earliest observations made were on electrons, since, being the lightest particles, these can most easily be accelerated to velocities near those of light. For example, an electron reaches a speed of $0.8c$ after passing through an electric field with a potential difference of about 300,000 volts, which is now commonplace in any modern laboratory. At this speed its kinetic energy is about twice as large as it would be in non-relativistic mechanics and the difference in its behaviour very marked.

Earlier experiments carried out with much lower voltages, and in which the deflection of electrons in electric and magnetic fields was measured accurately, showed up the variable mass and therefore confirmed relativistic mechanics. Perhaps the most striking demonstration of relativistic behaviour results from a collision of a fast particle with one at rest. Supposing an electron has been accelerated somehow to a velocity near that of light and in passing through matter collides with a stationary electron. Since the two particles have the same mass and since usually in such collisions no mechanical energy is converted into any other form . . . we should conclude therefore that after the collisions the two electrons would move in directions at right angles to each other. On the other hand, relativistic mechanics says that both electrons should still travel forward, their lines of motion making only a small angle with each other, this angle becoming less and less as the velocity of the incident electron approaches that of light. We shall later discuss modern techniques which have made it possible to photograph the track of a single electron. Figure 8

Fig. 8. Track of a fast electron colliding with a nearly stationary one.

shows a sketch of such a photograph, in which a fast electron is seen to be suddenly deflected from its path, and the track of a second electron originates from this point. All tracks are curved, since a magnetic field was applied in order to determine the speed of the electrons. After the collision the tracks form a very small angle with each other. This picture is a sketch of a photograph obtained by Dr. F. C. Champion.

Another drastic example is provided by modern particle accelerators, whose principle and purpose we shall discuss later. In these we have particles moving either along a straight line or (under the influence of a magnetic field) in a circle, and electric impulses are applied to them at suitable times. The performance of such machines depends therefore on knowing the speed of the particles so that the electric impulse can be arranged to be applied just as a particle passes a suitable point. Now there exist, for example, machines in which electrons can be given an energy of 300 MeV. By an MeV or million electron volts, we mean the energy which an electron would acquire on passing through an electric potential difference of one million volts. The rest energy of an electron is about $\frac{1}{2}$ MeV and therefore we are dealing with electrons whose energy is about 600 times its rest energy. A little arithmetic shows that this makes their velocity less than that of light only by a little over one part in a million. Without relativity, electrons of such energy should move with a speed of 35 times that of light, and it is evidently impossible to mistake one for the other.

These are only a few examples of the many tests to which relativistic mechanics has been subjected in recent years and the combined weight of these tests is sufficient to give us as

much confidence in the relativistic laws, for fast-moving particles, as we have in the validity of Newton's laws in the
domain of small velocities. . . .

Principle of Equivalence

The ground which we have covered in this chapter belongs
to the so-called "special" relativity. Before we leave the subject of relativity we should mention briefly the wider subject
of "general" relativity, though we shall not be able to do it
justice. . . .

Any acceleration produces the same effect as gravity. Imagine that we were in a closed box like the cage of a lift,
which was placed somewhere in free space a long way from
the earth or other bodies, so that there was no gravity. Then
all objects in this box would float around freely. If we pushed
out feet only slightly against the ground this would project
us upwards until we hit the ceiling. This state of affairs has
by now become familiar from books and films about space
travel. Now suppose we suddenly felt our weight again and
found objects falling to the floor. We could place two interpretations on this: either we could say that our box was really
the cage of a lift and that it had started accelerating upwards,
or we could say that the box was still stationary (or moving
with uniform velocity), but that we were now near the earth,
or near some planet, and there was gravity acting on us.

We could not settle the argument even if we had a small
window in the ceiling and saw through it that there was a cable
attached to the top of our cage, and that this cable was evidently in tension. This would still be expected on either view,
either to produce the acceleration, or to keep our cage suspended against the force of gravity.

Conversely, if we were travelling in an ordinary lift and the
supporting cable broke, so that the cage with all its contents
was falling freely, we would have the sensation of the absence
of weight, since the force of gravity would accelerate us at
just the same rate as the cage, so that no force was required
to keep our position relatively to the cage. We would therefore not know whether in fact our cage was falling or whether
the attractive force of the earth had suddenly ceased (this
doubt would of course last only until we hit the bottom of
the shaft). . . .

We conclude that accelerated motion does not leave the

laws of physics unchanged, but produces the same effect as a gravitational field. This "principle of equivalence," which is an essential feature of the general theory of relativity, depends essentially on the fact that the weights of two objects are exactly proportional to their masses, so that gravity gives them the same acceleration. The general theory of relativity is based on the view that an acceleration of all the landmarks to which we refer our observations is not merely in practice indistinguishable from the effect of a gravitational field, but that the two are essentially one and the same thing and cannot be distinguished in principle.

This does not mean to say that we could set up our laboratory in such a way that we could get rid of the whole gravitational attraction of the earth, for example, because this would require distributing observers in free-falling boxes all around the earth, and of course they would soon hit the earth, unless they had previously collided with each other. We can only in a small region of space and for a short time look at the motion of objects from the point of view of an observer for whom there exists no gravity.

General Relativity

If we tried to set up our description of nature in this way, doing our mechanics in a freely falling box we would come to the conclusion that space has curious properties. For example, suppose that our box arrives with high speed from large distances and the speed is so great that the attraction of the earth would not be sufficient to make it collide with the earth, but only deflect it through a small angle. In Figure 9, *E* shows

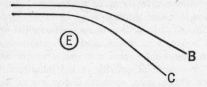

Fig. 9.   Two observers passing the earth.

the earth and the line marked *B* might be the path of our box. Suppose also that close to us a second box *C* with another observer was travelling parallel with us and at the same speed. Until we get near the earth, *C* would be at a constant distance from us, but being closer to the earth his box would be deflected more, as shown in the figure, and so afterwards we would be travelling in different directions and would be separating. We are both travelling freely and therefore experience no sensation of acceleration. On the ideas of general relativity we should therefore both be entitled to believe ourselves at rest, but we should then find the distance between us suddenly starting to increase without either of us apparently having started to move.

In other words, without answering the question about rest or motion and choosing as landmarks any objects which are allowed to move freely, we can detect the presence of a gravitational field and therefore the presence somewhere in the neighbourhood of heavy matter, by the way distances between different landmarks behave. This is the beginning of a description of a gravitational field in terms of the properties of space and time. It would require too much mathematics to pursue this description in greater detail, but I shall mention two conclusions which follow from it.

One is that in a gravitational field light does not travel in a straight line, but is deflected. This in fact could have been deduced already from the fact, proved in special relativity, that mass and energy are equivalent. A light ray carries energy and therefore mass, and if all mass is attracted by a heavy body like the sun or the earth, this should also hold for a light ray. However, the amount of deflection is not correctly given by this simple argument because the force of gravity on a moving object depends on its speed. This is unimportant for the slow speeds normally encountered in practice, but makes some difference when the speed is that of light. When light passes very close to the sun it should therefore be deflected by an amount which can be calculated, and therefore a star which we happen to see near the edge of the sun should appear slightly displaced from its normal position. It is of course not usually possible to see a star when it is close to the sun, but an opportunity for this arises during a total eclipse of the sun, when the sun's disc is covered by the moon so that its light does not make the stars invisible. Even then very accurate observation is needed to detect the displacement, since the expected amount is less then two seconds of arc, i.e. about the size of a penny seen from a distance of two miles. Nevertheless precise astronomical observations have made it possible to detect such displacements, and while astronomers still use

convenient solar eclipses to improve the accuracy of the result, it is certain that the displacement exists and is of roughly
the amount predicted by general relativity.

Another important conclusion is that the field of gravity
should affect the time scale, or more precisely that a clock
kept near a very massive star where there is a strong negative
(i.e. attractive) gravitational potential, should appear to us to
be slow. This, too, can be tested since the light from a hot star
contains radiation of very distinct colours, i.e. of certain
definite frequencies, which we know are due to the tendency
of some atoms in the star's atmosphere to set up electric oscillations with a definite frequency, much like a miniature radio
transmitter. Such atoms therefore may be regarded as standard
clocks, in that the period of vibration is a natural characteristic of the atom and not dependent on external circumstances.
Now it is known that from certain very massive stars these
characteristic rays do not have the same colours as those from
lighter stars, but their colour is more towards the red.

This agrees with the predictions of general relativity. We
must assume that from the point of view of an observer on
the surface of the star, the atomic "clock" would be correct,
but owing to the dependence of the time scale on gravitational
potential, to us watching from a distance, the clocks would
appear to be slow and the frequency of the light smaller, and
the colour of the light more in the red.

This influence of the gravitational potential on the behaviour of clocks is interesting also because of its connection with
a paradox which illustrates the close connection between the
ideas of special and general relativity.

It is sometimes objected that special relativity is inconsistent
in its prediction of time dilation. Suppose two observers pass
each other at high speed and make sure that when they pass
their clocks show the same time. Now if we regard ourselves
as standing still we would conclude that the clock of the other
observer must be slow. He in turn would say that our clock
should be slow because we were moving at high speed. Ad-
mittedly this cannot be checked as long as we are a long way
from each other, since we have seen that we cannot unambiguously compare the times of events taking place in distant
places. However, supposing later we meet the other observer
again, then we can compare our clocks and we shall know
whose clock is slow with respect to the other and therefore
who has been moving.

Within special relativity the correct answer to this objection is that as long as the two observers keep moving at uniform speed relatively to each other, they can never meet again. If they are to meet again at least one of them must either reverse his motion, or change direction, and therefore be subject to acceleration, so that he is no longer entitled to assume that he has been at rest all the time.

General relativity provides a more complete answer. Suppose for the sake of argument we have been moving without acceleration, whereas the other observer after travelling some distance away stopped and returned. In general relativity he would still be entitled to assume that he had been standing still all the time, but that there was a gravitational field present. This would account for the forces that he felt acting on him (which we would ascribe to his acceleration). Now gravitational fields, as we have seen, affect the clocks, and if this is applied in detail to the situation under discussion the answer is that again on both views the readings of the two clocks would compare in the same way. . . .

AUTOBIOGRAPHICAL

NOTES

ALBERT EINSTEIN

This selection consists of two pasages from the "Autobiographical Notes" in *Albert Einstein: Philosopher-Scientist*, edited by P. A. Schilpp, Library of Living Philosophers, 2nd ed., Open Court Publishing Company, La Salle, Ill., 1951.

. . . And now to the critique of mechanics as the basis of physics.

From the first point of view (confirmation by experiment) the incorporation of wave-optics into the mechanical picture of the world was bound to arouse serious misgivings. If light was to be interpreted as undulatory motion in an elastic body (ether), this had to be a medium which permeates everything; because of the transversality of the lightwaves in the main similar to a solid body, yet incompressible, so that longitudinal waves did not exist. This ether had to lead a ghostly existence alongside the rest of matter, inasmuch as it seemed to offer no resistance whatever to the motion of "ponderable" bodies. In order to explain the refraction-indices of transparent bodies as well as the processes of emission and absorption of radiation, one would have had to assume complicated reciprocal actions between the two types of matter, something which was not even seriously tried, let alone achieved.

Furthermore, the electromagnetic forces necessitated the introduction of electric masses, which, although they had no noticeable inertia, yet interacted with each other, and whose interaction was, moreover, in contrast to the force of gravitation, of a polar type.

The factor which finally succeeded, after long hesitation, to bring the physicists slowly around to give up the faith in the possibility that all of physics could be founded upon Newton's mechanics, was the electrodynamics of Faraday and Maxwell. For this theory and its confirmation by Hertz's experiments showed that there are electromagnetic phenomena which by their very nature are detached from every ponderable matter—namely the waves in empty space which consist of electromagnetic "fields." If mechanics was to be maintained as the foundation of physics, Maxwell's equations had to be interpreted mechanically. This was zealously but fruitlessly attempted, while the equations were proving themselves fruitful in mounting degree. One got used to operating with these fields as independent substances without finding it necessary to give one's self an account of their mechanical nature; thus mechanics as the basis of physics was being abandoned, almost unnoticeably, because its adaptability to the facts presented itself finally as hopeless. Since then there exist two types of conceptual elements, on the one hand, material points with forces at a distance between them, and, on the other hand, the continuous field. It presents an intermediate state in physics without a uniform basis for the entirety, which—although unsatisfactory—is far from having been superseded. . . .

Now for a few remarks to the critique of mechanics as the foundation of physics from the second, the "interior," point of view. In today's state of science, i.e., after the departure from the mechanical foundation, such critique has only an interest in method left. But such a critique is well suited to show the type of argumentation which, in the choice of theories in the future will have to play an all the greater role the more the basic concepts and axioms distance themselves from what is directly observable, so that the confrontation of the implications of theory by the facts becomes constantly more difficult and more drawn out. First in line to be mentioned is Mach's argument, which, however, had already been clearly recognized by Newton (bucket experiment). From the standpoint of purely geometrical description all "rigid" co-ordinate systems are among themselves logically equivalent. The equations of mechanics (for example this is already true of the law of inertia) claim validity only when referred to a specific class of such systems, i.e., the "inertial systems." In this the co-ordinate system as bodily object is without any significance. It is necessary, therefore, in order to justify the necessity of the specific choice, to look for something which lies outside of the objects (masses, distances) with which the theory is concerned. For this reason "absolute space" as originally determinative was quite explicitly introduced by Newton as the omnipresent active participant in all mechanical events; by "absolute" he obviously means uninfluenced by the masses and by their motion. What makes this state of affairs appear particularly offensive is the fact that there are supposed to be infinitely many inertial systems, relative to each other in uniform translation, which are supposed to be distinguished among all other rigid systems.

Mach conjectures that in a truly rational theory inertia would have to depend upon the interaction of the masses, precisely as was true for Newton's other forces, a conception which for a long time I considered as in principle the correct one. It presupposes implicitly, however, that the basic theory should be of the general type of Newton's mechanics: masses and their interaction as the original concepts. The attempt at such a solution does not fit into a consistent field theory, as will be immediately recognized.

How sound, however, Mach's critique is in essence can be seen particularly clearly from the following analogy. Let us imagine people construct a mechanics, who know only a very small part of the earth's surface and who also cannot see any stars. They will be inclined to ascribe special physical attributes to the vertical dimension of space (direction of the acceleration of falling bodies) and, on the ground of such a conceptual basis, will offer reasons that the earth is in most places horizontal. They might not permit themselves to be influenced by the argument that as concerns the geometrical properties space is isotrope and that it is therefore supposed to be unsatisfactory to postulate basic physical laws, according to which there is supposed to be a preferential direction; they will probably be inclined (analogously to Newton) to assert the absoluteness of the vertical, as proved by experience as something with which one simply would have to come to terms. The preference given to the vertical over all other spatial directions is precisely analogous to the preference given to inertial systems over other rigid co-ordination systems.

Now to [a consideration of] other arguments which also concern themselves with the inner simplicity, i.e., naturalness, of mechanics. If one puts up with the concepts of space (including geometry) and time without critical doubts, then there exists no reason to object to the idea of action-at-a-distance, even though such a concept is unsuited to the ideas which one forms on the basis of the raw experience of daily life. However, there is another consideration which causes mechanics, taken as the basis of physics, to appear as primitive. Essentially there exist two laws:

(1) the law of motion

(2) the expression for force or potential energy.

The law of motion is precise, although empty, as long as the expression for the forces is not given. In postulating the latter, however, there exists great latitude for arbitrary [choice], especially if one omits the demand, which is not very natural in any case, that the forces depend only on the co-ordinates (and, for example, not on their differential quotients with respect to time). Within the framework of theory alone it is entirely arbitrary that the forces of gravitation (and electricity), which come from one point are governed by the potential function $(1/r)$. Additional remark: it has long been known that this function is the central-symmetrical solution of the simplest (rotation-invariant) differential equation $\delta\phi = 0$; it would therefore have been a suggestive idea to regard this as a sign that this function is to be regarded as determined by a law of space, a procedure by which the arbitrariness in the choice of the law of energy would have been removed. This is really the first insight which suggests a turning away from the theory of distant forces, a development which—prepared by Faraday, Maxwell and Hertz—really begins only later on under the external pressure of experimental data.

I would also like to mention, as one internal asymmetry of this theory, that the inert mass occurring in the law of motion also appears in the expression for the gravitational force, but not in the expression for the other forces. Finally I would like to point to the fact that the division of energy into two essentially different parts, kinetic and potential energy, must be felt as unnatural; H. Hertz felt this as so disturbing that, in his very last work, he attempted to free mechanics from the concept of potential energy (i.e., from the concept of force). . . .

. . . The example I saw before me was thermodynamics. The general principle was there given in the theorem: the laws of nature are such that it is impossible to construct a *perpetuum mobile* (of the first and second kind). How, then, could such a universal principle be found? After ten years of reflection such a principle resulted from a paradox upon which I had already hit at the age of sixteen: If I pursue a beam of light with the velocity c (velocity of light in a vacuum), I should observe such a beam of light as a spatially oscillatory electromagnetic field at rest. However, there seems to be no such thing, whether on the basis of experience or according to Maxwell's equations. From the very beginning it appeared to me intuitively clear that, judged from the standpoint of such an observer, everything would have to happen according to the same laws as for an observer who, relative to the earth, was at rest. For how, otherwise, should the first observer know, i.e., be able to determine, that he is in a state of fast uniform motion?

One sees that in this paradox the germ of the special relativity theory is already contained. Today everyone knows, of course, that all attempts to clarify this paradox satisfactorily were condemned to failure as long as the axiom of the absolute character of time, viz., of simultaneity, unrecognizedly was anchored in the unconscious. Clearly to recognize this axiom and its arbitrary character really implies already the solution of the problem. The type of critical reasoning which was required for the discovery of this central point was decisively furthered, in my case, especially by the reading of David Hume's and Ernst Mach's philosophical writings.

One had to understand clearly what the spatial co-ordinates and the temporal duration of events meant in physics. The physical interpretation of the spatial co-ordinates presupposed a fixed body of reference, which, moreover, had to be in a more or less definite state of motion (inertial system). In a given inertial system the co-ordinates meant the results of certain measurements with rigid (stationary) rods. (One should always be conscious of the fact that the presupposition of the existence in principle of rigid rods is a presupposition suggested by approximate experience, but which is, in principle, arbitrary.) With such an interpretation of the spatial co-ordinates the question of the validity of Euclidean geometry becomes a problem of physics.

If, then, one tries to interpret the time of an event analogously, one needs a means for the measurement of the difference in time (in itself determined periodic process realized by a system of sufficiently small spatial extension). A clock at rest relative to the system of inertia defines a local time. The local times of all space points taken together are the "time," which belongs to the selected system of inertia, if a means is given to "set" these clocks relative to each other. One sees that *a priori* it is not at all necessary that the "times" thus defined in different inertial systems agree with one another. One would have noticed this long ago, if, for the practical experience of everyday life light did not appear (because of the high value of c), as the means for the statement of absolute simultaneity.

The presupposition of the existence (in principle) of (ideal, viz., perfect) measuring rods and clocks is not independent of each other; since a light signal, which is reflected back and forth between the ends of a rigid rod, constitutes an ideal clock, provided that the postulate of the constancy of the light-velocity in vacuum does not lead to contradictions.

The above paradox may then be formulated as follows. According to the rules of connection, used in classical physics, of the spatial co-ordinates and of the time of events in the transition from one inertial system to another the two assumptions of

(1) the constancy of the light velocity
(2) the independence of the laws (thus specially also of the law of the constancy of the light velocity) of the choice of the inertial system (principle of special relativity)

are mutually incompatible (despite the fact that both taken separately are based on experience).

The insight which is fundamental for the special theory of relativity is this: The assumptions (1) and (2) are compatible if relations of a new type ("Lorentz-transformation") are postuated for the conversion of co-ordinates and the times of events. With the given physical interpretation of co-ordinates and time, this is by no means merely a conventional step, but implies certain hypotheses concerning the actual behavior of moving measuring-rods and clocks, which can be experimentally validated or disproved.

The universal principle of the special theory of relativity is contained in the postulate: The laws of physics are invariant with respect to the Lorentz-transformations (for the transition from one inertial system to any other arbitrarily chosen system of inertia). This is a restricting principle for natural laws, comparable to the restricting principle of the non-existence of the *perpetuum mobile* which underlies thermodynamics.

First a remark concerning the relation of the theory to "four-dimensional space." It is a widespread error that the special theory of relativity is supposed to have, to a certain extent, first discovered, or at any rate, newly introduced, the four-dimensionality of the physical continuum. This, of course, is not the case. Classical mechanics, too, is based on the four-dimensional continuum of space and time. But in the four-dimensional continuum of classical physics the subspaces with constant time value have an absolute reality, independent of the choice of the reference system. Because of this [fact], the four-dimensional continuum falls naturally into a three-dimensional and a one-dimensional (time), so that the four-dimensional point of view does not force itself upon one as *necessary*. The special theory of relativity, on the other hand, creates a formal dependence between the way in which the spatial co-ordinates, on the one hand, and the temporal co-ordinates, on the other, have to enter into the natural laws.

Minkowski's important contribution to the theory lies in the following: Before Minkowski's investigation it was necessary to carry out a Lorentz-transformation on a law in order to test its invariance under such transformations; he, on the other hand, succeeded in introducing a formalism such that the mathematical form of the law itself guarantees its invariance under Lorentz-transformations. By creating a four-dimensional tensor-calculus he achieved the same thing for the four-dimensional space which the ordinary vector-calculus achieves for the three spatial dimensions. He also showed that the Lorentz-transformation (apart from a different algebraic sign due to the special character of time) is nothing but a rotation of the co-ordinate system in the four-dimensional space.

First, a remark concerning the theory as it is characterized above. One is struck [by the fact] that the theory (except for the four-dimensional space) introduces two kinds of physical things, i.e., (1) measuring rods and clocks, (2) all other things, e.g., the electro-magnetic field, the material point, etc. This, in a certain sense, is inconsistent; strictly speaking measuring rods and clocks would have to be represented as solutions of the basic equations (objects consisting of moving atomic configurations), not, as it were, as theoretically self-sufficient entities. However, the procedure justifies itself because it was clear from the very beginning that the postulates of the theory are not strong enough to deduce from them sufficiently complete equations for physical events sufficiently free from arbitrariness, in order to base upon such a foundation a theory of measuring rods and clocks. If one did not wish to forego a physical interpretation of the co-ordinates in general (something which, in itself, would be possible), it was better to permit such inconsistency—with the obligation, however, of eliminating it at a later stage of the theory. But one must not legalize the mentioned sin so far as to imagine that intervals are physical entities of a special type, intrinsically different from other physical variables ("reducing physics to geometry," etc.).

We now shall inquire into the insights of definite nature which physics owes to the special theory of relativity.

(1) There is no such thing as simultaneity of distant events; consequently there is also no such thing as immediate action at a distance in the sense of Newtonian mechanics. Although the introduction of actions at a distance, which propagate with the speed of light, remains thinkable, according to this theory, it appears unnatural; for in such a theory there could be no such thing as a reasonable statement of the principle of conservation of energy. It therefore appears unavoidable that physical reality must be described in terms of continuous functions in space. The material point, therefore, can hardly be conceived any more as the basic concept of the theory.

(2) The principles of the conservation of momentum and of the conservation of energy are fused into one single principle. The inert mass of a closed system is identical with its energy, thus eliminating mass as an independent concept.

Remark. The speed of light c is one of the quantities which occurs as "universal constant" in physical equations. If, however, one introduces as unit of time instead of the second the time in which light travels 1 cm, c no longer occurs in the equations. In this sense one could say that the constant c is only an *apparently* universal constant.

It is obvious and generally accepted that one could eliminate two more universal constants from physics by introducing, instead of the gram and the centimeter, properly chosen "natural" units (for example, mass and radius of the electron).

If one considers this done, then only "dimension-less" constants could occur in the basic equations of physics. Concerning such I would like to state a theorem which at present cannot be based upon anything more than upon a faith in the simplicity, i.e., intelligibility, of nature: there are no *arbitrary* constants of this kind; that is to say, nature is so constituted that it is possible logically to lay down such strongly determined laws that within these laws only rationally completely determined constants occur (not constants, therefore, whose numerical value could be changed without destroying the theory). . . .

The special theory of relativity owes its origin to Maxwell's equations of the electromagnetic field. Inversely the latter can be grasped formally in satisfactory fashion only by way of the special theory of relativity. Maxwell's equations are the simplest Lorentz-invariant field equations which can be postulated for an anti-symmetric tensor derived from a vector field. This in itself would be satisfactory, if we did not know from quantum phenomena that Maxwell's theory does not do justice to the energetic properties of radiation. But how Maxwell's theory would have to be modified in a natural fashion, for this even the special theory of relativity offers no adequate foothold. Also to Mach's question: "How does it come about that inertial systems are physically distinguished above all other co-ordinate systems?" this theory offers no answer.

That the special theory of relativity is only the first step of a necessary development became completely clear to me only in my efforts to represent gravitation in the framework of this theory. In classical mechanics, interpreted in terms of the field, the potential of gravitation appears as a *scalar* field (the simplest theoretical possibility of a field with a single component). Such a scalar theory of the gravitational field can easily be made invariant under the group of Lorentz-transformations. The following program appears natural, therefore: The total physical field consists of a scalar field (gravitation) and a vector field (electromagnetic field); later insights may eventually make necessary the introduction of still more complicated types of fields; but to begin with one did not need to bother about this.

The possibility of the realization of this program was, however, dubious from the very first, because the theory had to combine the following things:

(1) From the general considerations of special relativity theory it was clear that the *inert* mass of a physical system increases with the total energy (therefore, e.g., with the kinetic energy).

(2) From very accurate experiments (specially from the torsion balance experiments of Eötvös) it was empirically known with very high accuracy that the gravitational mass of a body is exactly equal to its *inert* mass.

It followed from (1) and (2) that the *weight* of a system depends in a precisely known manner on its total energy. If the theory did not accomplish this or could not do it naturally, it was to be rejected. The condition is most naturally expressed as follows: the acceleration of a system falling freely in a given gravitational field is independent of the nature of the falling system (specially therefore also of its energy content).

It then appeared that, in the framework of the program sketched, this elementary state of affairs could not at all or at any rate not in any natural fashion, be represented in a satisfactory way. This convinced me that, within the frame of the special theory of relativity, there is no room for a satisfactory theory of gravitation.

Now it came to me: The fact of the equality of inert and heavy mass, i.e., the fact of the independence of the gravitational acceleration of the nature of the falling substance, may be expressed as follows: In a gravitational field (of small spatial extension) things behave as they do in a space free of gravitation, if one introduces in it, in place of an "inertial system," a reference system which is accelerated relative to an inertial system.

If then one conceives of the behavior of a body, in reference to the latter reference system, as caused by a "real" (not merely apparent) gravitational field, it is possible to regard this reference system as an "inertial system" with as much justification as the original reference system.

So, if one regards as possible, gravitational fields of arbitrary extension which are not initially restricted by spatial limitations, the concept of the "inertial system" becomes completely empty. The concept, "acceleration relative to space," then loses every meaning and with it the principle of inertia together with the entire paradox of Mach.

The fact of the equality of inert and heavy mass thus leads quite naturally to the recognition that the basic demand of the special theory of relativity (invariance of the laws under Lorentz-transformations) is too narrow, i.e., that an invariance of the laws must be postulated also relative to non-linear transformations of the co-ordinates in the four-dimensional continuum.

This happened in 1908. Why were another seven years required for the construction of the general theory of relativity? The main reason lies in the fact that it is not so easy to free oneself from the idea that co-ordinates must have an immediate metrical meaning. The transformation took place in approximately the following fashion.

We start with an empty, field-free space, as it occurs—related to an inertial system—in the sense of the special theory of relativity, as the simplest of all imaginable physical situations. If we now think of a non-inertial system introduced by assuming that the new system is uniformly accelerated against the inertial system (in a three-dimensional description) in one direction (conveniently defined), then there exists with reference to this system a static parallel gravitational field. The reference system may thereby be chosen as rigid, of Euclidian type, in three-dimensional metric relations. But the time, in which the field appears as static, is *not* measured by *equally constituted* stationary clocks. From this special example one can already recognize that the immediate metric significance of the co-ordinates is lost if one admits non-linear transformations of co-ordinates at all. To do the latter is, however, *obligatory* if one wants to do justice to the equality of gravitational and inert mass by means of the basis of the theory, and if one wants to overcome Mach's paradox as concerns the inertial systems.

If, then, one must give up the attempt to give the co-ordinates an immediate metric meaning (differences of co-ordinates = measurable lengths, viz., times), one will not be able to avoid treating as equivalent all co-ordinate systems, which can be created by the continuous transformations of the co-ordinates.

The general theory of relativity, accordingly, proceeds from the following principle: Natural laws are to be expressed by equations which are covariant under the group of continuous co-ordinate transformations. This group replaces the group of the Lorentz-transformations of the special theory of relativity, which forms a sub-group of the former.

This demand by itself is of course not sufficient to serve as point of departure for the derivation of the basic concepts of physics. In the first instance one may even contest [the idea] that the demand by itself contains a real restriction for the physical laws; for it will always be possible thus to reformulate a law, postulated at first only for certain co-ordinate systems, such that the new formulation becomes formally universally covariant. Beyond this it is clear from the beginning that an infinitely large number of field-laws can be formulated which have this property of covariance. The eminent heuristic significance of the general principles of relativity lies in the fact that it leads us to the search for those systems of equations which are *in their general covariant* formulation the *simplest ones possible*; among these we shall have to look for the field equations of physical space. Fields which can be transformed into each other by such transformations describe the same real situation.

The major question for anyone doing research in this field is this: Of which mathematical type are the variables (functions of the co-ordinates) which permit the expression of the physical properties of space ("structure")? Only after that: Which equations are satisfied by those variables?

The answer to these questions is today by no means certain. The path chosen by the first formulation of the general theory of relativity can be characterized as follows. Even though we do not know by what type field-variables (structure) physical space is to be characterized, we do know with certainty a special case: That of the "Field-free" space in the special theory of relativity. Such a space is characterized by the fact that for a properly chosen co-ordinate system the expression

$$ds^2 = dx_1{}^2 + dx_2{}^2 + dx_3{}^2 - dx_4{}^2 \qquad (1)$$

belonging to two neighboring points, represents a measurable quantity (square of distance), and thus has a real physical meaning. Referred to an arbitrary system this quantity is expressed as follows:

$$ds^2 = g_{ik} dx_i dx_k \qquad (2)$$

whereby the indices run from 1 to 4. The g_{ik} form a (real) symmetrical tensor. If, after carrying out a transformation on field (1), the first derivatives of the g_{ik} with respect to the co-ordinates do not vanish, there exists a gravitational field with reference to this system of co-ordinates in the sense of the above consideration, a gravitational field, moreover, of a very special type. Thanks to Riemann's investigation of n-dimensional metrical spaces this special field can be invariably characterized:

(1) Riemann's curvature-tensor R_{iklm}, formed from the coefficients of the metric (2) vanishes.

(2) The orbit of a mass-point in reference to the inertial system (relative to which (1) is valid) is a straight line, therefore an extremal (geodetic). The latter, however, is already a characterization of the law of motion based on (2).

The *universal* law of physical space must now be a generalization of the law just characterized. I now assume that there are two steps of generalization:

(a) pure gravitational field

(b) general field (in which quantities corresponding some-how to the electromagnetic field occur, too).

The instance (a) was characterized by the fact that the field can still be represented by a Riemann-metric (2), i.e., by a symmetric tensor, whereby, however, there is no representation in the form (1) (except in infinitesimal regions). This means that in the case (a) the Riemann-tensor does not vanish. It is clear, however, that in this case a field-law must be valid, which is a generalization (loosening) of this law. If this law also is to be of the second order of differentiation and linear in the second derivations, then only the equation, to be obtained by a single contraction

$$0 = R_{kl} = g^{im}R_{iklm}$$

came under consideration as field-equation in the case of (a). It appears natural, moreover, to assume that also in the case of (a) the geodetic line is still to be taken as representing the law of motion of the material point.

It seemed hopeless to me at that time to venture the attempt of representing the total field (b) and to ascertain field-laws for it. I preferred, therefore, to set up a preliminary formal frame for the representation of the entire physical reality; this was necessary in order to be able to investigate, at least preliminarily, the usefulness of the basic idea of general relativity. This was done as follows.

In Newton's theory one can write the field-law of gravitation thus:

$$\Delta\phi = 0$$

(ϕ = gravitation-potential) at points, where the density of matter, ρ, vanishes. In general one may write (Poisson equation)

$$\Delta\phi = 4\pi k\rho \cdot (\rho = \text{mass-density}).$$

In the case of the relativistic theory of the gravitational field R_{ik} takes the place of $\Delta\phi$. On the right side we shall then have to place a tensor also in place of ρ. Since we know from the special theory of relativity that the (inert) mass equals energy, we shall have to put on the right side the tensor of energy-density—more precisely the entire energy-density, insofar as it does not belong to the pure gravitational field. In this way one gets the field-equations

$$R_{ik} - \tfrac{1}{2} g_{ik} R = -k T_{ik}.$$

The second member on the left side is added because of formal reasons; for the left side is written in such a way that its divergence disappears identically in the sense of the absolute differential calculus. The right side is a formal condensation of all things whose comprehension in the sense of a field-theory is still problematic. Not for a moment, of course, did I doubt that this formulation was merely a makeshift in order to give the general principle of relativity a preliminary closed expression. For it was essentially not anything *more* than a theory of the gravitational field, which was somewhat artificially isolated from a total field of as yet unknown structure.

If anything in the theory as sketched—apart from the demand of the invariance of the equations under the group of the continuous co-ordinate-transformations—can possibly make the claim to final significance, then it is the theory of the limiting case of the pure gravitational field and its relation to the metric structure of space. For this reason, in what immediately follows we shall speak only of the equations of the pure gravitational field.

The peculiarity of these equations lies, on the one hand, in their complicated construction, especially their non-linear character as regards the field-variables and their derivatives, and, on the other hand, in the almost compelling necessity with which the transformation-group determines this complicated field-law. If one had stopped with the special theory of relativity, i.e., with the invariance under the Lorentz-group, then the field-law $R_{ik} = 0$ would remain invariant also within the frame of this narrower group. But, from the point of view of the narrower group there would at first exist no reason for representing gravitation by so complicated a structure as is represented by the symmetric tensor g_{ik}. If, nonetheless, one found sufficient reasons for it, there would then arise an immense number of field-laws out of quantities g_{ik}, all of which are covariant under Lorentz-transformations (not, however, under the general group). However, even if, of all the conceivable Lorentz-invariant laws, one had accidentally guessed precisely the law which belongs to the wider group, one would still not yet be on the plane of insight achieved by the general principle of relativity. For, from the standpoint of the Lorentz-group two solutions would incorrectly have to be viewed as physically different from each other, if they can be transformed into each other by a non-linear transformation of co-ordinates, i.e., if they are, from the point of view of the wider field, only different representations of the same field.

One more general remark concerning field-structure and the group. It is clear that in general one will judge a theory to be the more nearly perfect the simpler a "structure" it postulates and the broader the group is concerning which the field-equations are invariant. One sees now that these two demands get in each other's way. For example: according to the special theory of relativity (Lorentz-group) one can set up a covariant law for simplest structure imaginable (a scalar field), whereas in the general theory of relativity (wider group of the continuous transformations of co-ordinates) there is an invariant field-law only for the more complicated structure of the symmetric tensor. We have already given *physical* reasons for the fact that in physics invariance under the wider group has to be demanded:[1] from a purely mathematical standpoint I can see no necessity for sacrificing the simpler structure to the generality of the group.

The group of the general relativity is the first one which demands that the simplest invariant law be no longer linear or homogeneous in the field-variables and in their differential quotients. This is of fundamental importance for the following reason. If the field-law is linear (and homogeneous), then the sum of two solutions is again a solution; as, for example, in Maxwell's field-equations for the vacuum. In such a theory it is impossible to deduce from the field equations alone an interaction between bodies, which can be described separately by means of solutions of the system. For this reason all theories up to now required, in addition to the field equations, special equations for the motion of material bodies under the influence of the fields. In the relativistic theory of gravitation, it is true, the law of motion (geodetic line) was originally postulated independently in addition to the field-law equations. Afterwards, however, it became apparent that the law of motion need not (and must not) be assumed independently, but that it is already implicitly contained within the law of the gravitational field.

The essence of this genuinely complicated situation can be

visualized as follows: A single material point at rest will be represented by a gravitational field which is everywhere finite and regular, except at the position where the material point is located: there the field has a singularity. If, however, one computes by means of the integration of the field-equations the field which belongs to two material points at rest, then this field has, in addition to the singularities at the positions of the material points, a line consisting of singular points, which connects the two points. However, it is possible to stipulate a motion of the material points in such a way that the gravitational field which is determined by them does not become singular anywhere at all except at the material points. These are precisely those motions which are described in first approximation by Newton's laws. One may say, therefore: The masses move in such fashion that the solution of the field-equation is nowhere singular except in the mass points. This attribute of the gravitational equations is intimately connected with their non-linearity, and this is a consequence of the wider group of transformations.

Now it would of course be possible to object: If singularities are permitted at the positions of the material points, what justification is there for forbidding the occurrence of singularities in the rest of space? This objection would be justified if the equations of gravitation were to be considered as equations of the total field. [Since this is not the case], however, one will have to say that the field of a material particle may the less be viewed as a *pure gravitational field* the closer one comes to the position of the particle. If one had the field-equation of the total field, one would be compelled to demand that the particles themselves would *everywhere* be describable as singularity-free solutions of the completed field-equations. Only then would the general theory of relativity be a *complete* theory.

Notes

[1] To remain with the narrower group and at the same time to base the relativity theory of gravitation upon the more complicated (tensor-) structure implies a naïve inconsequence. Sin remains sin, even if it is committed by otherwise ever so respectable men.

THE FOUR-DIMENSIONAL
WORLD

Moritz Schlick

Reprinted from Chapter 7 of *The Philosophy of Nature*, translated by Amethe von Zeppelin, Philosophical Library, New York, 1949.

As against our preceding argument to the effect that it is meaningless to speak of a spatial deformation in which all objects participate, it could be objected that although the distortion may not result in perceptible changes in magnitude, it may nevertheless give rise to changes in physical behaviour, provided that the latter are not excluded by special assumptions concerning the values of the constants of nature either before or after the deformation. But if we ask how these constants are to be discovered, we find that this is achieved by establishing a connection between various coincidences—a connection which involves the *simultaneous* occurrence of the coincidences. (The pointer of a galvanometer must, for instance, indicate a definite number, while the mercury column of a thermometer arrives simultaneously at a definite mark on the scale.) Thus, we need only add that the relations of simultaneity between neighbouring coincidences are not to be affected by the deformation, and we can, then, be quite certain that observation can yield no other constants or natural laws, and that the supposed change is not actual but only a change in the way of speaking. It follows, therefore, that the description of nature is solely a question of spatio-temporal coincidences. The meaning and scope of this proposition is most clearly demonstrated—according to Minkowski's method—by a graphical representation in which time is introduced as the fourth co-ordinate in addition to the other three spatial co-ordinates.

The world thus described is a four-dimensional continuum traversed by world-lines, each of which is the image of the motion of a point (material point, or packet of energy). Since the world-lines represent motions, a deformation of the four-dimensional model denotes a change in the states of motion, e.g. some kind of curved and irregular motion of a particle could succeed a state of rest or uniform rectilinear motion of the same particle. Now since—as we have seen—a deformation never describes a genuine change, but only a change in the way of speaking, it follows that whether we ascribe a position of rest or some kind of motion to a particle, is also only a change in the way of speaking. Actually, the type of motion of a particle depends solely on the choice of a system of reference.

The description in terms of world-lines is limited to objects which satisfy the condition of "genetic-identity"[1] or, in other words, to objects of which it can be said significantly that they have remained "the same" during a finite period of time. In this model, the coincidences are represented either by sections or by tangents of world-lines. A model of this type can be distorted in a completely arbitrary manner and, provided that the topological relations of the world-lines are not deranged, it is an equally adequate representation of reality.

The world-lines describe the motion of particles; but they must not be mistaken for the tracks of these particles. One may not, for example, say that a point traverses its world-line; or that the three-dimensional section which represents the momentary state of the actual present, wanders along the time-axis through the four-dimensional world. For a wandering of this kind would have to take place in time; and time is already represented within the model and cannot be introduced again from outside. Hence, those arguments, according to which the four-dimensional description of the world proves the unreality of time and depicts the world as rigid motionless existence, are absurd. On the other hand, there is much to be said for designating the four-dimensional world as absolute in contradistinction to three-dimensional representations with added time specifications, for these latter contain more arbitrary features and in consequence are relative to the methods of description. In comparison with the four-dimensional description, these behave like the image in perspective of an object as compared with its plastic reproduction. The method of investigation of relativity theory (which could with more justification be called *absolute* theory) does not allow of any scope for subjectivity or arbitrariness on the part of observers. On the contrary, this theory demonstrates a greater objectivity than any of the former methods of description.

We distinguish between a kinematic and a dynamic meaning of the word "motion." In the first interpretation, motion is defined as change of position in time; and since data concerning position can only be given relatively to a frame of reference, motion according to definition is relative. In the case of the dynamic interpretation on the other hand, motion is defined by the way in which it occurs in the laws of nature; and in this context it might be quite possible that coincidences with a definite body (Neumann's body a) play a part in all the laws of motion. In that case, these bodies would have to be described as "being in a state of rest," and the motions referred to them as "absolute" because they would be distinguished in accordance with natural laws. Thus, when the general theory of relativity maintains that all motions, even in the dynamic sense, are relative, this is not a tautological, analytical proposition, but a statement of the fact that no frame of reference has a privileged position with respect to any arbitrary motions.

In the physics of Newton on the other hand, although no single frame or body was selected, a certain group of them was singled out—namely all those which were either in a position of rest relative to the fixed stars, or in a state of uniform, rectilinear motion. We call the latter group of systems of reference, Galilean or inertial systems; and relative to this group, the principle of inertia in the Galilean form as well as all the other laws of Newtonian physics are true. Later, the "aether of space" became the sole privileged frame of reference and lost this distinction when it transpired that all Galilean systems of reference are equally valid for all mechanical as well as for all electro-magnetic processes (propagation of light) (Einstein's special theory of relativity 1905). The impossibility of speaking of coincidences with aether particles and hence of their genetic-identity, showed that the concept of substance is inapplicable to the vacuum—a result of great philosophical significance.

Accelerated or curvilinear motions have an absolute character both in Newton's system and in the "restricted theory of relativity" of 1905. Only the "general theory of relativity" of 1915 abolished this absolute character and with it the privileged position of special groups of bodies, and thus led the way to a more satisfactory fulfilment of causal requirements. Neither in Newton's system nor in the special theory of relativity was any reason given for the special distinction of inertial systems.

In order to achieve this, Einstein had to try to formulate the law of inertia in such a way that no reference was made to a special group of bodies (an inertial system), but only to the actual configuration of existing bodies. An indication that this was possible consisted in an extremely striking fact which had been neglected by the physics of that date—namely that the measure of the inertia of a body (its inertial mass) is exactly equal to the quantity which measures the effects that are solely dependent on the configurations of bodies, the so-called gravitational effects. On the strength of this fact, Einstein succeeded in discovering a law which includes both gravitational and inertial phenomena (and showed that both are essentially the same). In consequence not only are all special reference frames eliminated, finite systems abolished, but a tremendous simplification of the world picture is achieved.

FOUNDATIONS OF THE GENERAL THEORY OF RELATIVITY

The new law of motion is a differential law—in other words it represents the motion of a particle as dependent, not on the configuration of the world, but solely on the measurable relations prevailing in its immediate neighbourhood which, in turn, are only indirectly conditioned by that configuration. Thus, the law of motion expresses a relation between the motion of a particle and the result of the measurements of time and lengths which are supposed to have been taken in its immediate neighbourhood and which are usually designated as propositions concerning the prevailing, "curvature of space" or "gravitational potential in that neighbourhood."

It cannot be sufficiently emphasized that this reference to the actual method of measurements constitutes the only way of understanding the meaning of the world picture as sketched above.

In order to establish coincidences, the genetic-identity of material points is assumed, at least for short intervals; and no strict distinction between real coincidence and close proximity can be made. Furthermore, all the experiences of coincidence are united by a continuous field of perception. The world of experience is thereby given a very special structure which can, perhaps, be formulated by saying that a particular physical significance is ascribed to the concept of immediate proximity; or that, in other words, a definite order of magnitude of lengths is actually privileged, within which it is not really possible to speak of any arbitrary deformations. The description of reality with the help of the four-dimensional schema is a result of the construction of physical space from psychological spaces (visual and tactile spaces, etc.). These latter are, however, by no means relative. In small regions of these spaces, both lengths and motions can be spoken of in an absolute sense and not as though founded on coincidences. In these regions, the application of Euclidean geometry is more than an arbitrary convention. Nevertheless, the question as to which geometries shall be constructed for greater or smaller magnitudes, that is, for atomic or astronomical dimensions, is no longer psychologically, but purely physically, determined.

NOTES

[1] The concept of "genetic-identity" was introduced by Kurt Lewin in his book, *Der Begriff der Genese in Physik, Biologie und Entwicklungsgeschichte: Eine Untersuchung zur vergleichenden Wissenschaftslehre*, published in 1922 by Springer in Berlin.

SPACE AND TIME

H. Minkowski

A translation of an address delivered at the 80th Assembly
of German Natural Scientists and Physicians, at Cologne,
September 21, 1908. Reprinted from *The Principle of Relativity*,
a collection of papers by Einstein and others translated
by W. Perrett and G. B. Jeffery, with notes by A. Sommerfeld,
Dover Publications, Inc., New York, 1923. First published by
Methuen & Co., Ltd., London.

The views of space and time which I wish to lay before you
have sprung from the soil of experimental physics, and
therein lies their strength. They are radical. Henceforth space
by itself, and time by itself, are doomed to fade away into
mere shadows, and only a kind of union of the two will preserve
an independent reality.

I

First of all I should like to show how it might be possible,
setting out from the accepted mechanics of the present day,
along a purely mathematical line of thought, to arrive at
changed ideas of space and time. The equations of Newton's
mechanics exhibit a two-fold invariance. Their form remains
unaltered, firstly, if we subject the underlying system of spatial
co-ordinates to any arbitrary *change of position*; secondly, if
we change its state of motion, namely, by imparting to it any
uniform translatory motion; furthermore, the zero point of
time is given no part to play. We are accustomed to look upon
the axioms of geometry as finished with, when we feel ripe for
the axioms of mechanics, and for that reason the two invariances
are probably rarely mentioned in the same breath. Each
of them by itself signifies, for the differential equations of
mechanics, a certain group of transformations. The existence
of the first group is looked upon as a fundamental characteristic
of space. The second group is preferably treated with
disdain, so that we with untroubled minds may overcome the
difficulty of never being able to decide, from physical phenomena,
whether space, which is supposed to be stationary,
may not be after all in a state of uniform translation. Thus

the two groups, side by side, lead their lives entirely apart. Their utterly heterogeneous character may have discouraged any attempt to compound them. But it is precisely when they are compounded that the complete group, as a whole, gives us to think.

We will try to visualize the state of things by the graphic method. Let x, y, z be rectangular co-ordinates for space, and let t denote time. The objects of our perception invariably include places and times in combination. Nobody has ever noticed a place except at a time, or a time except at a place. But I still respect the dogma that both space and time have independent significance. A point of space at a point of time, that is, a system of values x, y, z, t, I will call a *world-point*. The multiplicity of all thinkable x, y, z, t systems of values we will christen the *world*. With this most valiant piece of chalk I might project upon the blackboard four world-axes. Since merely one chalky axis as it is, consists of molecules all a-thrill, and moreover is taking part in the earth's travels in the universe, it already affords us ample scope for abstraction; the somewhat greater abstraction associated with the number four is for the mathematician no infliction. Not to leave a yawning void anywhere, we will imagine that everywhere and everywhen there is something perceptible. To avoid saying "matter" or "electricity" I will use for this something the word "substance." We fix our attention on the substantial point which is at the world-point x, y, z, t, and imagine that we are able to recognize this substantial point at any other time. Let the variations dx, dy, dz of the space co-ordinates of this substantial point correspond to a time element dt. Then we obtain, as an image, so to speak, of the everlasting career of the substantial point, a curve in the world, a *world-line*, the points of which can be referred unequivocally to the parameter t from $-\infty$ to $+\infty$. The whole universe is seen to resolve itself into similar world-lines, and I would fain anticipate myself by saying that in my opinion physical laws might find their most perfect expression as reciprocal relations between these world-lines.

The concepts, space and time, cause the x, y, z-manifold $t = 0$ and its two sides $t > 0$ and $t < 0$ to fall asunder. If, for simplicity, we retain the same zero point of space and time, the first-mentioned group signifies in mechanics that we may subject the axes of x, y, z at $t = 0$ to any rotation we choose about the origin, corresponding to the homogeneous linear transformations of the expression

$$x^2 + y^2 + z^2.$$

But the second group means that we may—also without changing the expression of the laws of mechanics—replace x, y, z, t by $x - \alpha t, y - \beta t, z - \gamma t, t$ with any constant values of α, β, γ. Hence we may give to the time axis whatever direction we choose towards the upper half of the world, $t > 0$. Now what has the requirement of orthogonality in space to do with this perfect freedom of the time axis in an upward direction?

To establish the connexion, let us take a positive parameter c, and consider the graphical representation of

$$c^2 t^2 - x^2 - y^2 - z^2 = 1.$$

It consists of two surfaces separated by $t = 0$, on the analogy of a hyperboloid of two sheets. We consider the sheet in the region $t > 0$, and now take those homogeneous linear transformations of x, y, z, t into four new variables x', y', z', t', for which the expression for this sheet in the new variables is of the same form. It is evident that the rotations of space about the origin pertain to these transformations. Thus we gain full comprehension of the rest of the transformations simply by taking into consideration one among them, such that y and z remain unchanged. We draw (Fig. 1) the section of this sheet by the plane of the axes of x and t—the upper branch of the hyperbola $c^2 t^2 - x^2 = 1$, with its asymptotes. From the origin O we draw any radius vector OA' of this branch of the hyperbola; draw the tangent to the hyperbola at A' to cut the asymptote on the right at B'; complete the parallelogram $OA'B'C'$; and finally, for subsequent use, produce $B'C'$ to cut the axis of x at D'. Now if we take OC' and OA' as axes of oblique co-ordinates x', t', with the measures $OC' = 1$, $OA' = 1/c$, then that branch of the hyperbola again acquires the expression $c^2 t'^2 - x'^2 = 1$, $t' > 0$, and the transition from x, y, z, t to x', y', z', t' is one of the transformations in question. With these transformations we now associate the arbitrary displacements of the zero point of space and time, and thereby constitute a group of transformations, which is also, evidently, dependent on the parameter c. This group I denote by G_c.

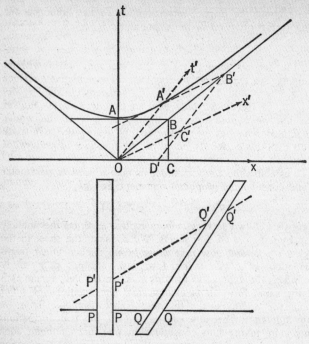

Fig. 1.

If we now allow c to increase to infinity, and $1/c$ therefore to converge towards zero, we see from the figure that the branch of the hyperbola bends more and more towards the axis of x, the angle of the asymptotes becomes more and more obtuse, and that in the limit this special transformation changes into one in which the axis of t' may have any upward direction whatever, while x' approaches more and more exactly to x. In view of this it is clear that group G_c in the limit when $c = \infty$, that is the group G_∞, becomes no other than that complete group which is appropriate to Newtonian mechanics. This being so, and since G_c is mathematically more intelligible than G_∞, it looks as though the thought might have struck some mathematician, fancy-free, that after all, as a matter of fact, natural phenomena do not possess an invariance with the group G_∞, but rather with a group G_c, c being finite and determinate, but in ordinary units of measure, *extremely great*. Such a premonition would have been an extraordinary triumph for pure mathematics. Well, mathematics, though it now can display only staircase-wit, has the satisfaction of being wise after the event, and is able, thanks to its happy antecedents, with its senses sharpened by an unhampered outlook to far horizons, to grasp forthwith the far-reaching consequences of such a metamorphosis of our concept of nature.

I will state at once what is the value of c, with which we shall finally be dealing. It is the velocity of the propagation of light in empty space. To avoid speaking either of space or of emptiness, we may define this magnitude in another way, as the ratio of the electromagnetic to the electrostatic unit of electricity.

The existence of the invariance of natural laws for the relevant group G_c would have to be taken, then, in this way:

From the totality of natural phenomena it is possible, by successively enhanced approximations, to derive more and more exactly a system of reference x, y, z, t, space and time, by means of which these phenomena then present themselves in agreement with definite laws. But when this is done, this system of reference is by no means unequivocally determined by the phenomena. It is still possible to make any change in the system of reference that is in conformity with the transformations of the group G_c and leave the expression of the laws of nature unaltered.

For example, in correspondence with the figure described above, we may also designate time t', but then must of necessity, in connexion therewith, define space by the manifold of the three parameters x', y, z, in which case physical laws would be expressed in exactly the same way by means of x', y, z, t' as by means of x, y, z, t. We should then have in the world no longer *space*, but an infinite number of spaces, analogously as there are in three-dimensional space an infinite number of planes. Three-dimensional geometry becomes a chapter in four-dimensional physics. Now you know why I said at the outset that space and time are to fade away into shadows, and only a world in itself will subsist.

II

The question now is, what are the circumstances which force this changed conception of space and time upon us?

Does it actually never contradict experience? And finally, is it advantageous for describing phenomena?

Before going into these questions, I must make an important remark. If we have in any way individualized space and time, we have, as a world-line corresponding to a stationary substantial point, a straight line parallel to the axis of t; corresponding to a substantial point in uniform motion, a straight line at an angle to the axis of t; to a substantial point in varying motion, a world-line in some form of a curve. If at any world-point x, y, z, t we take the world-line passing through that point, and find it parallel to any radius vector OA' of the above-mentioned hyperboloidal sheet, we can introduce OA' as a new axis of time, and with the new concepts of space and time thus given, the substance at the world-point concerned appears as at rest. We will now introduce this fundamental axiom:

The substance at any world-point may always, with the appropriate determination of space and time, be looked upon as at rest.

The axiom signifies that at any world-point the expression

$$c^2 dt^2 - dx^2 - dy^2 - dz^2$$

always has a positive value, or, what comes to the same thing, that any velocity v always proves less than c. Accordingly c would stand as the upper limit for all substantial velocities, and that is precisely what would reveal the deeper significance of the magnitude c. In this second form the first impression made by the axiom is not altogether pleasing. But we must bear in mind that a modified form of mechanics, in which the square root of this quadratic differential expression appears, will now make its way, so that cases with a velocity greater than that of light will henceforward play only some such part as that of figures with imaginary co-ordinates in geometry.

Now the impulse and true motive for assuming the group G_c came from the fact that the differential equation for the propagation of light in empty space possesses that group G_c.[1] On the other hand, the concept of rigid bodies has meaning only in mechanics satisfying the group G_∞. If we have a theory of optics with G_c, and if on the other hand there were rigid bodies, it is easy to see that one and the same direction of t would be distinguished by the two hyperboloidal sheets appropriate to G_c and G_∞, and this would have the further consequence, that we should be able, by employing suitable rigid optical instruments in the laboratory, to perceive some alteration in the phenomena when the orientation with respect to the direction of the earth's motion is changed. But all efforts directed towards this goal, in particular the famous interference experiment of Michelson, have had a negative result. To explain this failure, H. A. Lorentz set up an hypothesis, the success of which lies in this very invariance in optics for the group G_c. According to Lorentz any moving body must have undergone a contraction in the direction of its motion, and in fact with a velocity v, a contraction in the ratio

$$1 : \sqrt{1 - v^2/c^2}$$

This hypothesis sounds extremely fantastical, for the contraction is not to be looked upon as a consequence of resistances in the ether, or anything of that kind, but simply as a gift from above—as an accompanying circumstance of the circumstance of motion.

I will now show by our figure that the Lorentzian hypothesis is completely equivalent to the new conception of space and time, which, indeed, makes the hypothesis much more intelligible. If for simplicity we disregard y and z, and imagine a world of one spatial dimension, then a parallel band, upright like the axis of t, and another inclining to the axis of t (see Fig. 1) represent, respectively, the career of a body at rest or in uniform motion, preserving in each case a constant spatial extent. If OA' is parallel to the second band, we can introduce t' as the time, and x' as the space co-ordinate, and then the second body appears at rest, the first in uniform motion. We now assume that the first body, envisaged as at rest, has the length l, that is, the cross section PP of the first band on the axis of x is equal to $l \cdot OC$, where OC denotes the unit of measure on the axis of x; and on the other hand, that the second body, envisaged as at rest, has the same length l, which then means that the cross section $Q'Q'$ of the second band, measured parallel to the axis of x', is equal to $l \cdot OC'$. We now have in these two bodies images of two equal Lorentzian electrons, one at rest and one in uniform motion. But if we retain the original co-ordinates x, t, we must give as the extent of the second electron the cross section of its appropriate band parallel to the axis of x. Now since $Q'Q' = l \cdot OC'$, it is evident that $QQ = l \cdot OD'$. If dx/dt for the second band is equal to v, an easy calculation gives

$$OD' = OC\sqrt{1 - v^2/c^2},$$

therefore also $PP : QQ = 1 : \sqrt{1 - v^2/c^2}$. But this is the meaning of Lorentz's hypothesis of the contraction of electrons in motion. If on the other hand we envisage the second electron as at rest, and therefore adopt the system of reference x' t', the length of the first must be denoted by the cross section $P'P'$ of its band parallel to OC', and we should find the first electron in comparison with the second to be contracted in exactly the same proportion; for in the figure

$$P'P' : Q'Q' = OD : OC' = OD' : OC = QQ : PP.$$

Lorentz called the t' combination of x and t the local time of the electron in uniform motion, and applied a physical construction of this concept, for the better understanding of the hypothesis of contraction. But the credit of first recognizing clearly that the time of the one electron is just as good as that of the other, that is to say, that t and t' are to be treated identically, belongs to A. Einstein.[2] Thus time, as a concept unequivocally determined by phenomena, was first deposed from its high seat. Neither Einstein nor Lorentz made any attack on the concept of space, perhaps because in the above-mentioned special transformation, where the plane of x', t' coincides with the plane of x, t, an interpretation is possible by saying that the x-axis of space maintains its position. One may expect to find a corresponding violation of the concept of space appraised as another act of audacity on the part of the higher mathematics. Nevertheless, this further step is indispensable for the true understanding of the group G_o, and when it has been taken, the word *relativity-postulate* for the requirement of an invariance with the group G_c seems to me very feeble. Since the postulate comes to mean that only the four-dimensional world in space and time is given by phenomena, but that the projection in space and in time may still be undertaken with a certain degree of freedom, I prefer to call it the *postulate of the absolute world* (or briefly, the world-postulate).

III

The world-postulate permits identical treatment of the four co-ordinates x, y, z, t. By this means, as I shall now show, the forms in which the laws of physics are displayed gain in intelligibility. In particular the idea of acceleration acquires a clear-cut character.

I will use a geometrical manner of expression, which suggests itself at once if we tacitly disregard z in the triplex x, y, z. I take any world-point O as the zero-point of space-time. The cone $c^2t^2 - x^2 - y^2 - z^2 = 0$ with apex 0 (Fig. 2) consists of two parts, one with values $t < 0$, the other with values $t > 0$. The former, the front cone of O, consists, let us say, of all the world-points which "send light to O," the latter, the back cone of O, of all the world-points which "receive light from O." The territory bounded by the front cone alone, we may call "before" O, that which is bounded by the back

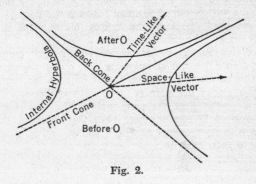

Fig. 2.

cone alone, "after" O. The hyperboloidal sheet already discussed

$$F = c^2t^2 - x^2 - y^2 - z^2 = 1, t > 0$$

lies after O. The territory between the cones is filled by the one-sheeted hyperboloidal figures

$$-F = x^2 + y^2 + z^2 - c^2t^2 = k^2$$

for all constant positive values of k. We are specially interested in the hyperbolas with O as centre, lying on the latter figures. The single branches of these hyperbolas may be called briefly the internal hyperbolas with centre O. One of these branches, regarded as a world-line, would represent a motion which, for $t = -\infty$ and $t = +\infty$, rises asymptotically to the velocity of light, c.

If we now, on the analogy of vectors in space, call a directed length in the manifold of x, y, z, t a vector, we have to distinguish between the time-like vectors with directions from O to the sheet $+F = 1$, $t > 0$, and the space-like vectors with directions from O to $-F = 1$. The time axis may run parallel to any vector of the former kind. Any world-point between the front and back cones of O can be arranged by means of the system of reference so as to be simultaneous with O, but also just as well so as to be earlier than O or later than O. Any world-point within the front cone of O is necessarily always before O; any world-point within the back cone of O necessarily always after O. Corresponding to passing to the limit, $c = \infty$, there would be a complete flattening out of the wedge-shaped segment between the cones into the plane manifold $t = 0$. In the figures this segment is intentionally drawn with different widths.

We divide up any vector we choose, e.g. that from O to x, y, z, t, into the four components x, y, z, t. If the directions of two vectors are, respectively, that of a radius vector OR from O to one of the surfaces $\mp F = 1$, and that of a tangent RS at the point R of the same surface, the vectors are said to be normal to one another. Thus the condition that the vectors with components x, y, z, t and x_1, y_1, z_1, t_1 may be normal to each other is

$$c^2 t t_1 - x x_1 - y y_1 - z z_1 = 0.$$

For the measurement of vectors in different directions the units of measure are to be fixed by assigning to a space-like vector from O to $-F = 1$ always the magnitude 1, and to a time-like vector from O to $+F = 1$, $t > 0$ always the magnitude $1/c$.

If we imagine at a world-point P (x, y, z, t) the world-line of a substantial point running through that point, the magnitude corresponding to the time-like vector dx, dy, dz, dt laid off along the line is therefore

$$d\tau = \frac{1}{c} \sqrt{c^2 dt^2 - dx^2 - dy^2 - dz^2}.$$

The integral $\int d\tau = \tau$ of this amount, taken along the world-line from any fixed starting-point P_0 to the variable end-point P, we call the proper time of the substantial point at P. On the world-line we regard x, y, z, t—the components of the vector OP—as functions of the proper time τ; denote their first differential coefficients with respect to τ by $\dot{x}$, $\dot{y}$, $\dot{z}$, $\dot{t}$; their second differential coefficients with respect to τ by $\ddot{x}$, $\ddot{y}$, $\ddot{z}$, $\ddot{t}$; and give names to the appropriate vectors, calling the derivative of the vector OP with respect to τ the velocity vector at P, and the derivative of this velocity vector with respect to τ the acceleration vector at P. Hence, since

$$c^2\dot{t}^2 - \dot{x}^2 - \dot{y}^2 - \dot{z}^2 = c^2,$$

we have

$$c^2\dot{t}\ddot{t} - \dot{x}\ddot{x} - \dot{y}\ddot{y} - \dot{z}\ddot{z} = 0,$$

i.e. the velocity vector is the time-like vector of unit magnitude in the direction of the world-line at P, and the acceleration vector at P is normal to the velocity vector at P, and is therefore in any case a space-like vector.

Now, as is readily seen, there is a definite hyperbola which has three infinitely proximate points in common with the world-line at P, and whose asymptotes are generators of a "front cone" and a "back cone" (Fig. 3). Let this hyperbola

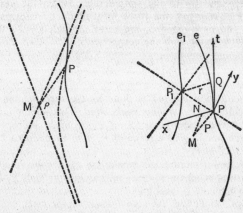

Fig. 3.   Fig. 4.

be called the hyperbola of curvature at P. If M is the centre of this hyperbola, we here have to do with an internal hyperbola with centre M. Let ρ be the magnitude of the vector MP; then we recognize the acceleration vector at P as the vector in the direction MP of magnitude c^2/ρ.

If $\ddot{x}$, $\ddot{y}$, $\ddot{z}$, $\ddot{t}$ are all zero, the hyperbola of curvature reduces to the straight line touching the world-line in P, and we must put $\rho = \infty$.

IV

To show that the assumption of group G_c for the laws of physics never leads to a contradiction, it is unavoidable to undertake a revision of the whole of physics on the basis of this assumption. This revision has to some extent already been successfully carried out for questions of thermodynamics and heat radiation,[3] for electromagnetic processes, and finally, with the retention of the concept of mass, for mechanics.[4]

For this last branch of physics it is of prime importance to raise the question: When a force with the components X, Y, Z parallel to the axes of space acts at a world-point P (x, y, z, t), where the velocity vector is $\dot{x}$, $\dot{y}$, $\dot{z}$, $\dot{t}$, what must we take this force to be when the system of reference is in any way changed? Now there exist certain approved statements as to the ponderomotive force in the electromagnetic field in the cases where the group G_c is undoubtedly admissible. These statements lead up to the simple rule: When the system of reference is changed, the force in question transforms into a force in the new space co-ordinates in such a way that the appropriate vector with the components iX, iY, iZ, iT, where

$$T = \frac{1}{c^2}\left(\frac{\dot{x}}{i}X + \frac{\dot{y}}{i}Y + \frac{\dot{z}}{i}Z\right)$$

is the rate at which work is done by the force at the world-point divided by c, remains unchanged. This vector is always normal to the velocity vector at P. A force vector of this kind, corresponding to a force at P, is to be called a "motive force vector" at P.

I shall now describe the world-line of a substantial point with constant mechanical mass m, passing through P. Let the velocity vector at P, multiplied by m, be called the "momentum vector" at P, and the acceleration vector at P, multiplied by m, be called the "force vector" of the motion at P. With these definitions, the law of motion of a point of mass with given motive force vector runs thus: *The force vector of motion is equal to the motive force vector.*[5] This assertion comprises four equations for the components corresponding to the four axes, and since both vectors mentioned are *a priori* normal to the velocity vector, the fourth equation may be looked upon as a consequence of the other three. In accordance with the above signification of *T*, the fourth equation undoubtedly represents the law of energy. Therefore the component of the momentum vector along the axis of *t*, multiplied by *c*, is to be defined as the kinetic energy of the point mass. The expression for this is

$$mc^2 \frac{dt}{d\tau} = mc^2 / \sqrt{1 - v^2/c^2},$$

i.e., after removal of the additive constant mc^2, the expression $\frac{1}{2}mv^2$ of Newtonian mechanics down to magnitudes of the order $1/c^2$. It comes out very clearly in this way, how the energy depends on the system of reference. But as the axis of *t* may be laid in the direction of any time-like vector, the law of energy, framed for all possible systems of reference, already contains, on the other hand, the whole system of the equations of motion. At the limiting transition which we have discussed, to $c = \infty$, this fact retains its importance for the axiomatic structure of Newtonian mechanics as well, and has already been appreciated in this sense by I. R. Schütz.[6]

We can determine the ratio of the units of length and time beforehand in such a way that the natural limit of velocity becomes $c = 1$. If we then introduce, further, $\sqrt{-1}\, t = s$ in place of *t*, the quadratic differential expression

$$d\tau^2 = -dx^2 - dy^2 - dz^2 - ds^2$$

thus becomes perfectly symmetrical in *x, y, z, s*; and this symmetry is communicated to any law which does not contradict the world-postulate. Thus the essence of this postulate may be clothed mathematically in a very pregnant manner in the mystic formula

$$3 \cdot 10^5 \, km = \sqrt{-1} \, secs.$$

V

The advantages afforded by the world-postulate will perhaps be most strikingly exemplified by indicating the effects proceeding from a point charge in any kind of motion according to the Maxwell-Lorentz theory. Let us imagine the world-line of such a point electron with the charge e, and introduce upon it the proper time τ from any initial point. In order to find the field caused by the electron at any world-point P_1, we construct the front cone belonging to P_1 (Fig. 4). The cone evidently meets the world-line, since the directions of the line are everywhere those of time-like vectors, at the single point P. We draw the tangent to the world-line at P, and construct through P_1 the normal P_1Q to this tangent. Let the length of P_1Q be r. Then, by the definition of a front cone, the length of PQ must be r/c. Now the vector in the direction PQ of magnitude e/r represents by its components along the axes of x, y, z, the vector potential multiplied by c, and by the component along the axis of t, the scalar potential of the field excited by e at the world-point P. Herein lie the elementary laws formulated by A. Liénard and E. Wiechert.[7]

Then in the description of the field produced by the electron we see that the separation of the field into electric and magnetic force is a relative one with regard to the underlying time axis; the most perspicuous way of describing the two forces together is on a certain analogy with the wrench in mechanics, though the analogy is not complete.

I will now describe the ponderomotive action of a moving point charge on another moving point charge. Let us imagine the world-line of a second point electron of the charge e_1, passing through the world-point P_1. We define P, Q, r as before, then construct (Fig. 4) the centre of M of the hyperbola of curvature at P, and finally the normal MN from M to a straight line imagined through P parallel to QP_1. With P as starting-point we now determine a system of reference as follows: The axis of t in the direction PQ, the axis of x in direction QP_1, the axis of y in direction MN, whereby finally the direction of the axis of z is also defined as normal to the axes of t, x, y. Let the acceleration vector at P be $\ddot{x}$, $\ddot{y}$, $\ddot{z}$, $\ddot{t}$, the velocity vector at P_1 be $\dot{x}_1$, $\dot{y}_1$, $\dot{z}_1$, $\dot{t}_1$. The motive force vector exerted at P_1 by the first moving electron e on the second moving electron e_1 now takes the form

$$-ee_1\left(\dot{t}_1-\frac{\dot{x}_1}{c}\right)\Re,$$

where the components $\Re_x$, $\Re_y$, $\Re_z$, $\Re_t$ of the vector $\Re$ satisfy the three relations

$$c\Re_t-\Re_x=\frac{1}{r^2}, \qquad \Re_y=\frac{\ddot{y}}{c^2r}, \qquad \Re_z=0,$$

and where, fourthly, this vector $\Re$ is normal to the velocity vector at P_1, and through this circumstance alone stands in dependence on the latter velocity vector.

When we compare this statement with previous formulations[8] of the same elementary law of the ponderomotive action of moving point charges on one another, we are compelled to admit that it is only in four dimensions that the relations here taken under consideration reveal their inner being in full simplicity, and that on a three-dimensional space forced upon us *a priori* they cast only a very complicated projection.

In mechanics as reformed in accordance with the world-postulate, the disturbing lack of harmony between Newtonian mechanics and modern electrodynamics disappears of its own accord. Before concluding I will just touch upon the attitude of Newton's law of attraction toward this postulate. I shall assume that when two points of mass m, m_1 describe their world-lines, a motive force vector is exerted by m on m_1, of exactly the same form as that just given in the case of electrons, except that $+mm_1$ must now take the place of $-ee_1$. We now specially consider the case where the acceleration vector of m is constantly zero. Let us then introduce t in such a way that m is to be taken as at rest, and let only m_1 move under the motive force vector which proceeds from m. If we now modify this given vector in the first place by adding the factor $\dot{t}^{-1}=\sqrt{1-v^2/c^2}$, which, to the order of $1/c^2$, is equal to 1, it will be seen[9] that for the positions x_1, y_1, z_1, of m_1 and their variations in time, we should arrive exactly at Kepler's laws again, except that the proper times τ_1 of m_1 would take the place of the times t_1. From this simple remark it may then be seen that the proposed law of attraction combined with the new mechanics is no less well adapted to explain astronomical observations than the Newtonian law of attraction combined with Newtonian mechanics.

The fundamental equations for electromagnetic processes in ponderable bodies also fit in completely with the world-postulate. As I shall show elsewhere, it is not even by any means necessary to abandon the derivation of these fundamental equations from ideas of the electronic theory, as taught by Lorentz, in order to adapt them to the world-postulate.

The validity without exception of the world-postulate, I like to think, is the true nucleus of an electromagnetic image of the world, which, discovered by Lorentz, and further revealed by Einstein, now lies open in the full light of day. In the development of its mathematical consequences there will be ample suggestions for experimental verifications of the postulate, which will suffice to conciliate even those to whom the abandonment of old-established views is unsympathetic or painful, by the idea of a pre-established harmony between pure mathematics and physics.

NOTES

[1] An application of this fact in its essentials has already been given by W. Voigt, *Göttinger Nachrichten*, 1887, p. 41.

[2] A. Einstein, *Ann. d. Phys.*, 17, 1905, p. 891; *Jahrb. d. Radioaktivität und Elektronik*, 4, 1907, p. 411.

[3] M. Planck, "Zur Dynamik bewegter Systeme," *Berliner Berichte*, 1907, p. 542; also in *Ann. d. Phys.*, 26, 1908, p. 1.

[4] H. Minkowski, "Die Grundgleichungen für die elektromagnetischen Vorgänge in bewegten Körpern," *Göttinger Nachrichten*, 1908, p. 53.

[5] H. Minkowski, *op. cit.*, p. 107. Cf. also M. Planck, *Verhandlungen der physikalischen Gesellschaft*, 4, 1906, p. 136.

[6] I. R. Schütz, "Das Prinzip der absoluten Erhaltung der Energie," *Göttinger Nachrichten*, 1897, p. 110.

[7] A. Liénard, "Champ électrique et magnétique produit par une charge concentrée en un point et animée d'un mouvement quelconque," *L'Eclairage Electrique*, 16, 1898, pp. 5, 53, 106; E. Wiechert, "Elektrodynamische Elementargesetze," *Arch. Néerl.* (2), 5, 1900, p. 549.

[8] K. Schwarzwald, *Göttinger Nachrichten*, 1903, p. 132; H. A. Lorentz, *Enzykl. d. math. Wissensch.*, V, Art. 14, p. 199.

[9] H. Minkowski, *op. cit.*, p. 110.

THE PHILOSOPHICAL RETENTION
OF ABSOLUTE SPACE
IN EINSTEIN'S GENERAL THEORY
OF RELATIVITY

ADOLF GRÜNBAUM

This selection is an enlarged excerpt from Grünbaum's essay by the same title published in *The Philosophical Review*, Volume LXVI (1957), pages 525–534.

The literature of recent decades on the philosophy and history of science has nurtured and given wide currency to a myth concerning the present status of the dispute between the absolutistic and relativistic theories of space. In particular, that literature is rife with assertions that the post-Newtonian era has witnessed "the final elimination of the concept of absolute space from the conceptual scheme of modern physics"[1] by Einstein's general theory of relativity and that the Leibniz-Huygens polemic against Newton and Clarke has thus been triumphantly vindicated.[2] In this vein, Philipp Frank recently reached the following verdict on Einstein's success in the implementation of Ernst Mach's program for a *relativistic* account of the *inertial* properties of matter: "Einstein started a new analysis of Newtonian mechanics which eventually vindicated Mach's reformulation [of Newtonian mechanics]."[3]

I shall now show that the history of the general theory of relativity does *not* bear out at all the widespread view set forth in the quotations from Max Jammer and Philipp Frank. And it will then become apparent in what precise sense there is ample justification for Einstein's own admission of 1953 as follows: the supplanting of the concept of absolute space is "a process which is probably by no means as yet completed."[4]

Mach had urged against Newton that both translational and rotational inertia are intrinsically dependent on the large-scale distribution and relative motion of matter. Assuming the indefinite extensibility of terrestrial axes to form an unlimited Euclidean rigid system S_e, the rotational motion of the stars seemed to be clearly defined with respect to S_e. Unfortunately, however, the general theory of relativity (hereafter referred to as "GTR") was not entitled to make use of S_e: the linear velocity of rotating mass points increases with the distance from the axis of rotation, and hence the existence of a system S_e of unrestricted size would allow *local* velocities greater than that of light, in contravention of the requirement of the local validity of the special theory of relativity. But to deny, as the GTR therefore must, that S_e can extend even as far as the planet Neptune is to assert that the Machian concept of the *relative* motion of the earth and the stars is no more meaningful physically than the Newtonian bugaboo of the *absolute* rotation of a solitary earth in a space which is structured independently of any matter that it might contain accidentally and indifferently![5] Accordingly, the earth must be held to rotate *not* relative to the stars but with respect to the local "star-compass" formed at the earth by stellar light rays whose paths are determined by the local *metrical field*. At Einstein's hands, Mach's thesis underwent not only this modification but also the following generalization: Einstein found that *both* the geometry of material rods and clocks *and* the inertial behavior of particles and light in the context of that geometry are functionally related to the same physical quantities. Probably unaware at the time that Riemann had previously conjectured the dependence of the geometry of physical space on the action of matter via a different line of reasoning,[6] Einstein named his own organic fusion of Riemann's and Mach's ideas "Mach's principle."[7] And he sought to implement that principle by requiring that the metrical field given by the quantities g_{ik} be *exhaustively* determined by properties and relations of ponderable matter and energy specified by the quantities T_{ik}. On this conception a single test particle would have no inertia whatever if all other matter and energy were either annihilated or moved indefinitely far away. But when the problem of solving the nonlinear partial differential equations which connect the derivatives of the g_{ik} to the T_{ik} was confronted, it became apparent that, far from having been exorcised by the GTR, the ghost of Newton's absolute space is nothing less than a haunting incubus. For to obtain a solution of these equations, it is necessary to supply the boundary conditions "at infinity." And to assume, as is done in Schwarz-schild's solution, that there are certain preferred co-ordinate systems in which the g_{ik} have the Lorentz-Minkowski values at infinity is to violate Mach's principle in the following two-fold sense: (a) The boundary conditions at infinity then assume the role of Newton's absolute space, since it is *not* the influence of matter that determines what co-ordinate systems at infinity are the Galilean ones of special relativity; and (b) instead of being the *source* of the *total* structure of space-time, matter then merely *modifies* the latter's otherwise autonomously flat structure. In 1916 Einstein first attempted to avoid this most unwelcome consequence by reluctantly altering the above field equations through the introduction of the cosmological constant λ, which yielded a solution in which space was closed (finite). But this rather forced step did not provide an escape from the troublesome philosophical difficulties that had cropped up in the boundary conditions at infinity, since these difficulties reappeared when W. de Sitter showed that the now *modified* equations violated Mach's prin-ciple by allowing a universe essentially devoid of matter to have a definitely structured space-time.[8] And in 1951 the Machian hope of subordinating space-time ontologically to matter was further dashed when A. H. Taub showed that there are conditions under which the *unmodified* field equations yield *curved* space in the absence of matter.[9]

These results inescapably raise the question of whether the failure of the GTR to implement Mach's principle is to be regarded as an inadequacy on the part of that theory or as a basis for admitting that the GTR was right in philosophically retaining Newton's absolute space to a significant extent, thinly disguised under new structural trappings. Einstein's own attitude in his last years seems to have been one of unmourn-ing abandonment of Mach's principle. His reason appears to have been that although matter provides the *epistemological* basis for the metrical field, this fact must *not* be held to confer *ontological* primacy on matter over the field: matter is merely part of the field rather than its source.[10] This is indeed a very far cry from, nay the very antithesis of Professor Jammer's "final elimination of the concept of absolute space from the conceptual scheme of modern physics" (*op. cit.*, p. 2). In fact, Jammer himself quotes a recent passage from Einstein in which Einstein says that if the *space-time* field were removed, there would be no space (*op. cit.*, p. 172). Yet Jammer gives no indication whatever that this is a drastically different thesis from Einstein's earlier one that if all *matter* were annihilated, then metric space would vanish as well.[11]

It is now clear that the GTR cannot be said to have resolved the controversy between the absolutistic and relativistic conceptions of space in favor of the latter. Instead, the current state of knowledge supports the following summary assessment given in 1961 by the physicists C. Brans and R. H. Dicke:

> The . . . view that the geometrical and inertial properties of space are meaningless for an empty space, that the physical properties of space have their origin in the matter contained therein, and that the only meaningful motion of a particle is motion relative to other matter in the universe has never found its complete expression in a physical theory. This picture is . . . old and can be traced from the writings of Bishop Berkeley[12] to those of Ernst Mach.[13] These ideas have found a limited expression in general relativity, but it must be admitted that, although in general relativity spatial geometries are affected by mass distributions, the geometry is not uniquely specified by the distribution. It has not yet been possible to specify boundary conditions on the field equations of general relativity which would bring the theory into accord with Mach's principle. Such boundary conditions would, among other things, eliminate all solutions without mass present.[14]

In view of the difficulties encountered by the attempt to incorporate Mach's principle into the GTR, Brans and Dicke have put forward a *modified* relativistic theory of gravitation which is "apparently compatible with Mach's principle."[15]

It remains to be seen whether this modification of the GTR turns out to be satisfactory in other respects.

NOTES

[1] Max Jammer, *Concepts of Space* (Cambridge, Mass.: Harvard University Press, 1954), p. 2.

[2] A very useful modern edition by H. G. Alexander of *The Leibniz-Clarke Correspondence* was published in 1956 by the Manchester University Press and by the Philosophical Library in New York.

3 Philipp Frank, *Philosophy of Science* (Englewood Cliffs, N.J.: Prentice-Hall, 1957), p. 153.

4 A. Einstein, Foreword to Max Jammer, *op. cit.*, p. xv.

5 For details, see H. Weyl, "Massenträgheit und Kosmos," *Naturwissenschaften*, XII (1924), 197. See also F. E. A. Pirani, "On the Definition of Inertial Systems in General Relativity," in *Bern Jubilee of Relativity Theory*, Suppl. IV of *Helvetica Physica Acta* (Basel, 1956), pp. 198-203.

6 Cf. Bernard Riemann, *Über die Hypothesen welche der Geometrie zu Grunde liegen*, H. Weyl, ed. (3d ed.; Berlin, 1923), pp. 3, 20. For an account of Riemann's reasoning, see A. Grünbaum, *Philosophical Problems of Space and Time* (New York: Knopf, 1963), chap. 1, sec. B, and chap. 15.

7 Albert Einstein, "Prinzipielles zur allgemeinen Relativitätstheorie," *Annalen der Physik*, LV (1918), 241.

8 The attempt to dispose of the difficulty at infinity by laying down the finitude of space as a *boundary condition* governing the solution of the *unmodified* field equations is unavailing for the purpose of rescuing Mach's principle, since such a speculative assumption involves a nonintrinsic connection between the overall structure of space and the properties of matter.

9 A. H. Taub, "Empty Space-Times Admitting a Three-Parameter Group of Motions," *Annals of Mathematics*, LIII (1951), 472.

10 For a discussion of the status of Einstein's program of field theory, see J. Callaway, "Mach's Principle and Unified Field Theory," *Physical Review*, XCVI (1954), 778. For an alternative theory of gravitation inspired by the aim of strict conformity to Mach's principle but incomplete in other respects, see D. W. Sciama, "On the Origin of Inertia," *Monthly Notices Roy. Astr. Soc.*, CXIII (1953), 35, and "Inertia," *Scientific American*, CXCVI (February, 1957), 99-109. Cf. also F. A. Kaempffer, "On Possible Realizations of Mach's Program," *Canadian J. Phys.*, 36 (1958), 151-159.

11 Jammer has since taken account of these criticisms on pp. vii and 195 of the revised Harper Torchbook edition of his book, published in New York in 1960.

12 G. Berkeley, *The Principles of Human Knowledge* (1710), pars. 111-117, and *De Motu* (1726).

13 E. Mach, *Conservation of Energy*, Note No. 1 (1872), reprinted by Open Court Publishing Company, LaSalle, Ill., 1911, and *The Science of Mechanics* (1883), reprinted by the same publisher, 1902, Chap. II, Sec. VI.

14 C. Brans and R. H. Dicke, "Mach's Principle and a Relativistic Theory of Gravitation," *Physical Review*, CXXIV (1961), 925.

15 *Ibid.*

IV

Recent
Philosophical
Analyses

OSTENSIBLE

TEMPORALITY

C. D. BROAD

Reprinted from parts of Chapter XXXV of *Examination of McTaggart's Philosophy*, Volume II, Part I, by C. D. Broad, Cambridge University Press, Cambridge, England, 1938.

Much the most puzzling set of temporal characteristics are those which are involved in facts of the following kind. An experience is at one time wholly in the future, as when one says "I am going to have a painful experience at the dentist's tomorrow." It keeps on becoming less and less remotely future. Eventually the earliest phase of it becomes present; as when the dentist begins drilling one's tooth, and one thinks or says "The painful experience which I have been anticipating has now begun." Each phase ceases to be present, slips into the immediate past, and then keeps on becoming more and more remotely past. But it is followed by phases which were future and have become present. Eventually the latest phase of this particular experience becomes present and then slips into the immediate past. There is the fact which one records by saying, "Thank God (on the theistic hypothesis) that's over now!" After that the experience as a whole retreats continually into the more and more remote past.

There is no doubt that the sentences which I have just been quoting record facts, and that such facts are of the very essence of Time. But it is, of course, quite possible that the grammatical form of these sentences is highly misleading. It may dispose people to take for granted a certain view of the structure and the elements of these facts, and this view may be mistaken and may lead to difficulties and contradictions.

The two aspects of duration and temporal relations are very closely interconnected, and it is in respect of them that there is a close analogy between Time and Space. I shall therefore class them together under the name of "the extensive aspect of temporal facts." The third feature is absolutely peculiar to

Time, and bears no analogy to any feature of spatial facts. I will call it "the transitory aspect of temporal facts." I will first take these two aspects separately, and will then consider the relations between them.

1·1. *The Extensive Aspect of Temporal Facts.* There is evidently a very close analogy between a person's mental history, taken as a whole, and a cord made up of shorter strands arranged in the following way. The shorter strands are all parallel to each other and to the axis of the cord. No strand stretches the whole length of the cord; the strands are of various lengths and the two ends of any one strand are in general at different positions, respectively, from the two ends of any other strand. Any short segment of the cord will contain segments of several overlapping strands; but two short segments of the cord at some distance apart may be composed of segments of wholly different strands. Some strands may be practically uniform in colour and texture throughout their length. Others may vary greatly in colour or texture from one end to the other. The former correspond to monotonous experiences, and the latter to variegated and exciting experiences.

This spatial analogy is valid and useful up to a point; but I will now indicate some important ways in which it breaks down. (i) The triadic relation "between" occurs both in a linear spatial series and in a temporal series. We can say both that Bletchley is between Euston and Rugby, and that the experience of writing this sentence is between the experience of eating my breakfast and that of eating my dinner. Nevertheless, there is a profound difference. Temporal betweenness is not fundamental; it is analysable into the relational product of a certain *dyadic* relation taken twice over. The fundamental facts are that eating my breakfast *preceded* writing the sentence, and that writing the sentence *preceded* eating my dinner. The triadic relational fact that writing the sentence is between eating my breakfast and eating my dinner is analysable into the conjunction of these two dyadic relational facts.

Now in the linear spatial series the exact opposite is the case. No doubt one can say that Euston is south of Bletchley and that Bletchley is south of Rugby, and one can compare this with my breakfast preceding my writing the sentence and the latter preceding my dinner. But there is a fundamental difference. The relation "south of" tacitly involves a reference to some third term beside those which are explicitly mentioned, viz., to the sun or to a compass-needle. But the relation "earlier than" is a genuinely dyadic relation which directly relates two experiences of the same person and contains no tacit reference to some third term.

We may sum this up as follows. In a linear spatial series there is no asymmetric dyadic relation intrinsic to the series. The only relation which does not involve a tacit reference to some term outside the series is the partly symmetrical and partly asymmetrical *triadic* relation of "betweenness." This is partly symmetrical because, if *B* is between *A* and *C*, then it is equally between *C* and *A*; and conversely. It is partly asymmetrical because, if *B* is between *A* and *C*, *C* cannot be between *A* and *B* and *A* cannot be between *B* and *C*. In the temporal series of experiences which constitutes a person's mental history there is a genuine dyadic relation which is intrinsic to the series and involves no reference to any term outside the latter. This is the relation "earlier than." It is the fundamental relation here, and *temporal* betweenness is definable in terms of it. In the temporal series there are two intrinsically opposite directions, earlier-to-later and later-to-earlier. In the linear spatial series there is no *intrinsic* direction. If direction is to be introduced, this must be done *extrinsically*, either by reference to motion along the line (and therefore to time), or by reference to the right and left hands of an external observer, or in some other way.

(ii) Spatial extension and the occurrence of spatial relations *presuppose* temporal duration and a certain determinate form of temporal relation. Shape and size are commonly ascribed to particulars which persist through periods of time and have histories of longer or shorter duration. Since, however, one and the same thing can have different determinate shapes and sizes at different times in its history, we have to divide its history into short successive phases during each of which its shape and size are sensibly constant. Thus we reach the limiting conception of "the shape and size of a certain thing at a certain moment." If the thing is very rigid and usually remains practically unchanged in shape and size over long periods, we often drop the reference to a particular moment and refer to the shape and size which it has at every moment throughout such a long period as "*the* shape and size of this thing." Again, if a thing is elastic, there may be a certain shape and size which it will automatically assume whenever it is free from external distorting or compressing forces. We sometimes refer to this as "*the* shape and size" or "the *natural* shape and size" of such a body, even though the body is at most moments in its history subject to external forces which distort or compress it. I think that it is clear from these remarks that the notions of shape and size, as applied to bodies, all involve a tacit or explicit reference to temporal characteristics.

We do not very often apply the notions of shape and size to events or processes, as distinct from material things. But we do, e.g., talk of a "long jagged flash of lightning." I think that we talk in this way only when the event or process is so short as to be sensibly instantaneous.

Lastly, we talk of spatial relations between two events only when each is sensibly instantaneous and the two are simultaneous with each other. And we talk of spatial relations between two material things only when the following conditions are fulfilled. The histories of the two things must go on parallel to each other in time. Then each history must be divided into successive instantaneous states, and we must consider the spatial relations between the two bodies at each pair of simultaneous instants in their respective histories. Thus there is a rather elaborate and complicated temporal relation implied in talking of spatial relations between bodies.

Now contrast all this with the extensive aspect of temporal facts. Temporal relations *directly* relate *events or processes*; they do not directly relate the continuants of which events and processes constitute the histories. Again, it is the events or processes which are temporally extended, i.e., which are longer or shorter in the temporal sense. The continuants, of which these events or processes constitute the histories, *endure through* periods of time. And the period through which a continuant endures is measured by the length of its history. Lastly, it is evident that a temporal whole may be composed of parts which do not temporally overlap but are completely successive to each other. Consider a variegated process, such as a single rendering of a certain tune on a piano. It can be regarded as a whole composed of adjoined phases, each of shorter duration, such that each phase wholly precedes one, and wholly follows another, phase of the same process.

I think that I have now made it plain that the unlikeness between spatial and temporal facts is almost as striking as the likeness, even when we confine ourselves to the purely extensive aspect. So far as I can see, all spatial illustrations even of the extensive aspect of temporal facts presuppose temporal notions. For our lines, etc., are all things which endure through certain periods and have longer or shorter temporal histories. This is no reason for refusing to use such illustrations, if we find them helpful. But it makes it certain that a point will be reached after which they can give us no further help.

1·2. *The Transitory Aspect of Temporal Facts.* We will now turn to the transitory aspect of temporal facts. Here there are two points to be considered, viz., (i) the characteristics of pastness, presentness, and futurity; and (ii) the fact that every event is continually changing in respect of these characteristics. It continually becomes less and less remotely future, then it becomes present, and then it continually becomes more and more remotely past.

The first remark to be made is concerned primarily with language, but it leads on to a conclusion which is not purely linguistic. In all the languages with which I am acquainted there are two different ways of recording such temporal facts as we are now considering. The most usual way is by means of differences of tense in inflected languages, or by means of a temporal copula, which can take three different forms, together with certain temporal adverbs. Thus I should most naturally say "I *had* my breakfast *lately*," "I *am* writing *now*," "I *shall be* eating my lunch *soon*," and so on. The other way is by means of a single uniform copula and temporal adjectives, which take three different forms, together with certain adverbs. Thus I might have said "Eating my breakfast is just past," "My writing is present," "Eating my lunch is slightly future," and so on. Such expressions are rather unnatural; but they are intelligible, and in some contexts they would be quite normal. Thus it sounds quite natural to say "The next glacial period is in the remote future." The various temporal copulas can be combined with the various temporal adjectives in many ways. Thus we can say "The invention of wireless broadcasting was still future when Queen Victoria died," "This spell of writing will be past (or over) when I am eating my lunch," and so on.

I come now to the point which is not purely linguistic. It is this. By using various forms of temporal adjective we may be able to reduce the number of forms of temporal copula needed in recording temporal facts to the single copula "is." We can, e.g., replace the sentence "I *was* eating my breakfast, I *am* writing, and I *shall be* eating my lunch" by the sentence

"Eating my breakfast *is* past, this spell of writing *is* present, and eating my lunch *is* future." But the "is" is the temporal copula "is now," which a person would use if he said of me "He is now writing"; it is not the non-temporal copula which would be used if one said "37 is a prime number" or "Scarlet is a determinate form of red."

The following considerations make this quite plain. Suppose that, on a certain occasion, I utter the sentence "The event *e* is present." And suppose that this utterance records a fact. If the word "is" in it were a non-temporal copula, *every* utterance by me of the same sentence would record the same fact, no matter whether it were earlier than, contemporary with, or later than this utterance of mine. But actually the only utterances of this sentence which would record the same fact as this utterance of mine would be those which are *contemporary with* my utterance. Earlier or later utterances of this sentence would simply be false; though an earlier utterance of the sentence "The event *e* is future" would be true, and a later utterance of the sentence "The event *e* is past" would also be true. Similar remarks would apply, *mutatis mutandis*, if we had taken as our example a true utterance of the sentence "The event *e* is future" or a true utterance of the sentence "The event *e* is past." It is clear then that there can be no question of getting rid altogether of temporal copulas, and replacing them by a single non-temporal copula and various temporal predicates. This point is highly relevant in connexion with McTaggart's argument against the reality of Time.

1·21. *Pastness, Presentness, and Futurity.* I will now make some remarks about the three temporal characteristics of pastness, presentness, and futurity. The first point to notice is this. If we regard them as three determinates under a single determinable, they are not on a level, as red, green, blue, etc., are when regarded as determinates under the determinable colour. Pastness and futurity are each capable of an infinite range of different degrees. Presentness is generally thought of as being incapable of variation in degree. (I am ignoring the doctrine of the "Specious Present" for the moment; I shall deal with it later.) Thus the three temporal characteristics are like the three determinates, hot, neutral, and cold, under the determinable of sensible temperature. If we want to put them on a level, we must take as our ultimate

determinates presentness, the various determinate degrees of
pastness, and the various determinate degrees of futurity.

When we do this, however, we are faced with the following
complication. Any experience has some duration; it is like a
line and not like a geometrical point. But neither presentness,
in the strict sense, nor any absolutely determinate degree of
pastness or futurity, can characterise a temporally extended
term. Such a term cannot be *present* as a whole. If it is *past*
as a whole, any earlier phase of it will have a greater degree
of pastness than any later phase; and, if it is *future* as a whole,
any earlier phase of it will have a less degree of futurity than
any later phase. Thus the notions of strict presentness and of
perfectly determinate degrees of pastness or futurity are in-
separably bound up with the notion of strictly instantaneous
terms, i.e., terms which have temporal position but no duration, and are analogous to geometrical points or unextended
particles. I propose to call such terms "event-particles." I think
that some event-particles are boundaries of events. And I
think that we are able to form the conception of event-
particles because we often prehend events as having boundaries. I will now try to explain what I mean by these cryptic
utterances.

I will begin by taking a spatial parallel. Suppose I draw a
line in blue ink on a sheet of white paper and look at it. Then
I shall prehend a long thin blue sensum surrounded by a
white sensum. Each of these is visibly extended in two dimensions; for the blue sensum, though thin, is visibly of some
width. But, in addition, I have an experience which I can
only describe as "prehending the common boundary of the
blue sensum and the white sensum." Such boundaries are, in
the strictest sense, *lines*, i.e., terms which are extended in one
and only one spatial dimension. Very likely the phrases which
I have used to describe this experience are in some respects
misleading; but I think that everyone will recognise the kind
of experience which I am recording by means of them. I
do not think that these "boundaries without breadth" are
coloured, and I should not be prepared to call them "visual
sensa." But anyone who prehends such visual sensa as I have
been describing will, *ipso facto*, prehend such a boundary. I
think that such experiences are the sensible basis of the notion
of lines without breadth. Suppose now that I draw on the
same sheet of paper a second ink-line which cuts the first. The

two lines, being of finite thickness, intersect in a small *blue area*. But their boundaries, being of only one dimension, intersect in *four colourless points* which are at the four corners of this area. Unless the lines are very thick these four points are, of course, very near together. The thinner the lines, the nearer together are the four points in which their boundaries intersect each other. As the lines approach indefinitely near to zero thickness, the four points of intersection approach indefinitely near to coalescing in a single point. We express this, briefly and inaccurately, by saying that two coloured lines without breadth would intersect in a certain point with position and no magnitude.

Let us now pass from the spatial analogy to the temporal problem which is our main business. Consider the following example. A uniform background of sound has been going on for some time, e.g., the faint hissing of my gas-fire. Suddenly the clock begins to strike. The striking-experience and the hissing-experience both have duration. But it seems to me that, in some important sense of "hearing," I also "hear" the boundary between hissing-without-striking and hissing-with-striking. It seems to me that this has absolutely determinate temporal position in my experience but no temporal extension. I do not think that it has an auditory quality, in the sense in which the hissing and the striking have auditory qualities, any more than the boundary between a blue sensum and a white sensum which surrounds it has a colour. I am inclined to think that the actual experience of temporal boundaries is the basis of the notion of event-particles.

Consider any process which is either qualitatively uniform or continuously varying in quality. An example of the first kind of process would be the hissing noise-process which is going on while I am hearing my gas-fire throughout the morning. An example of the second would be the noise-process which I hear when I stand on the platform of a railway station and an engine approaches whistling and runs past me. Here there is continuous variation both in loudness and in pitch. Such processes do not fall into successive temporally extended phrases audibly adjoined at boundaries. The first may be compared to a uniformly shaded red band; the second may be compared to a red band which varies continuously in shade from one end to the other. Now consider a third kind of auditory experience. Suppose that we have a toothed wheel which can be made to rotate for a period with uniform

velocity, and can be made to rotate at different times for the
same period with different uniform velocities. Suppose that a
card is held with its edge against the teeth of this wheel. Let
the experiments which I am about to describe all be conducted against the uniform auditory background of the hissing
of a gas-fire.

Let the wheel first be rotated fairly slowly with uniform
velocity for a short period. Then I shall hear a discontinuous
series of short qualitatively similar noises. Between each will
come a short phase of hissing. Each click will be adjoined at
its earlier and later ends to a phase of hissing, and the adjunc-
tion will audibly take place along a temporal boundary. Now
repeat the experiment for the same period with the wheel
being rotated uniformly but faster than before. There will be
more clicks and more phases of hissing; and each click will
be shorter and so too will be each intermediate phase of
hissing. Thus the later boundary of any click will now be
nearer in time to the earlier boundary of the next click. If
this experiment is repeated with the wheel rotating faster and
faster on each occasion, a point will be reached at which the
auditory experience changes in character. Instead of a discontinuous series of clicks, each with an audible boundary,
there will be a continuous noise-process. We can think of the
successive clicks as getting shorter and shorter and more and
more numerous, and we can think of the intermediate phases
of hissing as doing the same. Thus the later boundary of any
click approaches nearer and nearer in temporal position to the
earlier boundary of the next click. When the noise-process has
become continuous we can think of the successive clicks as
having approached to the limit of zero duration, and similarly
for the intermediate hissing phases; and we can think of the
later boundary of any click as having approached to the limit
of coalescing with the earlier boundary of the next click. Thus
we form the conception of a continuous uniform noise-process
as consisting of a *compact* series of qualitatively similar *event-
particles*, each with a different and absolutely determinate
temporal position and without any *duration*. Similarly, we
think of a continuously variable noise process as consisting
of a compact series of event-particles, each with its own absolutely determinate temporal position and sound-quality. The
sound-qualities of any two such event-particles will be more
and more alike the nearer together the event-particles are in
respect of temporal position; and, as the difference in temporal

position approaches zero as a limit, so the difference in sound-quality approaches zero as a limit.

I think that I have now given a fairly plausible account of the experiential basis of the notion of event-particles, and of the assumption that any process of finite duration can be regarded as consisting of a compact series of successive event-particles. It is evident that presentness, in the strict sense, and absolutely determinate degrees of pastness or futurity, belong only to event-particles and not to processes. But a process can be said to be past if its later boundary is past; and a process can be said to be more remotely past in proportion as its later boundary has a greater degree of pastness. Similarly, a process can be said to be future if its earlier boundary is future; and a process can be said to be more remotely future in proportion as its earlier boundary has a greater degree of futurity.

1·22. *Absolute Becoming.* We must now consider the other feature in temporal facts to which there is no spatial analogy, viz., temporal becoming. People have often tried to explain or to represent this in terms of qualitative change or motion. It seems to me quite evident that all such attempts are doomed to failure. Qualitative change and motion presuppose qualitative or substantial persistence, and both presuppose temporal becoming. It will be worth while to consider this point rather more fully.

Let us begin with the attempt to represent temporal becoming by means of motion. Here we are supposed to have a series of event-particles related by the relation of earlier and later. This may be represented by a straight line, which may be uniformly shaded if the process is to be qualitatively uniform, or may be coloured with a continuously variable shade from one end to the other if the process is to be one of continuous qualitative change. The characteristic of presentness is then supposed to move along this series of event-particles, in the direction from earlier to later, as the light from a policeman's bullseye might move along a row of palings.

The following fatal objections can at once be raised. (i) If anything moves, it must move with some determinate velocity. It will always be sensible to ask "How fast does it move?" even if we have no means of answering the question. Now this is equivalent to asking "How great a distance will it have traversed in unit time-lapse?" But here the series along which presentness is supposed to move is temporal and not spatial. In it "distance" *is* time-lapse. So the question becomes "How great a time-lapse will presentness have traversed in unit time-lapse?" And this question seems to be meaningless.

(ii) Consider any event-particle in the series. At a certain moment this acquires presentness and then loses it again without delay. Before that moment it was future, afterwards it is past. Now the acquisition and the loss of presentness by this event-particle is itself an event-particle of the second order, which happens to the first-order event-particle. Therefore every first-order event-particle has a *history* of indefinite length; and, at a certain stage of this there is one outstanding second-order event-particle; viz., the acquisition and the immediately subsequent loss of presentness. Yet, by definition, the first-order event-particle which we have been considering has no duration, and therefore can have no history, in the time-series along which presentness is supposed to move.

The two considerations which I have just mentioned would seem to make the following conclusion inevitable. If there is any sense in talking of presentness moving along a series of events, related by the relation of earlier-and-later, we must postulate a *second* time-dimension in addition to that in which the series is spread out. An event which has zero duration, and therefore no history, in the first time-dimension, will yet have an indefinitely long duration and a history in the second time-dimension. Let e_1 and e_2 be two first-order event-particles, and let e_1 precede e_2 by t units of the first time-dimension. Suppose that the second-order event-particle which is e_1's acquirement of presentness precedes the second-order event-particle which is e_2's acquirement of presentness by t' units of the second time-dimension. Then the velocity with which presentness moves along the original series will be measured by the ratio t/t'. The numerical value of this ratio is of no importance; it could always be given the value of $1/1$ by a suitable choice of the units in which we measure time-lapses in the two dimensions. The important point is that, whatever may be the numerical value, the ratio cannot possibly represent a rate of change unless its denominator measures a *lapse of time* and its numerator measures something *other than* a lapse of time in the same time-dimension.

Now let e_1, e_2, e_3, etc., be a series of event-particles of the first order, succeeding each other in the first time-dimension. Consider the following set of second-order event-particles, viz., e_1's acquirement of presentness, e_2's acquirement of presentness, e_3's acquirement of presentness, and so on. These might be denoted respectively by the symbols $e^2{}_1$, $e^2{}_2$, $e^2{}_3$, etc. These will form a series of second-order event-particles which succeed each other in the second time-dimension. Now, just as e_1 was future, became present, and then became past, so $e^2{}_1$ (i.e., e_1's acquirement of presentness) was future, became present, and then became past. Again, just as e_1 became present before e_2 became present, so $e^2{}_1$ (i.e., e_1's acquirement of presentness) became present before $e^2{}_2$ (i.e., e_2's acquirement of presentness) became present. Lastly, just as e_1 had ceased to be present when e_2 had become present, so $e^2{}_1$ (i.e., e_1's acquirement of presentness) had ceased to be present when $e^2{}_2$ (i.e., e_2's acquirement of presentness) had become present. Thus the series of second-order event-particles, $e^2{}_1$, $e^2{}_2$, $e^2{}_3$, etc., in the second time-dimension, is precisely like the series of first-order event-particles, e_1, e_2, e_3, etc., in the first time-dimension, in all those respects which led people to say that presentness "moves along" the first-order series. Such people ought therefore to say, if they want to be consistent, that presentness "moves along" the second-order series too.

Now, if they do say this, we can show by exactly the same arguments as we used at the first stage that a *third* time-dimension must be postulated. Each second-order event-particle, such as $e^2{}_1$, must be supposed to endure indefinitely and to have a history in this third time-dimension. And the acquirement of presentness by $e^2{}_1$ will be a third-order event-particle in the history of $e^2{}_1$. It could be symbolised by $e^3{}_1$, which thus stands for "the acquirement of presentness by the acquirement of presentness by e_1." It is easy to see that the argument is quite general, and that there is no stage at which one could consistently stop in postulating further time-dimensions and events of a higher order.

It is a great merit of Mr. J. W. Dunne, in his two books *An Experiment with Time* and *The Serial Universe*, to have insisted on what is substantially the same fact as this. Unfortunately he persuades himself, by false analogies with infinite series which have limits, that the regress is harmless and that it is sensible to postulate what he calls "the Observer at infinity." Actually the series which we have been considering could not have a last term or an upper limit, and so the conception of "the Observer at infinity" is the contradictory notion of the last term or upper limit of a series which, from its nature, could have neither. It may be remarked that Mr.

Dunne's attempted explanation of the alleged fact of pre-cognition, which is highly ingenious, does not require an unending series of time-dimensions. Any reader who is inter-ested in this subject may be referred to my article, "Mr. Dunne's Theory of Time," in *Philosophy*, Vol. x, No. 38.

When one finds oneself launched on an endless series of this kind it is generally a sign that one has made a false move at the beginning. I think that it is easy to see what the false move is in this case. The phrase "to become present" is gram-matically of the same form as the phrase "to become hot" or "to become louder." We are therefore tempted to think that sentences like "This event became present" record facts of the same kind as those which are recorded by sentences like "This water became hot" or "This noise became louder." Now a very little reflection is enough to show that this is a mistake.

Any subject of which we can significantly say that it "became hot" must be a more or less persistent substance, which persisted and had temperature before and after the date at which it became hot. The determinate form of its temperature was coldness for an earlier period and hotness for a later period, and the two periods are adjoined phases in its history. Again, any subject of which we can significantly say that it "became louder" must be a more or less prolonged noise-process, which divides into an earlier phase of less loud-ness adjoined to a later phase of greater loudness. But a literally *instantaneous* event-particle can significantly be said to "become present"; and, indeed, in the strict sense of "present" *only* instantaneous event-particles can be said to "become present." To "become present" is, in fact, just to "become," in an absolute sense; i.e., to "come to pass" in the Biblical phraseology, or, most simply, to "happen." Sentences like "This water became hot" or "This noise became louder" record facts of *qualitative change*. Sentences like "This event became present" record facts of *absolute becoming*. Now it is clear that qualitative change involves absolute becoming, and it seems to me equally certain that absolute becoming is involved in mere continuance without qualitative change. It is therefore hopeless to expect to treat absolute becoming as if it were a particular case of qualitative change. The endless series of time-dimensions and of orders of events, which such an attempt involves, is the sign and the measure of its futility. I do not suppose that so simple and fundamental a notion as that of absolute becoming can be analysed, and I am quite certain that it cannot be analysed in terms of a non-temporal copula and some kind of temporal predicate. . . .

A simple way of stating the theory[1] is as follows. Take the sentence "It is now raining." A number of utterances may occur at different times, which are all alike enough in the relevant respects to count as utterances of this sentence. Now any one who utters this sentence seriously on any occasion means to express his belief that an occurrence of rain falling in his neighbourhood is simultaneous with this utterance of his. And anyone who hears and understands any such utterance will take it to mean that an occurrence of rain falling in the speaker's neighbourhood is simultaneous with this utterance. Thus any utterance U of the type-sentence (to use Ramsey's phrase) "It is raining now" means "An occurrence of rain in the neighbourhood of the speaker who utters U is simultaneous with this utterance." We may abbreviate this into "An occurrence of rain is spatio-temporally contiguous with the utterance U." Now both speaker and hearer actually prehend the utterance U, since one makes it and both hear it. So, finally, when a speaker utters the type-sentence "It is raining now," what he means is "An occurrence of rain is spatio-temporally contiguous with *this* utterance of *mine*." And what the hearer understands could be expressed by the hearer saying "An occurrence of rain is spatio-temporally contiguous with that utterance of his." Different utterances of the same type-sentence necessarily have different meanings. One will mean "An occurrence of rain is spatio-temporally contiguous with U_1." Another will mean "An occurrence of rain is spatio-temporally contiguous with U_2." If U_1 and U_2 be successive, it may well be that one expresses a true proposition and the other a false proposition, though both are utterances of the same type sentence "It is raining now."

The theory may be summed up as follows. Any utterance of a type sentence, which is of a certain grammatical form and contains the type word "now" or "present" or some equivalent, is understood by speaker and hearers to mean that an event of a certain kind is *simultaneous* with *this* utterance. Any utterance of a type sentence, which is of a certain grammatical form and contains the type-word "past" or some equivalent, is understood by speaker and hearers to mean that an event of a certain kind is *earlier than this* utterance. And the same holds *mutatis mutandis*, for any utter-

ance of a type-sentence which is of a certain grammatical
form and contains the type-word "future" or some equivalent.
Unless there were people who uttered type-sentences of these
kinds nothing would be past, present, or future; though events
would still be simultaneous or successive.

Can this theory be accepted?

(i) In the form in which I have stated it I do not think
that it can possibly be the right analysis of what a *speaker*
means when he utters such a type-sentence as "It is raining
now," even if it were the right analysis of what his *hearers*
understand on such an occasion. For this would involve that
the speaker is using the utterance to express a judgment
which he is making about the utterance itself. I am very doubtful whether this is possible at all; and I am fairly certain that,
when I make such an utterance, I am not making a judgment
about the utterance which I am making. This difficulty does
not arise about the hearers.

(ii) The objection just mentioned could be removed by a
slight modification of the theory. We might say that what the
speaker means by his utterance is that an occurrence of rain
is simultaneous with *this*, where *this* is some particular, other
than the utterance itself, which he prehends simultaneously
with making the utterance. The particular in question might
be one of his own experiences or some sensum which he is
sensing. His hearers will almost certainly not prehend this
particular, and therefore what they understand by the utterance cannot be exactly the same as what the speaker means
to express by it. As regards the hearers, we may suppose that
each interprets the utterance to mean that an occurrence of
rain in the speaker's neighbourhood is roughly simultaneous
with certain auditory sensa which that hearer is sensing, viz.,
those which are manifestations to *him* of this utterance of the
speaker. Let us take the theory in this amended form, and
consider whether it is adequate.

(iii) The first comment to be made is this. The theory professes to give an analysis of those temporal facts which are
expressed by sentences containing temporal copulas, like "is
now," "was," or "will be," or temporal adjectives, like "past,"
"present," or "future." When we look at the proposed analysis
we find that it substitutes sentences of the form "Such and
such an event is simultaneous with, or is earlier than, or is
later than, *this*"; where "this" is used as a logical proper name
for some particular which the speaker or the hearer is pre-

hending when he makes or hears the utterance. Now what kind of copula is the "is" in these substituted sentences? Is it a timeless copula, like the "is" in "3 is the immediate successor of 2" or in "13 is a prime number"? Or is it the temporal copula "is now"? Or is it some third kind of copula which logicians and metaphysicians have not clearly recognised and distinguished?

If it is the timeless copula, the theory has *prima facie* been successful. If it is the temporal copula "is now," the theory has certainly failed. If it is supposed to be some third kind of copula, we must await further information about it from supporters of the theory.

Now ... we do not say "The Battle of Hastings precedes (or *is* followed by) the Battle of Waterloo." We say "The Battle of Hastings precede*d* (or *was* followed by) the Battle of Waterloo." Again, we do not say, on getting up in the morning, "My lunch precede*s* (or *is* followed by) a meeting of the Faculty Board of Moral Science." We say, "My lunch *will* precede (or *will* be followed by) a meeting of the Faculty Board of Moral Science." Thus it seems *prima facie* that the copula in propositions which assert temporal relations between events is not the timeless copula which occurs in propositions about the qualities and relations of abstract objects like numbers. The copula seems *prima facie* to be the temporal copula "is now," "was," or "will be," as the case may be. According to the theory which we are discussing, an utterance of the type-sentence "It will rain" means "An occurrence of rain in this neighbourhood is later than *this*," where "this" is used by the speaker as a proper name for a certain particular which he prehends when he makes the utterance. But no one except a philosopher doing philosophy ever does talk in this way. What we say is "An occurrence of rain in this neighbourhood *will* follow (or *will* happen later than) *this*." So *prima facie* the proposed analysis has failed to analyse away the temporal copula "will." Similar remarks apply, *mutatis mutandis*, to the proposed analysis of statements of the form "My breakfast is past" or "I have had my breakfast." We are told that an utterance of such a type-sentence means "Eating my breakfast precedes *this*." But no one ever does talk in this way in real life. Instead we say "Eating my breakfast precede*d* this." And so, *prima facie*, the temporal copula has not been analysed away.

Of course it may be answered that this objection depends

simply on defects in the language that we speak. It may be so. But I am more inclined to think that the obvious artificiality and awkwardness of the sentences which express temporal facts, according to this analysis of them, are a sign that we are trying and failing to force temporal facts into the mould of non-temporal facts about abstract objects such as numbers. The theory seems to presuppose that all events, past, present, and future, in some sense "co-exist," and stand to each other timelessly or sempiternally in determinate relations of temporal precedence. But how are we to think of this "co-existence" of events? It seems to me that the events and their temporal relations are thought of either by analogy with *timeless* abstract objects, such as the integers in their order of magnitude, or by analogy with *simultaneous persistent* particulars, like points on a line in spatial order from left to right. Neither of these analogies will bear thinking out; yet I suspect that the theory is made to seem intelligible and adequate to its supporters by the fact that these irrelevant analogies are always hovering about at the back of their minds.

(iv) It remains to make one more comment on the theory under discussion. It seems to me that the theory leaves altogether out of account the transitory aspect of Time. According to it, "past," "present," and "future," as used by a person at any moment, always denominate relational properties, in which the relation is "earlier than," "simultaneous with," or "later than," respectively, and the relatum is some particular which the speaker is prehending or some experience which he is having at that moment. Supposing this to be true, the transitory aspect of Time consists in the fact that the relatum is never the same on two different occasions on which these words are used.

Consider, e.g., that series of successive experiences which constitutes my mental history from the cradle to the grave. On the theory which we are discussing, there is no question of events "becoming" or "passing away." In some sense of "is," there "is" timelessly or sempiternally all that there ever has been or will be of the series. The qualitative changes that take place in the course of my experience are supposed to be completely analysable into the fact that different terms of this series differ in quality, as different segments of a variously coloured string differ in colour. But this leaves out the fact that at any moment a certain short segment of the series is marked out from all the rest by the quality of presentedness;

that at any two different moments the short segments thus marked out are different segments, though they may partially overlap if the two moments are near enough together; and that the relatum at any moment is, or is contained in, that short segment which has presentedness at that moment. Thus change has to be postulated in a sense not contemplated by the theory, viz., the steady movement of the quality of presentedness along the series in the direction from earlier to later. If we try to deal with this kind of change in the way in which the theory deals with the qualitative changes that take place in the course of my experience, we shall be committed to making each term in the original series a term in a second series in a second time-dimension. We shall have events of the first order. In fact we shall be landed in the endless series of time-dimensions and orders of events which I mentioned in Sub-section 1·22. . . . And this seems to me to be a most serious difficulty.

NOTE

[1] That "sentences which contain the words 'past,' 'present,' or 'future,' or their equivalents, can be translated without loss of meaning into sentences which do not contain these words or equivalents of them, but do contain the phrase 'earlier than' or some equivalent of it." From previous paragraph of original text.—ED.

TIME: A TREATMENT
OF SOME PUZZLES

J. N. FINDLAY

First published in *The Australasian Journal of Psychology and Philosophy*, Volume XIX, 1941.

The aim of this paper is to inquire into the causes of some of our persistent perplexities with regard to time and change. We do not propose to offer a solution for these difficulties, but rather to make clear how they have come to worry us. For we shall suggest that they have their origin, not in any genuine obscurity in our experience, but in our ways of thinking and talking, and we shall also suggest that the clear consciousness of this origin is the only way to cure them. It is plain that we do not, in any ordinary frame of mind, find time so hard to understand: we are in fact always competently dealing with what we may describe as "temporal situations." We are dealing with such situations whenever we say, without hesitation or confusion, that this lasted longer than that, that this took place at the same time as that, that this has just happened or that that will happen soon. We have no difficulty in showing other people what we mean by such forms of statement, nor in getting them to agree that we have used them truly and appropriately. Yet all these forms of statement, and the situations to which they refer, seem capable of creating the most intense perplexity in some people: people are led to say that time is "paradoxical," "contradictory," "mysterious," and to ask how certain things are "possible" whose actuality seems obvious. Thus it has been asked how it is "possible" for anything to reach the end of a phase of continuous change, or how it is "possible" for that which *is* the case ever to cease being the case, or how it is "possible" for the duration of any happening to have a length and a measure. In all such cases it seems reasonable to say that the burden of proof that there *is* a genuine problem or difficulty is on the person who feels it, and not on the person who refuses to depart from ordinary ways of speaking. And it certainly does seem odd that people who have always had to deal with changing objects and situations, and whose whole language is perfectly adapted to dealing with them, should suddenly profess to find time so very strange. If time is so odd, we may very well ask, in terms of what things more familiar and understandable shall we proceed to explain it or to throw light on its possibility? We may indeed regard it, as a strange disorder that people who have spent all their days "in time," should suddenly elect to speak as if they were casual visitors from "eternity." And it must be our business to cure them of this disorder through a clear awareness of its causes. There is indeed "a short way with puzzlers" who inquire into the "possibility" of perfectly familiar and understandable situations: we may simply point to some instance of the kind that perplexes them and say: "That's how it is possible for so-and-so to be the case." Thus if a man were to ask me "How is it possible that that which *is* the case should cease to be the case?" I might simply crook my finger and say "Now my finger is crooked," then straighten it and say "Now it has ceased to be crooked. And that's how it's possible for that which *is* the case to cease being the case."[1] But such an expedient, though perfectly proper in itself, and more than a man has a right to ask for in most cases, would not suffice to allay our questioner's perplexity, since he, presumably, is quite as familiar with ordinary usage as we are.

A treatment of the puzzles of time will also serve to illustrate a treatment which might be applied to many other questions and difficulties. For some people quite readily fall into a mood in which they feel that there is something mysterious and doubtful about things that they would normally regard as elementary and obvious. They are then led to ask questions which seem queer, because it is not in the least plain how one should set about answering them. Thus a man may wonder how it is possible for a number of distinct things to share in the same quality, or whether he really is the same person from year to year, or why *this* world exists rather than any other. Now in ordinary unreflective moods we should regard these questions as either unanswerable or not worth answering, but our questioner plainly wants an answer and he doesn't want an obvious answer. It is plain, in particular, that we couldn't remove our questioner's perplexity by "appealing to experience," by pointing to anything that both he and we could observe. For he *has* all the kinds of experience that could throw light on his problem, and yet he is puzzled. It seems clear that, where the simplest and most familiar instances of something occasion profound perplexity, we cannot hope to remove such perplexity, or even to allay it, by indefinitely accumulating other instances of the same kind, some of which would be strange and others highly complex. We are accordingly brought back to our supposition that there are some questions which beset us, not because there is anything genuinely problematic in our experience, but because the ways in which we speak of that experience are lacking in harmony or are otherwise unsatisfactory. We are sometimes thrown into a mood of interrogation not because we are in quest of further facts, but because we are in quest of clearer, or less discordant, or merely different ways of verbally dealing with those facts. Such moods of questioning plainly have no answers, in any ordinary sense of "answer"; we may nevertheless hope to relieve them by becoming clearly conscious of the underlying needs that prompt them, and by deliberately adopting ways of talking that provide appeasement for those needs.

There are other reasons why there is interest in our difficulties with regard to time. These difficulties form a relatively self-contained group of puzzles, which do not seem to share their entrails with too many other philosophical problems. We can find time difficult without finding anything else difficult, but we couldn't be puzzled by matter or mind or knowledge, without being puzzled by practically everything else. Hence we can deal more cleanly with these temporal puzzles than with other issues; they provide, accordingly, a simpler paradigm of method. These puzzles are also important in that philosophical difficulties seem to flourish more readily in the temporal field than in almost any other. It would be safe to say that rapid change and the "nothingness of the past" are things which can always be relied on spontaneously to vex a large number of unsophisticated people, and so to constitute one of the standing mysteries of our universe. We have reason, of course, to suspect such generalizations; for we know nowadays that there is no way of ascertaining the philosophical reactions of unphilosophical common sense, except by testing and questioning large numbers of people.[2] But in the absence of such testing, vague experience certainly bears witness to the generality of such puzzlement.

We may now point to a circumstance which is certainly responsible for *some* of our difficulties with regard to time.

This is the fact that it is possible to persuade a man, by an almost insensible process, to use certain familiar locutions in ways which become, on the one hand, steadily wider and more general, or, on the other hand, steadily narrower and stricter. This persuasive process is only one of the many processes by which an able dialectician can twist or stretch or shift or tear apart the web of words with which we overlay our world. In doing so, he relies on the fact that the boundaries of linguistic usage are seldom clear, that there are always ranges of cases in which it is simply doubtful whether a given locution is or is not applicable, and that there are, in addition, a number of deep-seated tendencies in language which facilitate linguistic shifts in certain directions. In the particular case we are now considering there are, it is plain, words and phrases whose use very readily widens: it is easy to persuade a man that they really *ought* to be used in cases in which it has never before occurred to anyone to use them. And it is also plain that there are words and phrases whose use very readily narrows, so that we are easily persuaded to say that it was "wrong" or "improper" to use them in cases where we previously used them without hesitation. And it is possible for the adroit dialectician, by making repeated use of a big stick called "consistency," on the one hand, and another big stick called "strictness," on the other hand, to persuade us to use such forms of speech so widely that they apply to everything, or so narrowly that they apply to nothing: the result in either case is to turn a serviceable mode of speaking into one that is totally unserviceable. Good examples of these dialectical processes would be arguments which led us to use the term "know" so widely, on the one hand, that we might be said, like the monads of Leibniz, always to know everything, or so narrowly, on the other hand, that we might never be said to know anything. There is, of course, nothing in such an exaggerated width or narrowness of reference which *necessarily* leads to paradoxes or problems. If we persuade a man to use words in new ways, we disorganize his linguistic habits for the time being, but there is no reason why he should not rapidly build up a new set of habits, which will enable him to talk of ordinary situations as plainly and as promptly as before. But the trouble is that such a sudden change of usage *may* produce a temporary disorientation, it is like a cerebral lesion from which an organism needs to recover, and in the interval before recovery sets in, and new connections take the place of the old, a man may readily become a prey to serious confusions. For even after a man has been persuaded to use certain phrases in totally new ways in certain contexts, he may still hark back to old uses in other contexts: he may even try to incorporate both uses in the same context, thus giving rise to statements and questions which cannot be interpreted in either way of speaking.

Now in regard to time it is plain that there is a strong tendency in language to use terms connected with the "present" in an ever stricter manner, so that, if this tendency is carried to the limit, the terms in question cease to have *any* application, or, at best, a novel and artificial one. It is also plain that *some* of the problems of time are connected with this fact. We can readily be persuaded to use the present tense and the temporal adverb "now" (as well as the imperfect past and imperfect future tenses and the words "then," "at that time," etc.) in stricter and stricter ways; and if we yield completely to such pressure, our normal habits of speech will be disorganized. Our use of the present tense and of the temporal adverb "now" is not very strict in ordinary circumstances: we are prepared to say, even of happenings that last a considerable time, that they are happening *now*, e.g. we say "The National Anthem is now being sung," "The Derby is now being run," etc. Now the present tense and the temporal adverb "now" *might* have been the sort of speech-form that we tended to use more and more widely, so that we might easily have been persuaded to say "The history of England is now running its course," "The heat death of the Universe is now taking place." We might then have been persuaded to allow that, since a *whole* cannot be happening now, unless all its component *parts* are also happening now, John is now really signing Magna Carta, life on the earth is now really extinct, and so on. The problems that this way of speaking might occasion would certainly be serious. The natural development of the speech-forms we are considering does not, however, lie in this direction. We tend rather, if pressed, to use the present tense and the temporal adverb "now" more and more narrowly: thus if we had said that the National Anthem was being sung, and someone asked us "But what are they singing *just now*?" we should not widen our reference to cover the whole evening's concert, but narrow it to apply to some line or phrase or word or note of the National Anthem. Now since our tendencies lie in *this* direction, we can readily be persuaded to give up saying that anything which takes an appreciable time is happening now. We can be bullied into admitting that this is a "loose" and "inaccurate" way of talking. And it is possible to force us to grant that the really strict speaker would not use these forms of speech in the case of anything but a happening which was so short that it took *no time at all*. Thus we might force a man first to admit that nothing which was *past*, nothing which was *no longer there*, could possibly be said to be happening now. We might then press him to admit the additional principle that nothing of which a *part* lay in the past could properly be said to be happening now. We might then persuade him to grant, with regard to any happening that "takes time," that it doesn't happen "all at once," but that it has parts which happen one after the other, and that, when any *one* of these parts *is* happening, all the *other* parts either *have* happened or *will* happen. It then becomes easy to prove that no happening which takes time can properly be said to *be* taking place, and that the only parts of it of which such a thing could ever be rightly said, would be parts that took *no time at all*.[3]

In all these arguments we are being persuaded to apply linguistic principles which are established in the case of happenings of *fairly long duration*, to happenings of very short duration; we are not obliged, but can be readily pressed, to be "consistent" in this manner since there are no clear lines between the long and the short. But the result of yielding to this pressure is to turn a serviceable way of talking into one that has no use. For it is obvious that all the happenings that we can point to (in any ordinary sense of "point to") take time, and that pointing itself takes time, so that if the only happenings of which we may say "This is happening now" are happenings which take no time, there are no happenings which we can point to, of which we may say "This is happening now." Now this does not, of course, imply that a clear and useful meaning cannot be given to phrases and sentences which mention happenings that take no time: it is plain, in fact, that very clear and useful meanings *have* been given to them by a long succession of mathematicians and philosophers. But it is also plain that these new forms of diction may, at first, merely serve to disorganize existing speech-habits, and that, while this lasts, we may fail to give any clear or serviceable meaning to "happenings which take no time"; we may tend to talk of them as if they were happenings we could point to, in the same sense in which we can point to happenings which *do* take time, and we may further credit them unthinkingly with many of the properties of happenings which *do* take time. Such ways of talking, it is plain, must lead to many quite unanswerable questions.

After this preliminary consideration of *one* source of our temporal difficulties, we may turn to Augustine's problem in the eleventh book of the *Confessions*. This we may phrase as follows: "How can we say of anything that it lasts a long time or a short time? How can a time have length? And how can that length be measured?"[4] What was it, we may ask, that Augustine found so difficult in the length and measure of time? We may perhaps distinguish three aspects of his bewilderment, which might be grounds for anyone's bewilderment. He found it difficult, in the first place (we may suppose), to see how happenings which take *no* time could ever be "added up" to make the happenings which *do* take time.[5] This difficulty is not peculiar to our thought of time, but applies to space as well. It seems absurd to say that an accumulation of events, the duration of each of which is zero, should have, together, a duration that is more than zero. The matter might be put more strongly. We are inclined to say that, if the duration of events were reduced to zero, "there would be nothing left of them," they would "just be nothing," and we obviously could not hope to make something out of an accumulation of nothings.[6] We may regard this as one side of the Augustinian problem. A second slightly different side consists in the fact that the stages of any happenings that takes time are never there *together*. Now it seems absurd to say of a number of things which are never together, but always apart, that they can ever *amount* to anything, or form a *whole* of any kind: it would be as if one were to try to build a house with bricks that repelled each other, so that each one moved away when the next one was brought up to it. At such a rate, it would seem, one could build no house and no interval of time.[7] But Augustine's problem has a third side which seems to have worried him particularly: that if we measure an interval of time, we must be measuring something of which a vanishing section only has reality: all the other sections of it, which give it breadth and bulk, are either *not yet there* or *not there any longer*. Now it is hard to grasp how we can measure something which is no longer there, which is "past and gone," of which we are tempted to say that it is "simply nothing." And it is also hard to grasp how we can measure something which is not yet there, which is merely expected, which we are likewise tempted to describe as "nothing." It would be like trying to measure a building of which all but the tiniest fragment had been blasted by a bomb, or existed merely in a builder's blue-print. In such a situation we should have no building to measure, and it seems we should be in the same position with regard to lengths of time.[8]

We shall now briefly point to some ways—there are an indefinite number of such ways—in which we might avoid these Augustinian perplexities. We might, first of all, evade the whole argument by which we have been bludgeoned into saying that there are some events that take no time, and that only these are ever truly present. We might refuse to say, of certain happenings which are very short, that any of their parts lie in the past or future; we do not normally, in fact, make use of the past and future tenses in speaking of the parts of very short events contemporary with our utterance. Alternatively we might say that some sufficiently short events can be "present as wholes," though most of their parts are past or future; this too agrees with ordinary usage, for we say that many fairly long events are happening, though we should talk in the past or future tense of some of their remoter parts. Or again we might deny—as Whitehead in his doctrine of epochal durations has denied—that certain very brief events come into being *part by part*.[9] There is, in fact, no plain empirical meaning to be given to the supposition that all events come into being part by part, since there must necessarily be a limit to the division of events by human judgments or instruments. Or again we might choose to follow certain other trends of language, and to say, of certain brief events, that they "took no time at all," thereby excluding from the start the whole issue of divisibility into successive parts.[10] It does not, in fact, matter, in all this choice of diction, *what* we say, provided only that we truly please ourselves: the facts are there, we can see and show them, and it is for us to talk of them in ways which will neither perplex nor embarrass us. It is desirable, in our choice of words, that we should be consistent, but it is not desirable that we should make a fetish of consistency. Consistency in language is most necessary if it means that we shall not, in a given context, fall victims to linguistic conflicts, that we shall not try to say something, while striving at the same

time to unsay it.[11] Consistency is also very desirable if it means
that we shall be guided by the analogies of things in what we
say in *different* contexts; in the absence of *some* degree of
such consistency, all language would be arbitrary and communication impossible. But consistency is wholly undesirable
if it becomes a bogey, if it makes us say something in one
context merely because we have said it in some other, more
or less analogous context, and if it then leads us on to say
further things which bewilder and confuse us. For the analogies of things are varied and conflicting, and it is impossible,
without disrupting human language, to do justice to them all.

So far we have pursued a line which shakes the dialectic on
which the Augustinian problem is founded. By so doing we
avoid giving a sense to the phrase "events which take no time,"
and are not obliged to say that these alone are truly present.
Suppose however we are moved by this dialectic, or by some
consideration of scientific convenience, to admit this talk of
"momentary presents," how then shall we proceed to deal
with the various aspects of the Augustinian problem? As regards the first aspect, the building of a whole which has size
out of parts which have *no* size, we may simply point out that
it mixes up the familiar sense in which a pile of money is
built up out of coins, with the new sense in which a happening
which takes time may be built up out of happenings which
take no time. Because one couldn't amass a fortune out of zero
contributions, one tends to think one couldn't make a measurable duration out of parts with no duration. But the situations are quite different; no one has witnessed a lapse of time
being built up out of instants, as he can witness a pile of
money being built up out of coins, nor can the former be
imagined as the latter is imagined.[12] Hence if we wish to
speak of "happenings which take no time," we are quite free
to fix what may be said of them, and this means that we may
simply rule that events which take time *are* made up of events
which take no time. And once misleading pictures are avoided,
we shall find no problem in this. We may in the same way
dispose of the difficulties which spring from the tendency to
say that an event which took no time would "just be nothing."
Either we must restrain this inclination—to which we are not
in duty bound to yield—or be prepared to say that certain
parts of real temporal wholes are simply nothing, and that
mere nothing can at times have definite properties. This way

of talking would no doubt do violence to our habits, and abound in dangerous suggestions, but we should not, with a little practice, find it difficult.

The second aspect of the Augustinian problem involves a similar confusion. Because it would be absurd to say of certain wholes—houses, mountains or libraries, for instance—that they existed and were measurable, although their parts were never together, we think it would be absurd to say the same thing of happenings. But the fact that we shouldn't say that *some* of the things we call parts could constitute the things we call their wholes, unless they were present together, does not oblige us to say this in the case of *other* things we also call parts and wholes. For the sense in which the parts were parts, and the whole wholes, and the former made up the latter, might be ruled to be different in the two sets of cases: we might say we were dealing with two totally different *sorts* of parts and wholes. And we do in fact rule so; for we regard it as nonsense to say of an event that takes time, that its parts are present together. And we recognize the difference between the two sets of cases by talking of *coexistent* parts in the one set of cases, and of successive parts in the other: the successive parts of a whole are, in fact, just those parts of it that *don't* need to be together. But if we feel ourselves unconquerably opposed to calling something a whole whose parts are not together, we may simply rule that some things may have magnitude although they are not wholes. And other similar expedients will meet other possible difficulties.

As regards the third difficulty of Augustine, how we manage to measure something which is in part past, we may again suggest a number of alternatives. We might, in the first place, reject the analogy between the measurement of a coexistent whole like a house, which isn't there to be measured if any parts of it lie in the past, and the measurement of a successive whole like a happening, which *must* have parts in the past. Or we might follow certain other trends of language, and say that we have succession *in the present*, and that certain happenings which are not too long are able to be present as wholes and so to be measured directly. Other longer happenings might then be measured by means of the briefer and directly measurable happenings which entered into their remembered history. Or if it is the "nothingness of the past" that troubles us, we must remember that we are not compelled to say that the past is nothing: we may, if we like, credit it with existence or sub-

sistence or any other suitable status. For we are only worried
by the "nothingness of the past" because we think it will stop
us from finding out any facts about the past, just as the
nothingness of a bachelor's children stops us from asking for
their ages or appearance. But there are so many clear and
agreed ways of establishing what has happened in the immediate or remoter past, that it would be nonsense to put past
events in the position of a bachelor's children. So that if we
wish to say that they exist or subsist, there is no good reason
why we should not do so. But if the "existence" of the past
is going to suggest to us that we could by some device revive
or revisit the past, as we could revive a drowned man or revisit
Palermo, then it is perhaps better to go on saying that the past
is nothing, allowing meanwhile that there may be measurable
wholes which have certain parts that are nothing.

The puzzles of Augustine lead on very naturally to the
problems of Zeno, or rather to a certain very general difficulty
which seems to be involved in every one of Zeno's paradoxes.
This is our difficulty in seeing how anything can happen, if
before it happens something else must happen, and *before*
that happens something else must happen, and so on indefinitely. If we make time continuous and infinitely divisible, we
also feel obliged to say that before any happening is completed, an infinity of prior happens must have been completed, and this seems to mean that *no* happening can ever
be completed. We seem to be in the plight of a runner in a
torch-race, who wants to hand on his torch to another runner
A, but is told by A that he will only take it from B, who tells
him he will only take it from C, who tells him he will only
take it from D, and so on indefinitely. Or in the plight of a
man who wants to interview a Cabinet Minister, and who is
informed by the Minister that he must first discuss his business
with the Under-Secretary, who informs him he must first
discuss it with the Chief Clerk, etc., etc. Our runner obviously will never get rid of his torch, and our harassed peti-
tioner will obviously never see his Minister, and it looks as if
all happenings involve the same hopeless difficulty. The difficulty we are presenting is, of course, not identical with any
one of Zeno's historical puzzles: in all of these the difficulties
of duration are complicated by the introduction of change
and motion. But it is plain that all these puzzles could be so
restated as to deal with happenings without regard to whether
those happenings were changes or persistent states, and with-

out regard to whether they involved motion or not. A plum continuing to hang on a tree for a certain period affords, less dramatically, the same species of philosophical perplexity as an arrow in its flight. Moreover, when we strip Zeno's problem of its spatial and other wrappings, its significance becomes clearer. For it is not, essentially, a problem of space or quantity, but solely one of time: it is only because all motion is *successive*, because an infinity of positions must be passed *before* any subsequent position, that the possibility of such motion seems so utterly ruled out. If the infinite stages of a motion could be there all at once, as the parts of a piece of space are, we should feel no problem in their infinite number. It is therefore foolish to imagine that we can meet Zeno's puzzles by the modern theory of the continuum or by the facts of infinite convergent numerical series.[13] And the problem assumes its most vexing form if we allow that ordinary happenings have ultimate parts that take no time. For of such parts it seems most natural to say that none can be next to any other,[14] and once this is said it is hard to understand how any ultimate part can ever pass away or be replaced by any other. For before such a part can be replaced by any other similar part, it must first have been replaced by an infinity of other similar parts. Our admission seems to leave us with a world immobilized and paralyzed, in which every object and process, like the arrow of Zeno, stands still in the instant, for the simple reason that it has no way of passing on to other instants.

As before, we may deal with our difficulties in several different ways. We might, in the first place, deny that very short happenings are divisible as fairly long ones are divisible: the divisibility of *all* happenings is in any case without a definite meaning. This is the line followed by Professor Whitehead, who makes time flow in indivisible drops, and says that it is "sheer succession of epochal durations."[15] But, far less drastically, we might give to all this talk of instants and of infinite divisibility a sense consistent with the obvious facts of our experience, that things happen and that phases are outlived, that the world is not immobilized, and that we seldom have to cast about for ways of passing on to novel stages. For the infinite happenings that must first occur before a given thing can happen, are not like ordinary happenings we can see and show, of which it would be absurd to say that an infinite number ever were completed. They are happenings of a new sort to which a meaning must be arbitrarily given.

And since *we* have to give a meaning to these happenings, it is for us to see that they mean nothing which conflicts with our established ways of saying things. And once we strip them of pictorial vividness, we also strip them of their puzzling character. Our problem also vanishes when we note that even to be "desperately immobilized," to "cast about in vain for means to pass to other stages," would both, if they were anything, be states that lasted and took time. Our problem therefore takes for granted the very thing it finds so difficult.

We turn, in conclusion, from these Augustinian and Zenonian difficulties, to a different set of temporal puzzles, quite unconnected with our tendency to use the present tense in more exact and narrow ways. We shall consider briefly the very general wonderment which professes to find something "unintelligible" or "contradictory" in time and change. "How is it possible," we sometimes like to ask, "for all the solid objects and people around us to melt away into the past, and for a new order of objects and persons to emerge mysteriously from the future?" This kind of wonderment is most strongly stirred by processes of *rapid change*: we wonder at things which have no constant quality for any length of time however short, at things which only reach a state to leave it, and so forth. A similar perplexity besets us in regard to "truths" or "facts": we wonder how what *is* the case can ever cease to be the case, or how what was false *then* can come to be true *now*, and so on. This week the peaches in our garden are not ripe; next week we find them ripe; the following week they are no longer ripe, but rotten: in certain frames of mind we find this difficult. Our difficulty with regard to change may also be expressed in terms of "happenings" and their "properties" of "pastness," "presentness" and "futurity," the form in which this problem was propounded by McTaggart. We wonder how it comes about that happenings which are at first remotely future, should steadily become more nearly future, how in the end they manage to be present, and how from being present they become past, and how they go on, ever afterwards, becoming more and more remotely past. McTaggart has shown plainly that we cannot solve this problem (if it is a problem) by bringing in the "different times" at which events are present, past and future, since these themselves (whatever we may mean by them) have also to be present, past and future, and so involve the very difficulty they are called in to remove.

Now it is hard to see, if we remain in any ordinary, un-reflective state of mind, what is the problem that is being raised by those who say they can't see how what *is* the case at one time, is not the case at other times, or that they can't see how a happening that is future can ever come to be a happening that is past. As we observed at the beginning of this paper, it should be possible to remove such difficulties by pointing to some ordinary happening around us, a man diving, for instance, and saying, as it happened, "Now he's not yet diving," "Now he's diving," "Now he is no longer diving," or other similar phrases. And if a man were really puzzled by our usage in such situations, it would not take him very long to master it. We do not ordinarily have difficulty in knowing what to say of happenings as they pass, nor any tendency both to say and not to say the same thing in a given context, a kind of inconsistency that is seldom desirable. Occasionally, where change is rapid, we may find ourselves at a loss to say whether something is or is not yellow, or whether it is or was yellow: we may also have a tendency to say that it is both or neither. But all this only means we lack a settled and satis-factory way of talking about very swiftly changing things. But in the case of changes which are less rapid, we find ourselves quite free from conflict or confusion. *Before* an event occurs we say, if we have evidence that it is not yet happening, that it hasn't yet happened, but that it will happen, while if it *is* happening we say that it is now happening, that it hasn't ceased happening and that it isn't about to happen, and *after* it has happened we say that it has happened, that it is no longer happening and that it is not going to happen. Stated in words these semantic rules might seem circular, but taught in connection with a concrete situation they are wholly clear. And our conventions with regard to tenses are so well worked out that we have practically the materials in them for a for-mal calculus.[16] Where all is so desirably definite, what room is there for puzzles or perplexities?

To give an answer to this question, we must point to a cer-tain aspiration which all our language to some extent fulfils, and which we are at times inclined to follow to unreasonable lengths. We desire to have in our language only those kinds of statement that are *not* dependent, as regards their truth or falsity, on any circumstance in which the statement happens to be made. We do not wish a statement which we call "cor-rect" and "justified by fact" when made by one person, to be incorrect when made by another person, and to have to be superseded by some other statement. In the same way we do not wish a statement which we call "correct" when made in one place, to be incorrect when made in another place, and to have to be superseded by some other statement. And there are occasions when we feel the same sort of thing about the *time* at which a statement is made: if we are right in saying something at a certain time, then, we sometimes feel, we must be right in saying the same thing at all other times. This means that we object, in certain frames of mind, even to the easy, systematic changes of tense which statements have to undergo when they are transmitted from period to period. We might express our general aspiration by saying that we wish our statements to be independent of "extraneous circumstances" in regard to their truth or falsity: "the facts" must settle whether what we say is true, and nothing else must come into consideration. But such a way of talking would be gravely question-begging, for it depends on the sort of language we are speaking whether a circumstance is or is not extraneous. If we spoke a language in which the statements permitted in one place differed systematically from the statements permitted in another place, then it wouldn't, in that language, be an extraneous circumstance, as regards the truth or falsity of a statement, whether that statement was made here or there. And those who used the language would protest quite legitimately that "something was left out" by other languages which ignored all local circumstances of utterance. But the point is that we do *in part* say things which may be passed from man to man, or place to place, or time to time, without a change in their truth-value, and we look at things from *this* angle when we say that time, place and speaker are extraneous circumstances, and require our statements to ignore them.

Now the urge behind these austerities seems simply to be the urge towards more adequate communication, which is the fundamental impulse underlying language. We are prepared to sacrifice local and personal colour, or period flavour, in order that our statements may be handed on unaltered to other persons who are differently situated, or to ourselves in other situations. But it is not *this* sacrifice which gives rise to our perplexities: if we always spoke rigorously in the third person of everyone, ourselves included, if we avoided the adverbs "here" and "there," if we purged our language of tenses, and talked exclusively in terms of dates and tenseless participles, we should never be involved in difficulties. And for the purposes of science it is perhaps desirable that we should always talk in this manner. But our difficulty arises because we try to talk in this way but are also uneasy in doing so; we feel that something worth-while has been omitted, and try to combine our old way of talking with our new one. Thus McTaggart first offers us an order of events in which there are no differences of past, present and future, but only differences of earlier and later, in which every happening always stays the sort of happening it is, and always occupies the same position in the time-series: he then slides back into another way of talking in which events are present, past and future, and always *change* in these modalities. And his attempt to combine these ways of talking results in the unanswerable question: how can a single happening have the incompatible properties of being past and present and future? Whereas if we talk in the ordinary way we never have to say these things at once, and if we talk in an artificial, tenseless manner the question can't arise, since the modalities in question can't be mentioned. It is as if a man tried to retain the use of personal pronouns, such as "I," "you," "he," etc., in a language in which everything that could truly be said by one man could be truly said by every other man, and were then led to ask: "How can one and the same person be I and you and he?" And once we see the source of such perplexities, we should be easily rid of them.

NOTES

[1] The example given and the general method indicated was suggested by Professor Moore's proof that external objects exist. He proves that there are such objects by proving that there are two human hands, the latter being proved "by holding up his two hands, and saying as he makes a certain gesture with the right hand, 'Here is one hand,' and adding, as he makes a certain gesture with the left, 'and here is another' " (*Proof of an External World*, p. 25).

[2] See, e.g., Arne Ness's *Truth as conceived by those who are not professional philosophers*, (Oslo, 1938).

[3] The typical historical case of this argument is Augustine, *Confessions* (Bk. XI, Chap. XV, 19, 20). . . . [Reprinted on pp. 59-60 of this volume.—ED.]

⁴ The interest in Augustine as a case of philosophical puzzlement is due to Wittgenstein.

⁵ Augustine: "The present hath no space. Where then is the time which we may call long?" See above.

⁶ Augustine: "If, then, time present . . . comes into existence only because it passes into time past, how can we say that even this *is*, since the cause of its being is that it will cease to be?" (Bk. XI, Chap. XIV, 17.) (See p. 58 of this volume.—ED.)

⁷ Augustine: "Therefore neither is the year now current present; and if not present *as a whole* (our italics) then is not the year present." See above.

⁸ Augustine: "But in what 'length,' then do we measure passing time? Is it in the future, from which it passes over? But what does not yet exist cannot be measured. Or, is it in the present, through which it passes? But what has no length we cannot measure. Or is it in the past into which it passes? But what is no longer we cannot measure." (Bk. XI, Chap. XXI, 27). (See p. 63 of this volume.—ED.)

⁹ "Accordingly we must not proceed to conceive time as another form of extensiveness. Time is sheer succession of epochal durations. . . . The epochal duration is not realized *via* its successive divisible parts, but is given *with* its parts" (*Science and the Modern World*, p. 158).

¹⁰ *How* brief the happenings must be, of which we say any of these things, is of course a matter for arbitrary dicision.

¹¹ Unless, indeed, a linguistic conflict is deliberately used to express some personal reaction to reality, as has been done by some philosophers.

¹² Though a sense might be invented in which we could be said to witness or imagine the former.

¹³ This point is clearly brought out by Whitehead. See *Process and Reality*, p. 95.

¹⁴ Unless we choose to say that there is a finite number of ultimate parts in any happening, or other queerer things.

¹⁵ *Science and the Modern World*, quoted above.

¹⁶ The calculus of tenses should have been included in the modern development of modal logics. It includes such obvious propositions as that

x present $=$ (x present) present;

x future $=$ (x future) present $=$ (x present) future;

also such comparatively recondite propositions as that

(x). (x past) future; i.e. all events, past, present and future, *will* be past.

TIME AND LANGUAGE,
AND THE
PASSAGE OF TIME

NELSON GOODMAN

Reprinted from sections 2–3 of Chapter XI of *The Structure of Appearance*, Harvard University Press, Cambridge, Mass., 1951.

TIME AND LANGUAGE

In ordinary discourse we often indicate the time of events not by explicit description but by such a word as "now", "yesterday", "next week", "past", "later", or by the tense of a verb. As a result, we have quite unequivocal statements that nevertheless seem, paradoxically, to change in truth value. For example, when I say "The Red Sox now lead the American League", I am being quite definite; I am not saying that they lead at some unspecified time but am indicating the time unmistakably. What I utter is thus not an open statement like "x is yellow" but a closed statement that is either true or false. And yet although it be true when I first utter it, it may be false when I repeat it later.

The point is, of course, that we must be more careful to distinguish between a statement and other statements that resemble it. In the last example given, we have two statements, not one. Each of the utterances is a distinct, definite statement; and the two in fact have different truth values. These utterances may be exactly alike in sound pattern; but it is each utterance and not anything common to the two that constitutes a statement. Similarly, it is each of the utterances of "now"—not anything common to the two—that constitutes a word and refers to a certain time. In platonistic terms, the distinction between the general pattern or *type* of a word or sentence and its particular instances or *tokens* was drawn many

years ago by Peirce. Too often, however, those who have
noticed the distinction have looked upon it as a matter of
isolated academic interest, and assumed that thereafter one
need be concerned only with the types. More recently, we
have been forced to recognize that often—as in the example
above—it is the tokens that function as words or sentences;[1]
for we find different tokens of the same type naming and
affirming different things.

Indeed, it is the types that we can do without. Actual discourse, after all, is made up of tokens that differ from and
resemble each other in various important ways. Some are
"now" 's and others "very" 's just as some articles of furniture
are desks and others chairs; but the application of a common
predicate to several tokens—or to several articles of furniture
—does not imply that there is a universal designated by that
predicate. And we shall find no case where a word or statement needs to be construed as a type rather than as a token.
The exclusion of types not only does away with some excess
baggage but also results, I think, in clarifying our immediate
problem.

Obviously the term "token" is no longer appropriate. It is
both misleading and superfluous; for utterances and inscriptions are no longer to be regarded as mere samples but as the
actual words or statements themselves, and the linguistic universals from which they were to be distinguished are no
longer to be countenanced at all. Nevertheless, to emphasize
the fact that words and statements are utterances or inscriptions or etc.—i.e., *events* of shorter or longer duration—I
shall sometimes use such terms as "word-events", "noun-
events", " 'here'-events", " 'Paris'-events", and so on, even
though the suffix is really redundant in all these cases. "Paris"-
events, of course, are not events in Paris but certain utterances and inscriptions—namely, the "Paris" 's. A word-event
surrounded by quotes-events is a predicate applicable to utterances and inscriptions; and any

> " 'Paris' consists of five letters"

is short for any

> "Every 'Paris'-inscription consists of five letter-
> inscriptions".

Although each utterance and inscription is a separate word
(or statement or letter, etc.), the difference between two

words often has no practical importance. For most purposes, we need not distinguish among the several "Pisa"'s, all of which name the same thing, even though they differ widely in size, shape, color, sound, place, date, etc. On the other hand, we must carefully distinguish a "Pisa" from a "Paris". It is true that a given "Pisa" may be more like a given "Paris" than like some other "Pisa", just as a given mushroom may look more like a given toadstool than like some other mushroom; but in both cases we must discern just that overt difference that is correlated with a difference in appropriate use. In the case of "Pisa"'s and "Paris"'s, and in many other cases, some certain difference of shape or sound-pattern is the clue to a difference in what the words name.

Yet by no means every difference of extension is accompanied by a difference in shape or sound-pattern. The nominata of two "Paris"'s that are exactly alike in shape may be as different as those of a "Paris" and a "Pisa"; for some "Paris"'s name a city in France while others name a town in Maine. To note from its shape that a given word is a "Paris" thus is not enough. In order to determine which of two places the particular "Paris" in question names, we must look to the *context*—i.e., to the surrounding words and to certain attendant circumstances. Similarly, the various "this continent"'s name six or more different individuals; the various "John Smith"'s a still greater number; and the various "I"'s name vastly many different individuals.

For convenience, let us speak of words (or letters or statements, etc.) that are catalogued under a single label as *replicas* of one another,[2] so that any "Paris" (or any "I say") is a replica of itself and of any other "Paris" (or "I say"). Roughly speaking, a word is an *indicator* if (but, as will be made clear later in this section, not necessarily only if) it names something not named by some replica of the word. What has been said above will suggest that almost every name has a replica somewhere that names something different, and that therefore almost every name will be an indicator according to this criterion. But the distinction between indicators and nonindicators becomes effective when applied to a limited discourse. Within such a discourse there will normally be many names that are not indicators—although proper nouns as well as pronouns will still often be indicators, and pronouns will occasionally be nonindicators.

Among the commonest indicators are the personal indicators, the spatial indicators, and the temporal indicators. Of the personal indicators, an "I" or "me" normally refers to its own utterer; a "we" or "us" refers to the utterer and certain others determined by the context; a "you" applies to those addressed by the utterer, and so on. Characteristically, even though there is no variation in what a given personal indicator names, there is wide variation in what several replicas of that indicator name. Much more remains to be said, of course. For one thing, the person in question is sometimes, as in ghost-writing, the ostensible rather than the actual utterer; and the indicators appearing in a copy or transcription relate not to the actual maker of these inscriptions but to the maker of the original inscription or utterance. Furthermore, some indicators, like the "his"'s, not only name but serve a prepositional purpose as well. Again, some words of the same shape as indicators are actually not indicators at all but simply variables; a case in point is the "he" in an "If anyone disapproves, he may leave." Finally, an inscription sometimes divides temporarily into several different indicators; for example, if a given placard reading "I hate Hitler" is carried by different persons on three successive days, then the three day-parts of the enduring "I"-inscription name different persons. But all this is by way of subscript to the main point.

The location of a spatial indicator has to be taken into account in much the same way as the producer of a personal indicator. Some spatial indicators like the "here"'s name regions they lie in, while others like the "yonder"'s are discrete from the regions they name. In most cases, just what region a given indicator names depends partly upon its context, including such supplementary aids as pointing. Even among the "here"'s, one may refer to part of a room while others refer severally to a town, a county, a state, a continent, etc. Analogues of all the subsidiary remarks about personal indicators apply to spatial indicators. For example, a "here"-inscription in a personal letter normally refers to the place where it was written; a "here" in a delivered telegram refers rather to the place where the original was written or spoken; and if a "No Parking Here" sign is moved about, certain different temporal parts of the "Here" name different places.

But we are primarily concerned with the temporal indicators. Part of what is to be said concerning them is already evident from our glance at the personal and spatial indicators. The "now"'s, for example, behave much like the "here"'s:

each "now" names a period in which it lies, and the periods named by different "now"'s range from a moment to an era. Other terms, like the "yesterday"'s and the "soon"'s, name periods earlier or later than themselves; but in every case the time of a temporal indicator is relevant to what it names. We need hardly review the other points of analogy between temporal indicators and those already discussed, but certain temporal indicators require special attention.

In the first place, the "past"'s, "present"'s, and "future"'s lend themselves to frequent abuse in theoretical discourse. Most "present"'s function exactly like most "now"'s, naming some period they lie in; and the various "present"'s name many different periods of varying length, some of them remote from others. A "past," however, most often names *all* the time preceding—and a "future" all the time following—a certain period in which it lies. Thus the period named by a given "past" overlaps, and indeed includes or is included in, the period named by any other "past"; and the same holds for "future"'s. This fact, that what is once past is always thereafter past (and that what is once future was always theretofore future), creates an illusion of fixity and leads to treating the "past"'s (or the "future"'s, or even the "present"'s) as if, like the "Eiffel Tower"'s, all named the same thing. Metaphysicians have capitalized on this confusion for some very purple passages on The Past, The Present, and The Future. We must be careful to remember that nonsimultaneous "past"'s (or "present"'s or "future"'s) commonly name different even if not discrete periods.

Very often, however, temporal indication is accomplished in a sentence not by any word devoted solely or chiefly to that purpose but rather by the tense of the verb. A "Randy ran" tells us not only who did what but also when, i.e., prior to the period of production of the sentence itself. The "ran," besides specifying the action performed, serves also as a temporal indicator; nonsimultaneous "ran"'s ordinarily indicate different periods of time. Incidentally, verbs in some languages may also serve the third purpose of personal indication; for example, a *creo* in Spanish indicates its utterer so definitely that the pronoun *yo* is customarily omitted.

A verb in the present tense normally indicates a period within which the verb is produced, while a verb in the future tense normally indicates the period after its own production.

The interpretation of compound tenses and of combinations of tensed verbs with other temporal indicators sometimes requires care but is seldom really difficult. A

"Randy had been running"

tells us that the running took place prior to a moment—presumably further specified in the context—that is in turn prior to the time of production of the sentence itself. An isolated

"World War II was present",

however, tells us simply, as does a

"World War II is past",

that World War II is prior to the sentence in question. The "present" in an "is present" or a "was present" or a "will be present" in no way affects the temporal indication accomplished by the verb alone. On the other hand, an "is past" or an "is future" functions in the same way as, respectively, a "was" or a "will be". No exhaustive survey of such combinations need be attempted here; but it should be noted that some may result in virtually vacuous statements. For instance, a

"World War II was future"

—if unaccompanied by any context determining what prior moment is being affirmed to precede World War II—says only what may be said about any event that did not begin at the first moment of time. Likewise, of any event that does not run to the end of time, we may truly say that it will be past. Of course, a combination such as a "was future" or a "will be past" is usually set within a restrictive context.

In many statements the tense is merely grammatical, the verbs not actually functioning as temporal indicators. This is true more often than not in formal discourse. For one thing, generalizations are usually without effective tense; an

"All men have spines"

refers not only to all men contemporary with the statement but also to all who preceded or will follow it. In many singular statements also the verb, although in the present grammatical tense, is adequately translated by a purely tenseless symbol.

For example, where an "*a*" and a "*b*" are proper names, a given

> "*a* overlaps *b*"

may speak simply of the overlapping of the two individuals, without indicating anything about the date of their common part; that is, the sentence may just say that *a* O *b*.³ On the other hand, another

> "*a* overlaps *b*"

may have effective tense, being used to affirm not just that *a* and *b* have some common part but that they have some common part that is contemporary with the sentence itself. The context makes the difference.

Now one may say that two things overlap (tenseless) if and only if they did or do or will overlap; and "*a* O *b*" is implied by an effectively tensed "*a* overlaps *b*" or "*a* overlapped *b*" or "*a* will overlap *b*," while an "*a* O *b*" implies none of these but only such a disjunction as an "*a* overlaps *b*, or *a* overlapped *b*, or *a* will overlap *b*." But parallel principles do not hold for all other verbs, indeed, an "*a* D *b*" obviously is not implied by an effectively tensed "*a* is discrete from *b*" or "*a* was discrete from *b*" or "*a* will be discrete from *b*," but implies them all. Moreover, even though each verb that is effectively in, say, the past tense indicates a period preceding the verb, the relationship affirmed to obtain between such a period and other individuals referred to in a sentence varies considerably with different verbs. While an

> "*a* overlapped *b*"

places a common part of *a* and *b* within such a period, an

> "*a* was earlier than *b*"

places *a* within such a period; a

> "color *c* was at place *p*"

places the (color-spot) sum of *c* and *p* within such a period, and a

> "color *c* matched color *d*"

seems to place *c* and *d* (but not their sum, of course) at a moment within such a period. These examples will perhaps be sufficient warning against certain kinds of hasty generalization about tense.

Like some verbs, some replicas of other temporal indicators are not themselves indicators. For example, in a

> "We can know at a given time only what is past at that time or present at that time, not what is future at that time",

the "past", "present", and "future" name no times. Rather, the "is past at", the "is present at", and the "is future at" are tenseless two-place predicates that may respectively be translated by the tenseless predicates "is earlier than", "is at", and "is later than".

Effective tense does not by itself prevent a string of words from constituting a genuine statement. A tensed statement has as constant a truth value as a tenseless one; and a tenseless statement, no less than a tensed one, is an event in time. The difference is that tensed statements and other statements with indicators are not, so to speak, "freely repeatable." Now of course no term or statement is ever repeated in the way a quale is repeated; for a term or statement is a particular event and not a universal. On the other hand, nearly all terms and statements are much repeated in that they have many replicas. But a term or statement is said to be freely repeatable in a given discourse if all its replicas therein are also translations of it. Indicators and statements containing them are not freely repeatable.

Ordinarily, when we want to make continued or renewed use of a given term or sentence that occurs earlier in our discourse, we just repeat it, i.e., introduce a replica to take its place. If the term or sentence is freely repeatable, then for most purposes we need not distinguish between it and its replicas; we proceed as if all were numerically identical. But in the case of an indicator or a sentence containing one, where not all the replicas are translations, this is obviously dangerous. Often, indeed, no available replica of a given term or sentence is a translation of it, so that an inaccessible original cannot, in effect, be brought back into play by repeating it. For this reason, although indicators are of enormous practical utility, they are likely to be awkward for formal discourse. Various remedies may be applied. One lies in supplying a freely repeatable name (or description) of the indicator, or of the sentence containing it, and thereafter, instead of repeating the term or sentence, referring to it by means of a replica of this name. For example, a given "now" might be identified by any

> "The 937th word uttered by George Washington in 1776".

A later repetition of that "now" is not a translation of it; but any replica of this descriptive name is a translation of every other, and names just the particular "now" in question. And using such a name, we can readily arrive at a repeatable *translation* of the indicator; e.g., the "now" in question is translated by any

> "The period referred to by the 937th word uttered by George Washington in 1776";

or alternatively, if the period is a day, by any

> "The day on which George Washington uttered his 937th word in 1776".

Or we may seek a translation that contains no name of the indicator itself, but rather another name for what the indicator names. Thus a certain "here" is translated by any "Philadelphia"; and a certain "ran" is translated by any

> "runs [tenseless] on Jan. 7, 1948 at noon E.S.T."

Against such translations, it is sometimes urged that they do not really convey the content of the originals. A spoken

> "Randy is running now"

tells us that the action takes place at the very moment of speaking, while a

> "Randy runs [tenseless] on October 17, 1948, at 10 P.M., E.S.T."

does not tell us that the action takes place simultaneously with either utterance unless we know in addition that the time of the utterance is October 17, 1948, at 10 P.M. E.S.T. Since—the argument runs—we recognize the tenseless sentence as a translation of the tensed one only in the light of outside knowledge, we have here no genuine translation at all. But this seems to me no more cogent than would the parallel argument that "l'Angleterre" is not a genuine translation of

"England" because we recognize it as a translation only if we know that l'Angleterre is England.

A different question may arise from the auxiliary function of tensed verbs as indicators. If two tensed predicates are co-extensive but indicate different times, are they translations of one another? Do we demand that the two agree in what they apply to, or do we demand that they agree also in what they indicate? It is to be noted that ordinarily predicates that indicate different times differ also in extension; for to say that a tensed predicate indicates a time is a convenient way of saying that the application of a tensed predicate is restricted to individuals at that time. Nevertheless in some cases predicates that—according to the looser locution—"indicate different times" may agree in extension. A clear if unimportant example is that of a

> "stood still while walking"

and a simultaneous

> "will stand still while walking".

Since neither applies to anything, they are coextensive. The question whether they are translations of each other is quite analogous to the question whether

> "orders a centaur steak"

and

> "orders a unicorn steak"

are translations of each other. Both questions illustrate a general problem concerning the criteria for the use of "translation". That general problem lies outside my province here. I can only remark in passing that I think (1) that the appropriate criteria may vary considerably with the nature and purpose of the discourse and (2) that criteria much more stringent than simple coextensiveness can be formulated within the framework of extensionalism.[4]

THE PASSAGE OF TIME

We have still to deal with statements that seem most patently to reflect the temporal flow of events. One speaks of time passing, of events moving from the future into the present and on into the past, of things growing steadily older. How is such language to be interpreted?

To say that time passes seems to amount to saying that a moment of time progresses constantly in a future-toward-past direction. Yet obviously a time does not shift its position with respect to other times; it is identified with its position in the temporal series, and if any time moves then all move together. Now we have seen how a

> "Time t is future",

a later

> "Time t is present",

and a still later

> "Time t is past"

may all be true; and how the conjunction of the three might have as a translation any

> "Washington's 27th 'future' is earlier than time t; his 13th 'present' is at time t; and his 49th 'past' is later than time t",

(where each "is" is tenseless). The motion of time t ostensibly expressed here consists simply of the fact that t has different relationships of precedence to different verbal events. Again a

> "Time t was future, is now present, and will be past"

says merely that this utterance is at time t, is later than some earlier time, and earlier than some later time. On the other hand, a statement like

> "A time is at first future, then becomes present, then becomes past"

is quite a different matter. The final clause, for example, says neither that a time is earlier than this particular "past" nor that it is earlier than some "past" or other. The clause says rather that a time is past at some time or other; and this, as we have seen, just says that a time is earlier than some time or other. What the clause in question says thus does not depend on the time of its own or any other utterance. Indeed the whole sentence contains no actual indicators at all but is freely repeatable. Of course we may quite understandably want a translation of it free of words having many replicas that are actual indicators; and such a translation is readily provided:

> "A time is later than some time x, identical with some time later than x, and earlier than some still later time".

In the case of a

> "Time t is past and constantly recedes further into the past",

uttered at time s, the first "past" is an actual indicator while the second is not. The sentence says that time t is earlier than s; and that if q and r are times later than s, and q is later than r, then t precedes q by more than t precedes r. What has been said here of statements concerning times can easily be adapted to the interpretation of parallel statements concerning events.

So far I have not considered statements like

> "While it endures, a thing constantly grows older".

This again is normally a tenseless, freely repeatable statement, saying in effect that if two times r and s are within the period of duration of a thing, and r is earlier than s, then a larger part of that period precedes s than precedes r.

Thus are sentences that express the passage of time or the flow of events translated by sentences that merely describe relationship of precedence in the temporal series. The suggestion of flow or of passing or of ageing disappears; and just for this reason, it may be felt that we are missing something important about time. Most efforts to formulate just what is missed end in vague poetry or in hopeless confusion over temporal indicators. Yet I think that underlying these efforts there is a certain peculiarity of time that deserves attention. Strangely enough it turns out not that time is more fluid than (say) space but rather that time is more static.

We saw that the analogy between space and time is indeed

close. Duration is comparable to extent. A thing may vary in color in its different spatial or in its different temporal parts. A thing may occupy different places at one time, or the same place at different times, or may vary concomitantly in place and time. The relation between the period of time occupied by a thing during its entire existence and the rest of time is as fixed as the relation between the region the thing covers during its entire existence and the rest of space. And yet there is this difference: two things may approach and then recede from each other in space, may grow more and then less alike in color, shape, etc.; but two things never become nearer and then farther apart in time. The location or the color or the shape of a thing may change, but not its time.

This may seem to depend on a mere verbal accident. Why not simply generalize the use of "change" a little so that a thing changes in a given respect if different parts of the thing have different qualities of the kind in question? Because, it may be fairly answered, this ignores the distinction between a minute mobile thing that travels over a given region, and a spatially large thing that occupies a comparable region at a single instant. Each of the two things has parts that differ from one another in location; but according to ordinary usage, only the former undergoes change. By applying the term "change" in the one case but not the other, ordinary usage marks an important distinction.

In other words, change is concomitant variation in time and some other respect. Since time is always one of the variant factors in change, we speak of *change in* whatever is the other variant factor in the given case. Thus although there is no change that does not involve time, there is no change in time.

NOTES

[1] See, for example, my *Study of Qualities* (1940), pp. 594-623; and Reichenbach's *Elements of Symbolic Logic* (New York: Macmillan, 1947), pp. 284-298.

[2] This usage differs from that of C. S. Peirce, who speaks of inscriptions or utterances as *replicas* of a word type; see *Collected Papers of Charles Sanders Peirce* (Cambridge: Harvard University Press, 1932), Vol. II, p. 143.

[3] Note that Goodman uses "*a* O *b*" to mean that *a* overlaps (tenseless) *b* (including the case where one of *a* or *b* is contained

in the other). Similarly, "*a* D *b*" means that *a* is (tenseless) discrete from *b*. O and D are used here instead of Goodman's own signs because those cannot be reproduced typographically.—Ed.

⁴ For a further discussion, see my paper "On Likeness of Meaning," *Analysis*, 10 (1949), pp. 1-7, and the sequel, "On Some Differences about Meaning," *Analysis*, 13 (1953), pp. 90-96.

TIME

W. V. Quine

Reprinted from the first part of section 36 of *Word and Object*, by W. V. Quine, John Wiley & Sons, Inc., New York, and Technology Press of Massachusetts Institute of Technology (M.I.T. Press), Cambridge, Mass., 1960. References to section numbers in this selection are to Quine's book.

Our ordinary language shows a tiresome bias in its treatment of time. Relations of date are exalted grammatically as relations of position, weight, and color are not. This bias is of itself an inelegance, or breach of theoretical simplicity. Moreover, the form that it takes—that of requiring that every verb form show a tense—is peculiarly productive of needless complications, since it demands lip service to time even when time is farthest from our thoughts. Hence in fashioning canonical notations it is usual to drop tense distinctions.

We may conveniently hold to the grammatical present as a form, but treat it as temporally neutral. One does this in mathematics and other highly theoretical branches of science without deliberate convention. Thus from "Seven of them remained and seven is an odd number" one unhesitatingly infers "An odd number of them remained," despite the palpable fallacy of the analogous inference from "George married Mary and Mary is a widow." One feels the "is" after "seven" as timeless, unlike the "is" after "Mary," even apart from any artifice of canonical notation.

Where the artifice comes is in taking the present tense as timeless always, and dropping other tenses. This artifice frees us to omit temporal information or, when we please, handle it like spatial information. "I will not do it again" becomes "I do not do it after now," where "do" is taken tenselessly and the future force of "will" is translated into a phrase "after now," comparable to "west of here." "I telephoned him but he was sleeping" becomes "I telephone him then but he is sleeping then," where "then" refers to some time implicit in the circumstances of the utterance.

This adjustment lays inferences such as the above ones about seven and George conveniently open to logical inspection. The valid one about seven becomes, with present tenses read timelessly, "Seven of them then remain and seven is an odd number; therefore an odd number of them then remain." In this form the inference no longer has an invalid analogue about George and Mary, but only a valid one: "George marries before now Mary and Mary is a widow now; therefore George marries before now (one who is) a widow now." (Whether to write "marries before now" as here, or "then marries" in parallel to the example about seven, is merely a question whether to suppose that the sentences came on the heels of some reference to a specific past occasion. I have supposed so in the one example and not in the other.)

Such rephrasing of tense distorts English, though scarcely in an unfamiliar way; for the treating of time on a par with space is no novelty to natural science. Of the perplexities that are thus lessened, instances outside the domain of logical deduction are not far to seek. One is the problem of Heraclitus (§ 24). Once we put the temporal extent of the river on a par with the spatial extent, we see no more difficulty in stepping into the same river at two times than at two places. Furthermore the river's change of substance, at a given place from time to time, comes to be seen as quite on a par with the river's difference in substance at a given time from place to place; sameness of river is controverted no more on the one count than on the other.

The problem of Heraclitus was already under control in § 24, without help of the alignment of time with space; but intuitively the alignment helps. Similarly for perplexities of personal identity: the space-time view helps one appreciate that there is no reason why my first and fifth decades should not, like my head and feet, count as parts of the same man, however dissimilar. There need be no unchanging kernel to constitute me the same man in both decades, any more than there need be some peculiarly Quinian textural quality common to the protoplasm of my head and feet; though both are possible.[1]

Physical objects, conceived thus four-dimensionally in space-time, are not to be distinguished from events or, in the concrete sense of the term, processes.[2] Each comprises simply the content, however heterogeneous, of some portion of space-time, however disconnected and gerrymandered. What then distinguishes material substances from other physical objects is a detail: if an object is a substance, there are relatively few atoms that lie partly in it (temporally) and partly outside.

Zeno's paradoxes, if they can be made initially puzzling, become less so when time is looked upon as spacelike. Typical ones consist essentially in dividing a finite distance into infinitely many parts and arguing that infinite time must be consumed in traversing them all. Seeing time in the image of space helps us appreciate that infinitely many periods of time can just as well add up to a finite period as can a finite distance be divided into infinitely many component distances.

Discussion of Zeno's paradoxes, as of much else, is aided by graphing time against distance. Note then that such graphs are quite literally a treatment of time as spacelike.

Just as forward and backward are distinguishable only relative to an orientation, so, according to Einstein's relativity principle, space and time are distinguishable only relative to a velocity. This discovery leaves no reasonable alternative to treating time as spacelike. But the benefits surveyed above are independent of Einstein's principle.[3]

Tense, then, is to give way to such temporal qualifiers as "now," "then," "before t," "after t," and to these only as needed. These qualifiers may be systematized along economical lines, as follows.

Each specific time or epoch, of say an hour's duration, may be taken as an hour-thick slice of the four-dimensional material world, exhaustive spatially and perpendicular to the time axis. (Whether something is an epoch in this sense will depend on point of view, according to relativity theory, but its existence as an object will not.) We are to think of t as an epoch of any desired duration and any desired position along the time axis.[4] Then, where x is a spatiotemporal object, we can construe "x at t" as naming the common part of x and t. Thus "at" is taken as tantamount to the juxtapositive notation illustrated in the singular term "red wine" (§ 21). Red wine is red at wine.

We easily extend "at" to classes. Where z is mankind, z at t may be explained as the class $\hat{y}(Ex)(y = (x \text{ at } t) \text{ and } x \in z)$ of appropriate man stages.

We can treat the indicator words "now" and "then" on a par with "I" and "you," as singular terms. Just as the temporary and shifting objects of reference of "I" and "you" are people, those of "now" and "then" are times or epochs. "I now" and "I then" mean "I at now," "I at then"; the custom just happens to be to omit the "at" here, as in "red wine."[5]

"Before" can be construed as a relative term predicable of times. Such constructions as "x is eating y before t" and "x is eating y after t" then come through thus:

$(\exists u)(u$ is before t and x at u is eating $y)$,

$(\exists u)(t$ is before u and x at u is eating $y)$.

In this example I have used the progressive aspect "is eating," in preference to "eats," because what is concerned is the state and not the disposition; contrast "Tabby eats mice" (§ 28). Temporal qualifications apply to the latter as well, for there may have been a time when Tabby had no taste for mice, and a time may come when she will lose it. Thus we may say "Tabby now eats mice," "Tabby at t eats mice," as well as "Tabby at t is eating mice," but in the one case we report a phase in her evolving pattern of behavior while in the other we report an incident in her behavior.

NOTES

[1] Cf. Goodman, *Structure of Appearance*, p. 94.

[2] They are what Strawson (*Individuals*, pp. 56 f.) has dismissed as *process-things*, "not to be identified either with the processes which things undergo *or* with the things which undergo them. . . . I was concerned to investigate . . . the categories we actually possess; and the category of process-things is one we neither have nor need." He supports his distinctions with examples of usage. Given his concern with usage conservation, I expect he is in the right. But our present concern is with canonical deviations.

[3] Einstein's discovery and Minkowski's interpretation of it provided an essential impetus, certainly, to spatiotemporal thinking, which came afterward to dominate philosophical constructions in Whitehead and others. But the idea of paraphrasing tensed sentences into terms of eternal relations of things to times was clear enough before Einstein. See e.g. Russell, *Principles of Mathematics* (1903), p. 471. For further discussion of tense elimination see my *Elementary Logic*, pp. 6 f., 111-115, 155 ff.; Goodman, *Structure of Appearance*, pp. 296 ff.; Reichenbach, *Elements of Symbolic Logic*, pp. 284-298; Taylor [pp. 381-396 of this volume]; Williams, "The Sea Fight Tomorrow" [for full reference see the Bibliographical Notes.—ED.].

[4] The question of an instant, or epoch of no duration, is best

set aside now and subsumed under § 52 [the next selection in this volume].

[5] In *Individuals,* p. 216, Strawson argues against viewing "now" as a singular term. His argument is that "now" sets no temporal boundaries. One possible answer might be to defend vagueness; another would be to construe the temporal boundaries as those of the shortest utterance of sentential form containing the utterance of "now" in question. The latter answer is in our present spirit of artificial regimentation, and we must note that the Strawson passage has a different context. I even share, in a way, an ulterior doctrine that he is there engaged in supporting, for I think it is of a piece with my reflections on the primacy of unanalyzed occasion sentences in the theory of radical translation and of infant learning.

GEOMETRICAL

OBJECTS

W. V. QUINE

Reprinted from section 52 of *Word and Object*, John Wiley & Sons, Inc., New York, and Technology Press of Massachusetts Institute of Technology (M.I.T. Press), Cambridge, Mass., 1960. References to section numbers are to Quine's book.

Traditionally geometry was the theory of relative position. For Poincaré and others influenced by the pluralism of non-Euclidean geometries, the geometries were a family rather of uninterpreted theory-forms, called geometries only because of structural resemblances to Euclid's original theory of position. The question of the nature of the objects of geometries in the latter sense need not detain us, secured as it is against an answer. But meanwhile geometry also in something like the traditional sense continues as a handmaiden, by whatever name, of natural science. Its objects would appear to be points, curves, surfaces, and solids, conceived as portions of a real space that bathes and permeates the physical world. They are objects which we are tempted to admit on a par with physical objects as values of our variables of quantification, as when we say that Boston, Buffalo, and Detroit are cut by a great circle of the earth.

The objects of geometry can for some purposes be adequately explained away in the manner already contemplated for the ideal objects of mechanics; for we may think of points, curves, and geometrical surfaces as ideally small particles, ideally slender wires, and ideally thin sheets. This treatment conforms well enough to the pure universal statements of geometry—statements to the effect merely that any geometrical objects interrelated in such and such ways are interrelated in such and such further ways. But it ill fits the existential statements of geometry, which require there to be points, curves, surfaces, and solids everywhere.

May we then hold to the naïve view? Here we have a

dualistic theory of spatiotemporal reality, whose two sorts of objects, physical and geometrical, interpenetrate without conflict. There is no conflict simply because the physical laws are not extended to the geometrical objects.

But if such a plan is tolerable here, why could we not equally in § 51 have admitted the ideal objects of mechanics in a single spatiotemporal universe along with the full-fledged physical objects, simply exempting them from some of the laws? Is it just that these two categories would be intuitively too much alike to make the separation of laws seem natural? No. There is a more substantial reason why mass points and the like, as objects supplementary to the full-fledged bodies, should be less welcome than geometrical objects. No sense has been made of their date and location. Evidently, to judge by what is said of them, mass points and such ideal objects are supposed to be in space-time of some sort, ours or another; but just where is each? And, if we waive location, there supervenes a perplexity of identity: when do mass points (or frictionless surfaces, etc.) count as one and when as two? Talk of ideal objects in mechanics, significantly enough, tends not to turn on such questions. There is in this circumstance strong reason to define the ideal objects away—say along Weierstrassian lines of § 51—rather than to keep them and try to settle the perplexities of position or identity by multiplying artificialities. On the other hand geometrical objects raise no such evident problems of position or identity; they are positions outright.

But are we prepared to admit absolute positions, and therewith an absolute distinction between rest and motion? Is not motion relative rather, so that what would count from one point of view as an identical position twice over would count from another point of view as two distinct positions? No doubt. However, we can accommodate this relativistic scruple simply by adding a dimension and speaking of positions not in space but in space-time. Distinct point-instants are distinct absolutely, regardless of the relative movement of the point of view.

If motion is relative, then obviously the question whether a given spatiotemporal region (or aggregate of point-instants) is constant in shape through time, or whether its internal distances readjust through time, will depend on the relative movement of the point of view; and so will the question whether its shape at a time is spherical or elongated. But this is to say only that shape-at-a-time is relative to frames of reference; the geometrical objects whose shapes are concerned remain absolute aggregates of point-instants, however shaped and however specified.

Whether it was better to stay within the three spatial dimensions for our geometrical objects, or better to look beyond space into space-time for them, turned on whether it was wise to assume an absolute distinction between rest and motion. This question, in turn, is the question what theory will best systematize the data of physics. Thus we may fairly say that the question of the nature of the geometrical objects is, like the question of the nature of the elementary particles of physics, a question of physical theory. Granted, laboratory data only incline and do not constrain us in our geometrizing; but likewise they only incline and do not constrain us in our invention of a physical theory. Let fashion and terminology not mislead us into viewing geometry too differently from physics.

And the fact is that Einstein's physical theorizing included geometrical decisions also beyond the relativity of motion. Considerations of overall theoretical simplicity of physical theory induced him to settle for a non-Euclidean form of geometry, simpler though the Euclidean is when considered apart. Accepting then this four-dimensional non-Euclidean geometry along with relativity physics as the literal truth (by today's lights), we may view Euclidean geometry on a par with Newtonian physics as a convenient myth, simpler for some purposes but symbolic of that ulterior truth. The geometrical objects of Euclidean geometry then take on, relative to the "real" geometrical objects of the non-Euclidean "true" geometry of ideal objects, the status of manners of speaking, limit myths, explicable in principle by paraphrasing our sentences along Weierstrassian lines.

There remain also other geometries, other in various ways. There are the more abstract ones, culminating in topology, which treat of the geometrical objects in decreasingly specific detail. These geometries raise no further ontic problem, for their objects can be taken to be our same old geometrical objects; we can look upon these geometries as merely saying less about them.

And there remain geometries that are not just more abstract than, but actually contrary to, our "true" geometry of relativity physics. Shall we rate these as simply false? Or seek ways of reconstruing their words that would make them true after all, whether of our same old geometrical objects or of something else? We need do none of this; an uninterpreted theory-form can be worthy of study for its structure without its talking about anything. When it is brought into connection with the quantifiers of a broader scientific context in such a way as to purport to talk unfeignedly of objects of some sort, then it is time enough to wonder what the objects are.

Up to now I have defended geometrical objects not because I think it is best to admit them as part of the furniture of our universe, but only in order to exhibit relevant considerations. Meanwhile there remains, obviously, an objection to geometrical objects, on the score of economy of objects. Let us now consider how we may manage without them.

The only sentences we need try to paraphrase, for elimination of reference to geometrical objects, are those that cannot be facilely dismissed as gibberish of an uninterpreted calculus: those that contribute, rather, like sentences about the equator or the one about Boston, Buffalo, and Detroit, to discourse about the real world outside geometry. Now sentences about the equator can all probably be satisfactorily paraphrased into forms in which "equator" has the immediate context "nearer the equator than"; and these four words can be treated as a simple relative term or even defined away in terms of centrifugal force or mean solar elevation. The more serious cases are sentences which, like that about Boston, Buffalo, and Detroit, ostensibly call for a geometrical object as value of a variable of quantification.

But the reference to geometrical objects in such cases is an auxiliary merely to what we want to say about the movements and spatiotemporal relations of bodies; and we can hope to by-pass the geometrical objects by falling back on a relative term of distance or spatiotemporal interval, conceived as relating physical bodies and numbers. This course involves accepting numbers as objects along with bodies, but spares us assuming geometrical objects in addition. The elements are thus simplified. The practical convenience of geometrical objects can still be preserved by reinstating definitionally (cf. § 39) whatever idioms we analyze away.

The elimination can be systematized along the lines of analytical geometry. In its minimum essentials the idea is as follows, for our four-dimensional space-time. We pick five particle-events a, b, c, d, e, not quite at random. (The requirement is merely that they mark the vertices of a full-fledged four-dimensional "hyper-solid," rather than all lying in a plane or a three-dimensional solid.) We can think of the five as given by proper names, or, what comes to the same (cf. § 37), by general terms true uniquely of each. Now every point (or point-instant) in space-time is uniquely determined once we specify its "distance" (or interval: the analogue of distance in four-dimensional space-time) from each of the five. The position of a body in space-time is therefore determined by the distance of its various extremes from each of the five reference particle-events. The attribution of (four-dimensional) shapes to bodies can be paraphrased as attribution of appropriate arithmetical conditions to the classes of ordered quintuples of numbers that fix the bodies' boundaries. Correspondingly for the attribution of collinearity or other geometrical relations.

We could take the further step, if we wish, of nominally restituting geometrical objects by *identifying* points (actually point-instants) with the appropriate ordered quintuples of numbers, and identifying the rest of the geometrical objects with the classes of their constituent points in that sense. Whether to speak of geometrical objects as by-passed or as reconstrued is a matter of indifference.

The five-point type of coordinate system thus simply described would be prohibitively awkward in practice. The least of its inelegances is that it ill exploits its numerical resources. The compatible distances from the five points make up a quite special and not quickly recognizable class of quintuples. The more strictly Cartesian scheme, of fixing each point by its distance from each of various mutually perpendicular planes, is far superior: it gets by with quadruples of numbers instead of quintuples, it wastes no quadruples, and, above all, it correlates the important geometrical conditions with far simpler arithmetical conditions than our five-point method would do. Certainly one would want to set up a system of Cartesian coordinates. But its construction, given as starting point only a distance measure and selected reference particles, is a long story. The five-point method is a more readily describable one to the same theoretical effect, and suffices for conveying some concrete sense of what the elimination of geometrical objects means.

To the same end it may be worth while now to turn more specific still: to cut through the whole apparatus of systematic reference points and quadruples or quintuples of real numbers, and consider rather how some very definite geometrical remark about physical bodies, considered simply by itself, might be paraphrased into terms of distance without geometrical objects. Let us take the sentence to the effect that there is a line passing through the bodies A, B, and C, where B is the one in the middle.

A simple paraphrase that does not quite fill the bill is this: there are particles x, y, and z, respectively in A, B, and C, such that the distance from x to z *is* the sum of those from x to y and from y to z. The trouble with it is that it makes no allowance for gaps between the component particles (or particle-events) of a body. It does not allow for the possibility that every line through A, B, and C that hits particles of both A and C passes between particles of B, hitting none.

There is a way through this difficulty which may most readily be grasped in principle if we suppose that we are working in just two dimensions. So A, B, and C are now clusters of dots on a page; and we want to say in effect that there is a geometrical line through A, B, and C, without actually referring to any objects but the dots and their clusters, nor relating them otherwise than in point of distance. We suppose still that of these three clusters (if a line does cut through them) the middle one is B. What we want in effect to say, then, though within the allotted means, is that there are a dot x of A, a dot z of C, and dots y and y' of B (same or different), such that the geometrical line xz hits y or y' or passes between them. But xz hits y or y' or passes between them if and only if the area of the triangle xyz plus the area of the triangle $xy'z$ equals the area of the triangle xyy' plus the area of the triangle zyy'. But the area of a triangle is a known function f of the lengths of the sides. The following, then, is a formulation to our purpose, where "dxy" means "distance from x to y":

There are a dot x of A, a dot z of C, and dots y and y' of B such that $f(dxy, dyz, dxz) + f(dxy', dy'z, dxz) = f(dxy, dyy', dxy') + f(dzy, dyy', dzy')$.

SPATIAL AND TEMPORAL
ANALOGIES AND THE
CONCEPT OF IDENTITY

Richard Taylor

Presented in a symposium on "Space, Time, and Individuals" at a meeting of the American Philosophical Association, Eastern Division, December 27, 1955. First published in the *Journal of Philosophy*, Volume LII, 1955.

Few things have engendered more philosophical puzzlement than time. Unlike space, which has generally seemed above all simple and obvious, time has always been regarded by a great many philosophers and theologians as a dark subject of speculation, fundamentally enigmatic, even incomprehensible. It is also something concerning which men can become bewitched over statements which, on the slightest analysis, turn out to express the most trivial truisms—such as, "the past cannot be changed," "the future (or the past) is nothing," "time cannot be reversed," and so on.

I want to remove some of this mysteriousness by showing that temporal and spatial relations, contrary to much traditional thought, are radically alike; or, more precisely, that (1) terms ordinarily used in a peculiarly temporal sense have spatial counterparts and vice versa, and that accordingly (2) many propositions involving temporal concepts which seem obviously and necessarily true, are just as necessarily but not so obviously true when reformulated in terms of spatial relations; or, if false in terms of spatial concepts, then false in terms of temporal ones too.

Such a project is sometimes rejected as a "spatializing of time," but what I have in mind is no more a spatialization of time than a temporalization of space; if it is either, it is the other as well. Of course I am not the first to press the analogies between space and time,[1] but I believe they can be carried much farther than has been thought possible heretofore.

Basic concepts. A basic notion to be employed is that of *place*, which can be either spatial or temporal. "At Boston, Mass." designates a spatial place, "On May 1, 1955" a temporal one. A corollary is the notion of *distance*, which is likewise either spatial or temporal. New York and Boston are spatially distant from each other and from other things, while Plato and Kant are temporally so; but distances, of either kind, can of course be great or small. The allied notion of *being present*, incidentally, is in fact commonly used in both senses, as meaning "here" or "now" or both. Again, the concept of *length* or *extension* has a place in both contexts, though this is easily overlooked. Things can be spatially long or short, but so too they can have a long or brief duration, i.e., be temporally long or short.[2] Indeed, there is no reason why temporal dimension should not be included in any description of the shape of a thing. The notion of length, in turn, leads to that of *parts*, both spatial and temporal.[3] Distinctions between the spatial parts of things are commonplace, but it is no less significant to reason that things have temporal parts too, often quite dissimilar to each other—for instance, widely separated parts of a man's history, or narrowly separated temporal parts of a kaleidoscope. Again, the notion of *direction* has a use with respect to both spatial and temporal relations; one can, for instance, speak of the direction from past to future, from future to past, from north to south, and so on, none of which directions is any more or less genuine or intrinsic than the others. Finally, the concept of a *physical object* involves both space and time, since any such object has, for instance, both kinds of extension and both kinds of parts. Objects are often distinguished from *events*, on the assumption that the notion of the former is a fundamentally spatial one and that of the latter fundamentally temporal, but no such distinction is necessary and none will be made here. Any physical object is itself an event, i.e., endures and has a more or less interesting history; or, as Nelson Goodman has expressed it, "a thing is a monotonous event; an event is an unstable thing."[4]

Most of these terms are thus ambiguous, but I think the context will prevent misunderstanding in what follows. A spatial place, however, will usually be called just a "place" (abbreviated "*L*"), and a temporal one a "time" (abbreviated "*T*"). "Spatial part" and "temporal part" will sometimes be abbreviated as "*S*-part" and "*T*-part." Other abbreviations will be obvious.

Spatial and temporal analogies. My procedure now will be to state, in the form of objections, propositions which are commonly thought to be obviously true, and to express a radical difference between space and time. Concerning each such proposition I shall show, in the form of a reply, that no such difference is expressed, that is, that no such proposition is, under similar interpretations, true of time and false of space or vice versa. I shall begin with the simpler and easier propositions and conclude with the more difficult.

First objection. An object cannot be in two places at once, though it can occupy two or more times at only one place.

An object occupies two times in one place by remaining awhile where it is, or by being removed from its place and later returned, or by being annihilated and subsequently re-created at the same spot. But it seems plain that no object can be in two places at one time, and in particular that it cannot be "returned" to a time, for when one tries to imagine situations that might be so described he unavoidably finds himself thinking of at least two, perhaps similar, objects.

Reply. This statement seems to express a simple and obvious difference between space and time, but it does not. For it should be noted that an object is ordinarily said to be in one place at two times, only if it also occupies all the time in between, whether at that place or another. But with a similar proviso, an object can likewise be in two places at one time, namely, by occupying the space between them as well.[5] A ball, for instance, occupies two places at once, if the places be chosen as those of opposite sides; but in so doing, it also occupies all the places between. It is tempting to say that only *part* of the ball is in either place; but then, it is a different *temporal* part of an object which, at the same place, is in either of two times.

The situation analogous to that of an object which is removed from its place and later returned is more complicated, and will be considered more fully in answer to another objection (the seventh), but for now the analogy can be shown superficially as follows.

An object which occupies one place at two times by being returned to that place is one which fulfills the following description:

> O is at L_1 and T_1 and T_2; it temporally extends from T_1 to T_2, but is *then* (i.e., sometime in that interval) at places other than L_1.

The analogy of a thing being in two places at one time by being returned to that time would therefore be something fulfilling this description:

> O is at T_1 and L_1 and L_2; it spatially extends from L_1 to L_2, but is *there* (i.e., somewhere in that interval) at times other than T_1.

And an example of this would be some widespread physical disturbance, like a roll of thunder, existing simultaneously in two nearby towns, but at a different time, earlier or later, somewhere between them. One might want to insist that several objects or events are involved in this case, in contrast to the selfsame object involved in the former, but this would only betray a prejudice in the common notion of identity. For just as we can and ordinarily do say that moving about in space—i.e., acquiring and losing spatial relations with other things over a lapse of time—does not destroy the identity of a thing, we have equal reason to say that moving about in time—i.e., acquiring and losing temporal relations with other things over a lapse of space—does not destroy it either.

Finally, we might want to say that an object can be in the same place at two times without filling the time in between—which would, of course, simply amount to its being annihilated and then recreated at the same place. And it seems that we might be entitled in some situations to regard it as the *same* object at two times, or at least that there is no overwhelming reason for thinking that two wholly distinct though similar objects are involved; we do, for instance, sometimes speak of hearing the same note of a whistle twice over. But the analogy to this is exceedingly simple, viz., any object simultaneously at two places and nowhere between, such as, a billiard ball which is at once at both sides of a table. Here most people would want to insist that we have two wholly distinct balls, however similar to each other, simply on the ground that they stand in quite different spatial relations to other things. But in the former case, too, the object stands in quite different temporal relations to other things at the two times involved, so there is no significant difference in the two examples. It is perhaps arbitrary whether we say that there is an identity or a

diversity of things in either example, but it would be utterly
capricious to insist that there is an identity in the one case
but a diversity in the other, for the two situations are analogous. I would myself, however, say that the things are diverse
in both cases.

Second objection. But time, unlike space, is an essential ingredient of motion and change, of coming to be and passing
away.[6]

This would seem to follow from the *meanings* of "motion"
and "change." Further, it seems impossible to form any idea
of change, process, or mutation, abstracted from any concept
of time; spatial relations, on the other hand, are static in
essence.

Reply. This objection is necessarily true, only because "motion" and "change," as ordinarily used and defined, *mean*
temporal processes; "coming to be" and "passing away" *mean*
generation and destruction *in time*, and are appropriately
tensed. Such ordinary usage and definition do not, however,
preclude the possibility of a spatial kind of change, exactly
analogous to temporal change and no less significant to reason;
and, in fact, if spatial relations are substituted for temporal
ones in any description of motion or change, we find that we
do have a description of something real and familiar.

Ordinarily, to say that a thing *moves* is to say that it
occupies one place at one time and another place at another.
But it is, evidently, the same thing to say that it occupies one
time at one place and another time at another; so *this* kind of
change involves spatial relations just as much as, and in the
same way as, it involves temporal ones; it is neither more nor
less a temporal process than a spatial one.[7]

To say, further, that a thing *changes*, in a more general
sense of "change," means that it has an interesting history,
acquiring and losing properties while enduring—which means
simply that its temporal parts are dissimilar. But why may we
not say, analogously, that a thing may have an interesting
geography, acquiring and losing properties in its spatial extension, i.e., that at one and the same time its various spatial
parts are dissimilar? An example of temporal change would
be an object which, at any given place, is blue at one time
and red at another. An example of spatial change would be
an object—e.g., a wire—which, at any given time, is blue
at one end and red at the other, and perhaps various other
colors between. This sense of "change" is not, moreover, at

all unusual; it would make sense, for instance, to say of a wire, which was found to be red in one town and blue in another, that *somewhere* (not sometime) between the two towns it changes color, and such change might, like temporal change, be gradual or abrupt, i.e., occur over a long or brief interval. This manner of speaking would, I suggest, sound artificial only to one who insisted, quite arbitrarily, that temporal change is somehow more genuine than spatial change.

To say, finally, that something "comes to be" and then "passes away" means simply that there are two times, that the thing extends from one such time to the other, and does not extend beyond either of them.[8] The spatial analogy to this is too evident to require description.

Third objection. Things can change their spatial positions, but not their temporal ones, these being, once given, fixed eternally.

A thing which is north of a given object, for instance, can be switched to the south of it, but a thing which is at any time future to a given thing is everlastingly so, and can in no way be shifted to become past to it. It is precisely this which constitutes the irreversibility of time, and which has no parallel in space. And it is such considerations as these which should lead us to conclude that it is space, if anything, which is the more "fluid," and time "static."[9]

Reply. What has happened here is that the analogies have not been made complete, and that the notion of change has been used only in its temporal sense in both contexts. The statement, profoundly uttered, that things cannot change their positions in time derives its truth from, and in fact amounts to no more than, the utterly trivial statement that a thing cannot be in two times at once (at one time), and is comparable to the equally trivial statement that a thing cannot be in two places at one place.[10] Moreover, the claim that two things which are so related that the one is future to the other are *always* so related, and cannot at another time be truthfully represented as oppositely related, is surely true; but it is hardly more significant than a statement that two things which are so related that one is north of the other are *everywhere* so related, and cannot at another place be truthfully represented as oppositely related. It is true in Rhode Island, for instance, that I am now south of Boston; it is no less true in California.

But of course at another time I might be north of Boston. Can we find an analogy for this in temporal relations? I.e., can a description be given of something which is at one place past to another thing, and at another place future to it? It seems clear that it can, and that, accordingly, in any sense in which things can change positions in space, they can in a precisely analogous and equally significant way change them in time, with the result that time is neither more nor less "fluid" and neither more nor less "static" or "fixed" than space.

For note, first, that things can change their positions in space, *only* through a lapse of time, time being used up in moving from one place to another. A description of two things, A and B, which change their relative positions in space would thus be:

At T_1 A is north of B,
At T_2 A is south of B.

Analogously, a description of the temporal relations of two things changing would be:

At L_1 A is future to B,
At L_2 A is past to B.

And an example fitting this description is not hard to find. Let A, for example, be an earthquake, occurring gradually over an area which includes two towns, and let B be a stroke of a clock (any place in the world). Now it is possible that in one town, A is future to B, and in the other, past to B. This fulfills the description.

It is surely tempting to object here that the analogy is mistaken, on the ground that the earthquake occurring in one place is one event and that occurring at the other place another, so that it is not one and the same thing which is both past and future to another. But all this calls attention to is the fact that a lapse of space is required to make the example work— and we have already seen that, analogously, a lapse of time is absolutely required for things to change their spatial positions. Moreover, if one wanted to insist that the earthquake existing at two times in different places was in fact two earthquakes, we could just as well say, as we certainly would not, that any object existing at two places in different times was in fact two objects—that if something is moved from the south to the north of Boston, for instance, then it is no longer the same object, but a new and wholly different one. The reason we do not say this is that, in occupying different times and places, it occupies all the time in between (though not all at one place); but so too, the earthquake existing in two places and times occupies all the space in between (though not all at one time), and accordingly remains one identical thing throughout this lapse of space if any object moving in space remains one identical thing throughout its lapse of time. The analogy is, then, complete.

Fourth objection. But time is something moving, or flowing, in a fixed direction from future to past and at an unalterable rate; space, on the other hand, is everywhere the same and unchanging.[11]

It is for this reason that we speak meaningfully of the passage of time, of the continual recession of things past and the approach of things future. It is because of this, too, that people naturally think of time as like a great river, engulfing all things in its course. Nothing of the sort, however, is appropriate to the notion of space.

Reply. It seems quite certain that there is *no* sense in which time moves. For, in the first place, no given time can move with respect to any other time, without involving the absurdity of two different times temporally coinciding, just as no place can move with respect to any other place, without involving the absurdity of two places spatially coinciding. Neither the date nor the place of one's birth, for instance, can move in relation to the date or place of anything else.[12]

But if, on the other hand, one supposes that the *whole* of time moves, how can one express the direction of such motion? Towards what, away from what, or in relation to what can it move? We cannot say that time arises out of the future and moves into the past, nor that it unfolds itself from the past and moves into the future, for the past and the future are themselves part of the whole of time, and must move with it if time moves at all. Nor can we say that time moves in relation to the present, for there is no one place in time, called "the present," with respect to which anything can move. A constant temporal distance obtains between any time whatever and any other time that anyone calls "the present."

Of course there is a temptation to say that the present moves in some sense, since the expression "the present" never designates the same time twice over; a moment no sooner emerges from the future and becomes present than it lapses forever into an ever receding past. But this kind of statement, gravely asserted, says only that the word "now" never, i.e., at no time, designates more than one time, viz., the time of its utterance. To which we can add that the word "here" likewise nowhere, i.e., at no place, designates more than one place, viz., the place of its utterance.

Moreover, if it did make sense to speak of time, or any part of it such as the present, as moving, then it would make equal sense to speak of its ceasing to move, or moving more or less rapidly than it does, however persuaded one might be of the falsity of such suggestions. Yet such ways of speaking do not merely say what is false, they say nothing at all. When one tries, for instance, to think of time "standing still," what he in fact envisages is everything *but* time standing still—i.e., a cessation of change, over a period of time. Indeed, the notion of anything at all being at rest involves the idea of its being unchanged in some respect over an interval of time, which surely renders the idea of time's being at rest unintelligible. Similarly, the suggestion that the rate of time's flow might be accelerated or diminished involves the same difficulties as that of space being expanded or contracted, the former just bringing to mind the thought of things happening more or less rapidly than usual, and the latter the thought of all things in space becoming larger or smaller, or at greater or lesser distances from each other.

I conclude, then, that if time moves, all time moves together, just as, if space expands or contracts, it all does so together; but that in fact neither supposition is intelligible, and space and time are in this respect, again, alike.[13]

Fifth objection. While time may not then in any clear sense be moving, yet everything *in* time moves from the future through the present and on into the past.[14]

It is for this reason that, though things need not always be moving in space, nothing can pause or rest in time, but becomes past immediately upon having been present, and then becomes increasingly remote in the past. It is with this in mind that we speak of history unfolding, and of facing the future and leaving the past ever farther behind us. Such ways of speaking are too common and useful to be thought to express no truth at all.

Reply. It makes little difference whether one says that time moves, or that things move in time, and we have already considered the former suggestion. The following additional comments can, however, be made.

All that is meant in saying that anything or everything

"moves in time" in this sense is that it has temporal extension;[15] that it "moves" from the future through the present
and into the past means simply that some of its temporal parts
are earlier than others—which of course must be true in any
case, if we are to avoid the idea that two times might coincide.
Nor is everything moving, even in this sense, in the direction
of the future; we cannot say of things past, for instance, that
they are so moving, any more than we can say of something
south of us—say, some southern state—that it here stretches
towards the north. Of course of things past we can say that
they were first future, then present, then became past; but this
again says only that they have temporal extension, and that
they do not extend to the temporal present, i.e., do not reach
to "here" in time. The analogy to this is not something moving
through space in the usual way, but simply something spatially
extended which does not reach to the spatial present, such as,
again, some southern state.

That, moreover, any object O should pause or be at rest in
space means only that it occupies the whole of a given place
at two different times and at all times between; that is, that

O is at L_1 throughout the interval T_1 through T_2.

That, similarly, an object should pause or be at rest in time
means that it occupies the whole of a given time at two different places and at all places between; that is, that

O is at T_1 throughout the interval L_1 through L_2,

which applies to an enormous number of things, such as, for
instance, bridges and roads. Of course the object in our second description is very likely to occupy more time than is
designated by "T_1," especially if this be taken as representing
but an instant, within which time nothing at all can exist; but
so also is the object of our first description likely to occupy
more space than is designated by "L_1," especially if this be
taken as a spatial point, within which again nothing at all can
exist. Both the place of the first description and the time of
the second must, accordingly, be taken as having some size,
and it is not *necessary* that the first object exist anywhere
else—i.e., be any larger than L_1—nor that the second object
exist "anywhen" else—i.e., be temporally any larger than T_1.

Sixth objection. But two things can move closer together or
farther apart in space; they cannot do so in time.[16]

That things move closer together or farther apart in space is an everyday fact of experience. But the idea of increasing or diminishing the temporal intervals between two things—such as Caesar's birth and his death—involves obvious absurdities.

Reply. To say that two things can move closer together or farther apart in space only means that they can at different times be separated by smaller or larger spatial intervals. This is surely true, but it is necessary to notice that these differing spatial intervals do not separate the same temporal parts of such objects. The description of such a situation would thus be:

> At T_1 T-part$_1$ of object A and T-part$_1$ of object B are separated by a spatial interval x.

> At T_2 T-part$_2$ of object A and T-part$_2$ of object B are separated by a spatial interval y, larger or smaller than x.

To say, analogously, that two things can move closer together or farther apart in time means that they can at different places be separated by smaller or larger temporal intervals. That is:

> At L_1 S-part$_1$ of object A and S-part$_1$ of object B are separated by a temporal interval x.

> At L_2 S-part$_2$ of object A and S-part$_2$ of object B are separated by a temporal interval y, larger or smaller than x.

And this, too, describes a familiar kind of motion, though not so obviously. Consider, for example, two rolls of thunder, considered as aerial disturbances either heard or unheard, each existing in two nearby towns, separated in one town by an interval of one second, and in the other by an interval of two seconds. This situation fits our description. And while these different temporal distances obviously separate different spatial parts of the two things involved, we saw (what is easy to overlook) that in the former case it was different temporal parts of two objects which were at different times separated by different spatial distances. If, accordingly, anyone should want to insist that it is not the *same* two things which are at different places separated by different temporal intervals, simply on the ground that the things involved *are* at different places, and that they therefore in no sense *move* closer together in time, we can with equal reason insist that it is not the *same* two things which move closer together or farther apart in space, and that they therefore in no sense really *move* in space, simply on the ground that the things involved are at different times.

Now I think the reason this sort of analogy does not occur to one very readily is that we unhesitatingly make allowance for a lapse of time when we think of two things moving closer together or farther apart in space, whereas it is not obvious how a similar lapse of space will enable things to move closer together in time. We somehow feel, prior to reflection, that the temporal distance between things should change with no lapse of space in order for them to move closer together or farther apart in time; but in fact there is no more point to this impossible requirement than to requiring, as no one would, no lapse of time in the movement of things towards or away from each other in space. And this way of thinking, again, results from a prejudice concerning identity. We tend to think of an object as remaining one and the same throughout its temporal length—even, sometimes, in spite of discontinuity or gaps between its temporal parts—whereas a spatially extended thing is likely to be thought of as an amalgam of contiguous parts, and any spatial discontinuity between these is enough to destroy utterly the thought that it is one and the same thing throughout. It is not uncommon, for instance, to speak of the *same* sound, like the blast of a whistle, being heard several times over, but we are never tempted to say that the same object, like a billiard ball, is lying about in various parts of a room; we prefer to say that they are different, similar objects, for no other reason than that they are spatially discrete.

Seventh objection. A thing can move back and forth in space, though it cannot do so in time.[17]

That things move back and forth in space, reoccupying the places where they were, is, again, a common fact of experience; but it is difficult to see how anything could move back and forth in time, reoccupying a time now past, without moving backwards in time, and thus being in two times at once.

Reply. To speak of a thing "moving backwards" in time is but a misleading way of expressing the idea that, at times future to now, the thing occupies times past, which is plainly impossible but not very profound.[18] For in *this* sense it is equally impossible for a thing to move backwards in space—e.g., at a place north of here, to occupy a place south.

The real difficulty raised by this objection, however, is that of seeing how, if at all, it is possible for anything to move back and forth in time in a sense which is analogous to that in which things *do* most obviously move back and forth in space. But if the analogy is really carried out at every point, it can be seen that it does still hold for temporal as well as spatial relations.

An object that moves back and forth in space is one which is at one place at one time, at another place at another time, and in the first place at a third time, *without* occupying any two such places at once, i.e., without being so large that it fills both. Assume, then, a small ball which is in one town at noon, in another town at 1:00 o'clock, and back in the first town at 2:00 o'clock. Now it is obvious that one condition of its thus moving back and forth is that it has a considerable temporal length, long enough to reach from the first through the last of the times mentioned; if it did not extend to 2:00 o'clock, it might get to the neighboring town but it would not get back. (We might say that "there would not be time enough," but this is not right. It is not that time would run out, but that the object would run out in time.)

Having temporal extension, such an object also has temporal parts. It will be useful, then, to distinguish three such parts, occupying the times when the object is in either of the two towns; and to keep them distinct, let us give them different colors—say, blue, green, and yellow. This means that the ball starts out as blue, and becomes successively green and yellow as it moves forth and back.

The rather complicated history of this object can thus be fully described as follows:

> At T_1 T-part$_1$ of O (blue) is at L_1
> No part is at L_2
> T-part$_2$ (green) and T-part$_3$ (yellow) do not then exist.
>
> At T_2 T-part$_2$ of O (green) is at L_2
> No part is at L_1
> T-part$_1$ (blue) and T-part$_3$ (yellow) do not then exist.
>
> At T_3 T-part$_3$ of O (yellow) is at L_1
> No part is at L_2

T-part$_1$ (blue) and T-part$_2$ (green) do not then exist.

Now if there is any analogy to this, consisting of an object moving back and forth in time, it must be one fulfilling the following description, wherein temporal relations are substituted for spatial ones and vice versa:

> At L_1 S-part$_1$ of O is at T_1
> No part is at T_2
> S-part$_2$ and S-part$_3$ do not there exist.
> At L_2 S-part$_2$ of O is at T_2
> No part is at T_1
> S-part$_1$ and S-part$_3$ do not there exist.
> At L_3 S-part$_3$ of O is at T_1
> No part is at T_2
> S-part$_1$ and S-part$_2$ do not there exist.

Can anything fulfill this description? Evidently it can, though it is a bit shocking to discover it. Consider, for instance, an aerial disturbance such as a whistle blast, existing nonsimultaneously in three nearby towns, A, B, and C, B being located between the other two. At T_1 the disturbance exists in A and C but not in B, and at T_2 it is heard at neither A nor C but is at B. And if it is desired to distinguish this object into spatial parts in some obvious way, as the first was distinguished into temporal ones, we can suppose that it has a different pitch in each town, each pitch thus identifying one spatial part. This completes the analogy, though two further remarks are needed.

The first is, that it would not be strange for people in all three towns to say they heard the same blast. A philosopher might want to argue that they heard three different sounds, in view of their differences in pitch and the difference in their spatial and temporal locations; but there would then be the *same* reason for saying, as one ordinarily would not, that it was three different balls which were involved in our first example. Differences of color and pitch, moreover, were introduced only to make it easier to distinguish between parts, and are otherwise quite unessential.

Secondly, the latter example might seem to involve no real temporal *movement*, though there is no doubt that the ball in the first example moves in space. But the fact that the ball is spatially in linear motion is expressed by this statement: that it is spatially quite small but temporally large—i.e., is a small ball that lasts quite awhile—and that over an interval of time it occupies a space greater than its own spatial dimension. But we find, analogously, that the following statement is true of the object in our second example: that it is temporally quite small and spatially large—i.e., is a brief blast which covers a large area—but that over an interval of space it occupies a time greater than its own temporal dimension. And this statement can, accordingly, be taken as expressing the fact that this object moves—indeed, in this case, moves back and forth—in time.

NOTES

1 See in particular Donald Williams, "The Myth of Passage," *Journal of Philosophy*, Vol. XLVIII (1951), pp. 457-472, and Nelson Goodman, *The Structure of Appearance* (Harvard, 1951), concluding chapter [See pp. 356-369 of this volume.—ED.]

2 *Cf.* Goodman, *op. cit.*, p. 285.

3 *Ibid.*, pp. 285, 301.

4 *Ibid.*, p. 285. *Cf.* C. D. Broad, *Scientific Thought* (London, 1923), p. 54: "By an *event* I am going to mean anything that endures at all, no matter how long it lasts or whether it be qualitatively alike or qualitatively different at adjacent stages in its history."

5 *Cf.* Goodman, *op. cit.*, p. 301.

6 *Cf.* Goodman, *ibid.*

7 *Cf.* Williams, *op. cit.*, p. 463.

8 This applies to absolute generation and destruction only (Aristotle's γένεσις καὶ ψθορά) as distinct from coming to be or ceasing to be with respect to this or that property (κίνησις), for which the description could be modified in an obvious way.

9 As Goodman in fact does conclude, *op. cit.*, pp. 301-302.

10 *Cf.* Williams, *op. cit.*, p. 463. Goodman says (*op. cit.*, p. 301) that "a minimal spatially changing (moving) compound not merely occupies two places but occupies them at different times. . . . Analogously, a minimal temporally changing compound would have not merely to occupy two times but to occupy them at different times," which is of course absurd. But if this were a real analogy, the last word of this statement would be "places" rather than "times," in which case there would be no absurdity at all.

11 John Wild develops this point in "The New Empiricism and Human Time," *Review of Metaphysics,* Vol. VII (1954). He says "everything in the world of nature seems to be engulfed in an irreversible flux of time which cannot be quickened or retarded, but

flows everywhere at a constant rate" (p. 543), whereas "space as such is fixed" (p. 541).

[12] Cf. Goodman, op. cit., p. 298.

[13] Much more thoroughgoing arguments than mine on this point are to be found in Goodman, op. cit., concluding chapter, and Williams, op. cit., pp. 461 ff.

[14] John Wild, in addition to holding (I believe) that time "flows everywhere at a constant rate," holds also that everything in time flows at a constant rate, for he says that "time . . . determines every being in time to flow with it in a single direction at a constant rate" (ibid., italics supplied).

[15] Cf. Williams, op. cit., p. 463: "Each of us proceeds through time only as a fence proceeds across a farm: that is, parts of our being, and the fence's, occupy successive instants and points, respectively."

[16] Cf. Goodman, op. cit., p. 301.

[17] Cf. Goodman, ibid.

[18] Cf. Williams, ibid.

TIME, IRREVERSIBLE
PROCESSES, AND THE
PHYSICAL STATUS OF BECOMING

Adolf Grünbaum

Reprinted with some minor changes and omissions from sections III and IV of Grünbaum's essay, "Carnap's Views on the Foundations of Geometry," in *The Philosophy of Rudolf Carnap*, edited by P. A. Schilpp, Library of Living Philosophers, Open Court Publishing Company, La Salle, Ill., 1964.

III. Time and Irreversible Processes

Although the serial relation "later than" itself does have *a* direction in virtue of being asymmetric, the set of states ordered by it does *not* have *a* direction, since that set is likewise ordered by the converse relation "earlier than." But in a universe containing irreversible processes, the set of temporally ordered states exhibits a special *difference* of structure between the *two opposite directions*. Thus, when we speak of the anisotropy of time, this must *not* be construed as equivalent to making assertions about "*the*" direction of time. J. J. C. Smart and Max Black have correctly pointed out (see note 38) that reference to "the" direction of time is inspired by the notion that time "flows." In particular, as we shall see in Section IV, Reichenbach's use of this term rests on his incorrect supposition that there is a physical basis for *becoming* in the sense of the shifting of a physically defined "now" along *one* of the two physically distinguished directions of time. In speaking of the anisotropy of physical time, we intend to refer only to the static directional difference between earlier and later and thus make no commitment whatever to a transient division of time into the past and the future by a present whose "advance" would define "*the*" direction of time. In fact, we shall argue in Section IV that the concept of becoming has no significant application outside human conscious awareness. Nevertheless, we shall find it desirable, after having entered this explicit *caveat*, to use the locution "the direction of time" as a synonym not only for "the future direction" in *psychological* time but also for "the one of two physically distinguished directions of time which our theory calls 'positive.'"

Our analysis of the logical relations between symmetric causality, open time, extrinsic vs. intrinsic seriality of time, and anisotropy of time requires us to reject the following statement by Reichenbach:

In the usual discussions of problems of time it has become customary to argue that only irreversible processes supply an asymmetrical relation of causality, while reversible processes allegedly lead to a symmetrical causal relation. This conception is incorrect. Irreversible processes alone can define a direction of time; but reversible processes define at least an [serial] order of time, and thereby supply an asymmetrical relation of causality. The reader is referred to the discussion of the relation *to the left of* (...). The correct formulation is that only irreversible processes define a unidirectional causality.[1]

Reichenbach notes that while the causal processes of classical mechanics and special relativity are reversible, the temporal order affirmed by these "reversible" theories is serial. He then infers that (a) the causal relation in a reversible world must be asymmetric, and (b) in an irreversible world, we require a temporal relation which is not "merely" serial but also "unidirectional," as well as a causal relation which is both asymmetric and unidirectional. But he overlooks that in the reversible worlds, the seriality of time is extrinsic, because the assignment of the lower of two real numbers as the temporal name to *one* of two causally connected events does *not* express any objective asymmetry on the part of the causal relation itself.

We are now ready to examine in detail the physical basis of the anisotropy of time.

Our problem now is whether entropy, whose values are given by real numbers, succeeds, unlike the causality of reversible processes, in conferring anisotropy on open time by intrinsically defining a serial ordering in the class of states of a closed system.

In its original, non-statistical form, the second law of thermodynamics tells us that the entropy of a *closed* system, not already in thermodynamic equilibrium, always increases with time. This statement is synthetic in an obvious sense, *if* the direction of increasing time is defined independently of the entropy-increase either by reference to the continuous matter-energy accretion (as distinct from energy *dispersion*) postulated by the "new cosmology"[2] or—in the spatially limited and cosmically brief career of man—by reliance on the subjective sense of time flow in human consciousness. We shall see, however, that contemporary information theory and thermodynamics *explain* several important features of man's subjective sense of time on the basis of the participation of his organism in the entropic lawfulness of physical nature. And, being unwilling to base the empirical content of so earthy a law as that of Clausius on a highly speculative cosmology, we reject both of these criteria. Instead, while postponing statistical considerations, we can follow Eddington and use the second law of thermodynamics itself to give a coordinative definition of the positive direction of time,[3] thereby inviting the query, often raised in the literature, how that law then avoids being a mere tautology. If we restrict ourselves to a *single* closed system and say that of two given entropy states, the state of greater entropy will be said to be "later than" the state of smaller entropy, then indeed we have merely given a coordinative definition. But just as with other coordinative definitions of empirical science,[4] this definition is prompted by the empirical fact that it does not give rise to ambiguities or contradictions, when *different* closed systems are used. For—statistical modifications being temporarily ignored—there is concordance in the behavior of all closed systems: given any two such systems A and B, but not in thermodynamic equilibrium, if an entropy state S_{A_j} of A is simultaneous with a state S_{B_j} of B, then there is no case of a state S_{A_k} being simultaneous with a state S_{B_i}, such that $S_{A_k} > S_{A_j}$ while $S_{B_i} < S_{B_j}$.[5]

Since the relation "larger than" for real numbers is serial, the entropic definition of "later than" just given renders the seriality of time,[6] once its openness is assured by suitable boundary conditions, openness and seriality being attributes concerning which the causal theory of time had to be non-committal. (Time is said to be topologically "open"—rather than closed—if its states form a system of *betweenness* having the formal properties of the triadic relation of betweenness ordering the points of an open—as opposed to a closed—line.) But, as Eddington neglected to point out, that theory of time played an essential role in our entropic definition of "later than" by furnishing coordinative definitions for the concepts of "temporally between" and "simultaneous," which are needed to give meaning to the second law of thermodynamics. For this law uses the concept of "closed system," which Reichenbach defines as a system "not subject to differential forces,"[7] these being forces whose presence is correlated with *changes* of varying degree in different kinds of materials. But the absence of a change at any given instant t means the constancy of a certain value (or values) *between* the termini of a time interval containing the instant t (no *anisotropy* of time being assumed). Thus the concept of closed system presupposes the ordinal concept of temporal betweenness. We see therefore that the meaning of temporal betweenness is presupposed by the statement of the second law of thermodynamics, as is the concept of simultaneity, since the law makes reference to the entropy of an *extended* system *at a certain time*, and implicitly, to the *simultaneous* entropy states of several systems. . . .

Bridgman raises an objection to Eddington's definition of time direction: "How would one go to work in any concrete case to decide whether time were flowing forward or backward? If it were found that the entropy of the universe were decreasing, would one say that time was flowing backward, or would one say that it was a law of nature that entropy decreases with time?"[8] But under what circumstances would it be found that the entropy of the universe "is decreasing"? Remembering that we are concerned with a non-statistical study of closed systems, this situation would arise in the purely hypothetical case in which the direction of increasing entropy among physical systems is not *also* the direction of memory or information increase among biological organisms, "higher" memory states corresponding to *lower* entropy states of physical systems. In other words, the direction of entropy increase and the future direction of psychological time would then be *counter-directed*. Since in our actual world the production of memory traces depends on an entropy increase in the overall external environment, as we shall see, Bridgman's hypothetical situation is hardly an argument against Eddington, who was concerned with the physical basis of the temporal anisotropy of our *actual* world and not of other logically possible worlds. But even if the situation depicted by Bridgman were to arise, it would certainly not invalidate Eddington's claim that (i) the behavior of the class of closed physical systems defines a structural *difference* in time direction via the difference in material content between the relation "y is a higher entropy state than x" and its converse, and (ii) the direction of increasing entropy can be called the direction of time increase. Eddington did not contend that entropically characterized time "is flowing forward" in the sense of a transiency or becoming, since he makes a special point of emphasizing that the shifting "now," so familiar from psychological time, eludes conceptual rendition as an attribute of physical processes.[9] But by his very unfortunate choice of the name "time's arrow" for the anisotropy of physical time, he ironically invited the very misunderstanding which he had been at pains to prevent, viz., that he was intending to offer a thermodynamic basis for the "undirectional flow" of psychological time. There can be no problem of physical time flowing backward rather than forward, since it does *not* do any flowing at all. And in the context of *psychological* time, the locution "flow backward" is self-contradictory. Hence if Bridgman's hypothetical situation of counterdirectedness could actually materialize, then we would say that the entropy is decreasing with increasing psychological time and *not* that time is flowing backward.

More fundamentally, if the situation envisioned by Bridgman did arise, we would hardly survive long enough to be troubled by it. Poincaré and Costa de Beauregard have explained, in a qualitative way, why prediction and action would become impossible under the circumstances posited by Bridgman: two bodies initially at the same temperature would acquire different temperatures, while we would be unable to anticipate which of these bodies will become the warmer one.[10] Friction would no longer be a retarding influence but would set stationary bodies into motion in unpredictable directions. The most carefully conceived plan of action would precipitate enormous catastrophes, since unstable equilibria would now be the rule rather than the exception. . . .

It is natural to ask why it is that in so many cases involving irreversible processes, we seem to be far more reliably informed concerning the past than concerning the future. This question is raised by Schlick, who points out that human footprints on a beach enable us to infer that a person was there in the past but not that someone will walk there in the future. His answer is that "the structure of the past is inferred not from the extent to which energy has been dispersed but from the spatial arrangement of objects."[11] And he adds that the spatial traces, broadly conceived, are always produced in accord with the entropy principle. Thus, in the case of the beach, the kinetic energy of the person's feet became dispersed in the process of arranging the grains of sand into the form of an imprint, which owes its (relative) persistence to the fact that the pedal kinetic energy lost its organization in the course of being imparted to the sand. To be sure, Schlick's claim that the process of leaving a trace occurs in accord with the entropy principle is quite true, but he fails to stress the crucial point at which the entropy principle is invoked in the retrodictive inference. Initially considering the beach itself as a closed system not far from equilibrium, we are informed by the discovery of footprints that the degree of order possessed by the grains of sand is higher and hence the entropy is lower than it should be, if the beach had actually been an isolated system. The *non*-statistical entropy principle, which precludes the beach's having evolved isolatedly from an earlier state of randomness to its present state of greater organization, therefore enables us to infer that the beach must have been an *open* system whose increase in order was acquired at the expense of an at least equivalent decrease of organization in the system with which it interacted (the stroller, who is metabolically depleted).

Our *statistical* analysis below will show, however, that the true conclusion of this inference *cannot* be made to rest on the statistically untenable premise that the entropy increase with time is monotonic in a permanently closed system. It will turn out that the justification for this inference derives from the fact that (i) most systems which we now encounter in an isolated state of low entropy, behaving as if they might remain isolated, were not in fact permanently closed in the past, and (ii) in the case of such *temporarily* isolated or "branch" systems we *can* reliably infer a portion of the past from a present ordered state, an inference which is *not* feasible, as we shall see in detail, on the basis of the statistical version of the second law of thermodynamics as applied to a single, *permanently* closed system, and (iii) the assumption that a transition from an earlier high entropy state to a present low one is overwhelmingly *im*probable, which is the basis for this inference, refers to the frequency of such transitions within a space-ensemble of branch systems, each of which is considered at *two* different times; this improbability does *not* refer to the time-sequence of entropy states of a single, permanently-closed system.

Hence, in the case of the beach whose sand forms a smooth surface except for one place where it is in the shape of a human footprint, we know with high probability that instead of having evolved *isolatedly* from a prior state of uniform smoothness into its present uneven configuration according to the *statistical* entropy principle for a permanently-closed system, the beach was an *open* system in *interaction* with a stroller. And we are aware furthermore that if there is some quasi-closed wider system containing the beach and the stroller, as there often is, the beach achieved its ordered low entropy state of bearing the imprint or interaction-indicator at the expense of an at least compensatory entropy increase in that wider system comprising the stroller: the stroller increased the entropy of the wider system by scattering his energy reserves in making the footprint.

We see that the sandy footprint shape is a genuine indicator and not a randomly achieved form resulting from the unperturbed chance concatenations of the grains of sand. The imprint thus contains information in the sense of being a veridical indicator of an interaction. Now, in all probability the entropy of the imprint-bearing beach-system increases after the interaction with the stroller through the smoothing action of the wind. And this entropy increase is parallel, in all probability, to the direction of entropy increase of the majority of branch systems. Moreover, we saw that the production of the indicator by the interaction is likely to have involved an entropy increase in some wider system of which the indicator is a part. Hence, *in all probability the states of the interacting systems which do contain the indicators of the interaction are the relatively higher entropy states of the majority of branch systems as compared to the interaction state. Hence the indicator states are the relatively later states.* And by being both *later* and indicators, these states have *retrodictive* significance, thereby being traces, records or memories. And due to the high degree of retrodictive univocity of the low entropy states constituting the indicators, the latter are veridical to a high degree of specificity. . . .

As already mentioned, our affirmation of the temporal asymmetry of recordability of interactions must be qualified by dealing first with the exceptional case of pre-recordability of those interactions which are veridically predicted by human beings (or computers). For any event which could be predicted by a scientist could also be "pre-recorded" by that scientist in various forms such as a written entry on paper asserting its occurrence at a certain later time, an advance drawing, or even an advance photograph based on the pre-drawing. By the same token, artifacts like computers can pre-record events which they can predict. A comparison between the written, drawn or photographic pre-record (i.e., recorded prediction) of, say, the crash of a plane into a house and its post-record in the form of a caved-in house, and a like comparison of the corresponding pre- and post-records of the interaction of a foot with a beach will now enable us to formulate the essential differences in the conditions requisite to the respective production of pre-records and post-records as well as the usual differences in make-up between them.

The production of at least one retrodictive indicator or post-record of an interaction such as the plane's crash into the house requires only the occurrence of that interaction (as well as a moderate degree of durability of the record). The retrodictive indicator states in the system which interacted with an outside agency must, of course, be distinguished from the *epistemic use* which human beings may make of these physical indicator states. And our assertion of the sufficiency of the interaction for the production of a post-record allows, of course, that the *interpretation* of actual post-records by humans as bona fide documents of the past requires their use of theory and not just the occurrence of the interaction. In contrast to the sufficiency of an interaction itself for its (at least short-lived) post-recordability, no such sufficiency obtains in the case of the pre-recordability of an interaction: save for an overwhelmingly improbable freak occurrence, the production of even a single pre-record of the coupling of a system with an agency external to it requires, as a necessary condition, *either* (a) the use of an appropriate theory by symbol-using entities (humans, computers) having suitable information, *or* (b) the pre-record's being a partial effect of a cause that also produces the pre-recorded interaction, as in the barometric case to be dealt with below. And in contexts in which (a) is a necessary condition, we find the following: since pre-records are, by definition, veridical, this necessary condition cannot *generally* also be sufficient, unless the predictive theory employed is deterministic *and* the information available to the theory-using organism pertains to a closed system.

In addition to differing in regard to the conditions of their production, pre-records generally differ from post-records in the following further respect: unless the pre-record prepared by a human being (or computer) *happens* to be part of the interacting system to which it pertains, the pre-record will not be contained in states of the interacting system which it concerns but will be in some other system. Thus, a pre-record of the crash of a plane into a house in a heavy fog would generally *not* be a part of either the house or the plane, although it can happen to be. But in the case of *post*-recording, there will always be at least one post-record, however short-lived, in the interacting system itself to which that post-record pertains.

Our earlier example of the footprint on the beach will serve to illustrate more fully the asymmetry between the requirements for the production of a pre-record and of a post-record. The pre-recording of a *later* incursion of the beach by a stroller would require extensive information about the motivations and habits of people not now at the beach and also knowledge of the accessibility of the beach to prospective strollers. This is tantamount to knowledge of a large system which is *closed*, so that all relevant agencies can safely be presumed to have been included in it. For otherwise, we would be unable to guarantee, for example, that the *future* stroller will *not* be stopped en route to the beach by some agency not included in the system, an eventuality whose occurrence would deprive our pre-record of its referent, thereby destroying its status as a veridical indicator. In short, in the case of the footprint, which is a post-record and *not* a pre-record of the interaction of a human foot with the beach, the interaction itself is *sufficient* for its post-recording (though not for the extended *durability* of the record once it exists), but *not* for its pre-recording and prediction. Since a future interaction of a potentially open system like the beach is *not* itself sufficient for its pre-recordability, open systems like beaches therefore do not themselves exhibit pre-records of their own future interactions. Instead—apart from the second species of pre-recordability to be considered presently—pre-recordability of interactions of potentially open systems requires the mediation of symbol- and theory-using organisms or the operation of appropriate artifacts like computers. And such pre-recordability can obtain successfully only if the theory available to the pre-recording organism is deterministic and sufficiently comprehensive to include all the relevant laws and boundary conditions governing the pertinent closed systems.

The second species of exceptions to the asymmetry of recordability is exemplified by the fact that a sudden drop in the pressure reading of a barometer can be an advance indicator or "pre-record" of a subsequent storm. To be sure, it is the immediately *prior* pressure change in the spatial vicinity of the barometer and only that particular prior change (i.e., the *past* interaction through pressure) which is recorded numerically by a given drop in the barometric reading, and *not* the pressure change that *will* exist at that same place at a *later* time. To make the predictions required for a pre-recording of the pressure changes which will exist at a given space point at later times (i.e., of the corresponding future interactions), comprehensive meteorological data pertaining to a large region would be essential. *But* it *is* possible in this case to base a rather reliable prediction of a future storm on the present sudden barometric drop. The latter drop, however, is, in fact, a *bona fide* advance indicator *only because* it is a partial effect of the very comprehensive cause which also produces (assures) the storm. Thus, it is the fulfillment of the *necessary condition* of having a causal ancestry that overlaps with that of the storm which is needed to confer the status of an advance indicator on the barometric drop. In contrast to the situation prevailing in the case of *post*-recordability, the existence of this necessary condition makes for the fact that the future occurrence of a storm is *not sufficient* for the existence of an advance indicator of that storm in the form of a sudden barometric drop at an earlier time.

An analogous account can be given of the following cases, which Mr. F. Brian Skyrms has suggested to me for consideration: situations in which *human intentions* are highly reliable advance indicators of the events envisaged by these intentions. Thus, the desire for a glass of beer, coupled with the supposed presence of the conditions under which beer and a glass are obtainable produces as a partial effect the intent to get it. And, *if* external conditions permit (the beer is available and accessible), and, furthermore, if the required internal conditions materialize (the person desiring the beer remains able to go and get it), then the intent will issue in the obtaining and drinking of the beer. But in contrast to the situation prevailing in the case of retrodictive indicators (post-records), the future consumption of the beer is *not* a *sufficient condition* for the existence of its probabilistic advance indicator in the form of an intention.[12] . . .

. . . Our present entropic account of the circumstances under which the past can be inferred from the present while the future cannot, as well as of the circumstances when only the converse determination is possible enables us to specify the conditions of validity for the following statements by Reichenbach: "Only the totality of all causes permits an inference concerning the future, but the past is inferable from a partial effect alone" and "one can infer the total cause from a partial effect, but one cannot infer the total effect from a partial cause."[13] A partial effect produced in a system while it is open permits, on entropic grounds, an inference concerning the earlier *interaction* event which was its cause: even though we do not know the *total* present effect, we know that the part of it which is an ordered, low entropy state was (most probably) preceded by a still lower entropy state and that the *diversity* of the interactions associated with such a very *low* interaction entropy state is relatively *small*, thereby permitting a rather specific assertion about the past.

Thus, the asymmetry of inferability arises on the macro-level in the absence of knowledge of the microscopic state of the *total* (closed) system at a given time and is made possible by the relative retrodictive univocity of local low entropy states which result from interactions. We are therefore in possession of the answer to the question posed by J. J. C. Smart when he wrote: "Even on a Laplacian view, then, we still have the puzzling question 'Why from a limited region of space can we deduce a great deal of the history of the past, whereas to predict similar facts about the future even a super-human intelligence would have to consider initial conditions over a very wide region of space?'"[14] . . .

Our analysis of entropy so far has been in the macroscopic

context of thermodynamics and has taken no adequate account of the important questions which arise concerning the serviceability of the entropy criterion for the definition of a time direction, when the entropy law is seen in the statistical light of both classical and quantum mechanics. These questions, which we must now face, derive from the attempt to deduce the phenomenological irreversibility of classical thermodynamics from principles of statistical mechanics asserting that the motions of the microscopic constituents of thermodynamic systems are completely reversible.

As is well-known, in the form of Boltzmann's H-theorem, the statistical version of the phenomenological entropy law affirms that the increase of entropy with time is overwhelmingly probable by virtue of the approach of the particles to their equilibrium distribution.[15] But soon after Boltzmann's enunciation of his theorem, it was felt that there is a logical hiatus in a deduction which derives the overwhelming probability of macroscopic irreversibility from premises attributing complete reversibility to micro-processes. For according to the principle of dynamical reversibility, which is integral to these premises, there is, corresponding to any possible motion of a system, an equally possible reverse motion in which the same values of the coordinates would be reached in the reverse order with reversed values for the velocities.[16] Thus, since the probability that a molecule has a given velocity is independent of the sign of that velocity, separation processes will occur just as frequently in the course of time as mixing processes. J. Loschmidt therefore raised the *reversibility objection* to the effect that for any behavior of a system issuing in an increase of the entropy S with time, it would be equally possible to have an entropy decrease.[17] A similar criticism was presented in the *periodicity objection*, based on a theorem by Poincaré[18] and formulated by Zermelo.[19] Poincaré's theorem had led to the conclusion that the long-range behavior of an isolated system consists of a succession of fluctuations in which the value of S will decrease as often as it increases. And Zermelo asked how this result is to be reconciled with Boltzmann's contention that if an isolated system is in a state of low entropy, there is an overwhelming probability that the system is actually in a microscopic state from which changes in the direction of higher values of S will ensue.[20]

These logical difficulties were resolved by the Ehrenfests.[21]

They explained that there is no incompatibility between (i) the assertion that *if* the system is in a low entropy state, then, *relative* to that state, it is highly probable that the system will soon be in a higher entropy state, and (ii) the contention that the system plunges down from a state of high entropy to one of lower entropy as frequently as it ascends entropically in the opposite direction, thereby making the *absolute* probability for these two opposite kinds of transition *equal*. The compatibility of the equality of these two absolute probabilities with a *high* relative probability for a future transition to a higher entropy becomes quite plausible, when it is remembered that (i) the low entropy states to which the high relative probabilities of subsequent increase are referred are usually at the low point of a trajectory at which changes back to higher values are initiated, and (ii) the Boltzmann *H*-theorem therefore does not preclude such a system's exhibiting decreases and increases of *S* with equal frequency. The time variation of the entropy, embodying these two claims compatibly, can be visualized as an entropy staircase curve.[22]

Boltzmann's *H*-theorem can thus be upheld in the face of the reversibility and periodicity objections, but only if coupled with a very important *proviso*: the affirmation of a high probability of a future entropy increase must *not* be construed to assert a high probability that present low entropy values were *preceded* by *still lower* entropies *in the past*. For the relative probability that a *low* entropy state was *preceded* by a state of *higher* entropy is just as great as the relative probability that a low state will be *followed* by a higher state. . . .

. . . Reichenbach has argued convincingly that although there is no contradiction between the high relative probabilities of Boltzmann's *H*-theorem and the equality among the absolute probabilities of the reversibility and periodicity objections, the aforementioned time-symmetry of the statistical results on which these objections are based shows decisively that the entropic behavior of a single, permanently-closed system does not confer anisotropy on time.[23] For this time-symmetry precludes the inference that the *lower* of two given entropy states is the *earlier* of the two: a low entropy state is *preceded* by a high state no less frequently than it is *followed* by a high state. And . . . it is entirely meaningful to assert that the system is in the *same* macro-state and has the *same* entropy S_k at *different* times t and t', since a difference in the underlying *micro*-states assures the non-identity of the states of the system at times t and t'.[24]

Thus, the H-theorem fails to provide a basis for the anisotropy ("arrow") of time. . . .

. . . Schrödinger's perceptive guiding idea that the attempt to characterize phenomenological time entropically without running afoul of the reversibility and periodicity objections can succeed only if we regard the entropy law as an assertion about *at least two temporarily* closed systems was developed independently by Reichenbach.[25] And the *valid core*—but only the valid core—of Reichenbach's version of this idea seems to me to provide a basis for an entropic criterion of a statistical anisotropy of physical time. Believing that Reichenbach's account requires modification in order to be satisfactory, I shall now set forth what I consider to be a correct elaboration of his principal conception.

We must first describe certain features of the physical world having the character of initial or boundary conditions within the framework of the theory of statistical mechanics. The sought-after basis of a statistical anisotropy of time will then emerge from principles of statistical mechanics relevant to these *de facto* conditions.

The universe around us exhibits striking disequilibria of temperature and other inhomogeneities. In fact, we live in virtue of the nuclear conversion of the sun's reserves of hydrogen into helium, which issues in our reception of solar radiation. As the sun dissipates its reserves of hydrogen via the emission of solar radiation, it may heat a terrestrial rock embedded in snow during the daytime. At night, the rock is no longer exposed to the sun but is left with a considerably higher temperature than the snow surrounding it. Hence, at night, the warm rock and the cold snow form a quasi-isolated subsystem of either our galactic or solar system. And the relatively low entropy of that subsystem was purchased at the expense of the dissipation of the sun's reserves of hydrogen. Hence, *if* there is some quasi-closed system comprising the sun and the earth, the branching off of our subsystem from this wider system in a state of low entropy at sunset involved an entropy increase in the wider system. During the night, the heat of the rock melts the snow, and thus the entropy of the rock-snow system increases. The next morning at sunrise, the rock-snow subsystem merges again with the wider solar system. Thus, there are subsystems which branch off from the wider solar or galactic system in a state of relatively low entropy, remain quasi-closed for a *limited* period of time, and then merge again with the wider system from which they had been separated. Following Reichenbach, we have been using the term "branch system" to designate this kind of subsystem.[26] . . .

Thus, our environment abounds in branch-systems whose initial relatively low entropies are the products of their earlier coupling or interaction with outside agencies of one kind or another. This rather constant and ubiquitous formation of a branch-system in a relatively low entropy state resulting from interaction often proceeds at the expense of an entropy increase in some wider quasi-closed system from which it originated. And the *de facto*, nomologically contingent occurrence of these branch systems has the following *fundamental consequence*, at least for our region of the universe and during the current epoch: among the quasi-closed systems whose entropy is relatively low and which behave as if they might remain isolated, the vast majority have not been and will not remain permanently closed systems, being branch systems instead.

Hence, upon encountering a quasi-closed system in a state of fairly low entropy, we know the following to be overwhelmingly probable: the system has *not* been isolated for millions and millions of years and does *not* just *happen* to be in one of the infrequent but ever-recurring low entropy states exhibited by a permanently isolated system. Instead, our system was formed not too long ago by branching off after an interaction with an outside agency. For example, suppose that an American geologist is wandering in an isolated portion of the Sahara desert in search of an oasis and encounters a portion of the sand in the shape of "Coca Cola." He would then infer that, with overwhelming probability, a kindred person had interacted with the sand in the recent past by tracing "Coca Cola" in it. The geologist would not suppose that he was in the presence of one of those relatively low entropy configurations which are assumed by the sand particles spontaneously but very rarely, if beaten about by winds for millions upon millions of years in a state of effective isolation from the remainder of the world.

There is a further *de facto* property of branch systems that concerns us. For it will turn out to enter into the temporally asymmetrical statistical regularities which we shall find to be exhibited in the entropic behavior of these systems. This property consists in the following *randomness* obtaining *as a matter of nomologically contingent fact* in the incidence of the W_1 micro-states belonging to the initial macro-states of a *space*-ensemble of branch-systems each of which has the same initial entropy $S_1 = k \log W_1$: For each class of *like* branch-systems having the *same* initial entropy value S_1, the micro-states constituting the identical initial macro-states of entropy S_1 are *random samples* of the set of all W_1 micro-states yielding a macro-state of entropy S_1.[27] This attribute of randomness of micro-states on the part of the initial states of the members of the *space*-ensemble will be recognized as the counterpart of the following attribute of the micro-states of one single, permanently-closed system: there is equi-probability of occurrence among the W_1 micro-states belonging to the *time*-ensemble of states of equal entropy $S_1 = k \log W_1$ exhibited by one single, permanently-closed system.

We can now state the statistical regularities which obtain as a consequence of the *de facto* properties of branch systems just set forth, when coupled with the principles of statistical mechanics. These regularities, which will be seen to yield a temporally asymmetric behavior of the entropy of *branch*-systems, fall into two main groups as follows.[28]

Group 1. In most space-ensembles of quasi-closed branch-systems each of which is initially in a state of non-equilibrium or relatively *low* entropy, the majority of branch systems in the ensemble will have *higher* entropies *after* a given time t. But these branch systems simply did not exist as quasi-closed, distinct systems at a time t *prior to* the occurrence of their initial, branching off states. Hence, not existing then as such, the branch systems did in fact *not* also exhibit the same higher entropy states at the *earlier* times t, which they would indeed have done then had they existed as closed systems all along. In this way, the space-ensembles of branch-systems do *not* reproduce the entropic time-symmetry of the single, permanently-closed system. And whatever the behavior of the components of the branch systems prior to the latter's "birth," that behavior is irrelevant to the entropic properties of branch systems as such.

The increase after a time t in the entropy of the overwhelming majority of branch systems of initially low entropy —as confirmed abundantly by observation—can be made fully intelligible. To do so, we note the following property of the *time*-ensemble of entropy values belonging to a single, permanently-closed system and then affirm that property of space-ensembles of branch systems: since *large* entropic down-grades or decreases are *far less* probable (frequent) than moderate ones, the *vast majority* of *non*-equilibrium entropy states of a permanently-closed system are located either at or in the immediate temporal vicinity of the *bottom* of a *dip* of the one-system entropy curve. In short, the vast majority of the *sub*-maximum entropy states are on or temporally very near the *upgrades* of the one-system curve. The application of this result to the space-ensemble of branch-systems whose initial states exhibit the aforementioned *de facto* property of *randomness* then yields the following: among the initial low entropy states of these systems, the vast majority lie at or in the immediate temporal vicinity of the bottoms of the one-system entropy curve at which an upgrade begins.

Group 2. A decisive *temporal asymmetry* in the statistics of the temporal evolution of branch-systems arises from the fur-ther result that in most space ensembles of branch systems each of whose members is initially in a state of *equilibrium* or very *high* entropy, the vast majority of these systems in the ensemble will *not* have *lower* entropies *after* a finite time *t*, but will still be in equilibrium. For the aforementioned ran-domness property assures that the vast majority of those branch systems whose initial states are equilibrium states have maximum entropy values lying somewhere *well within* the plateau of the one-system entropy curve, rather than at the extremity of the plateau at which an entropy *decrease* is initiated.[29]

We see therefore that in the vast majority of branch sys-tems, either one end of their finite entropy curves is a point of low entropy and the other a point of high entropy, or they are in equilibrium states at both ends as well as during the inter-vening interval. And it is likewise apparent that the statistical distribution of these entropy values on the time axis is such that the vast majority of branch systems have the *same direc-tion of entropy increase* and hence also the same opposite direction of entropy decrease. Thus, the statistics of entropy increase among branch systems assure that in most space ensembles the vast majority of branch systems will increase their entropy in *one* of the two opposite time directions and decrease it in the other: in contradistinction to the entropic time-symmetry of a single, permanently-closed system, the probability within the space-ensemble that a low entropy state *s* at some given instant be *followed* by a higher entropy state *S* at some given later instant is much *greater* than the probability that *s* be *preceded* by *S*. In this way the entropic behavior of branch systems confers the *same* statistical anisotropy on the vast majority of all those cosmic epochs of time during which the universe exhibits the requisite disequilibrium and contains branch systems satisfying initial conditions of "randomness."[30]

Let us now call the direction of entropy increase of a *typical representative* of these epochs the direction of "later," as indeed we have done from the outset by the mere assignment of higher time numbers in that direction but *without* prejudice to our findings concerning the issue of the anisotropy of time. Then our results pertaining to the entropic behavior of branch systems show that the directions of "earlier than" and "later than" are not merely opposite directions bearing decreasing and increasing time coordinates respectively but are statistically *anisotropic* in an objective physical sense.

The achievements of this entropic criterion of temporal anisotropy are as follows: (i) it provides an empirical justification for interpreting present ordered states as veridical traces of actual past interaction events, a justification which the entropic behavior of a single, permanently-closed system was incompetent to furnish, as we saw, and (ii) it explains why the subjective (psychological) and objective (physical) directions of positive time are parallel to one another by noting that man's own body participates in the entropic lawfulness of space-ensembles of physical branch systems in the following sense: man's memory, just as much as all purely physical recording devices, accumulates "traces," records or information. And as we saw earlier, the direction of that accumulation is dictated by the statistics of branch systems. Thus, Spinoza was in error when he wrote Oldenburg that "*tempus non est affectio rerum sed merus modus cogitandi.*" Contrary to Watanabe's conception of man's psychological time sense as *sui generis*, we see that the future direction of psychological time is parallel to that of the accumulation of traces (increasing information) in interacting systems, and hence parallel to the direction defined by the positive entropy increase in the branch systems.

But processes characterized by temporally asymmetric entropy changes are not the sole source of the anisotropy of time. Reichenbach's account took no cognizance of processes in nature which are *irreversible* yet do *not* involve any entropy increase. That there are such processes has recently been emphasized by K. R. Popper, who cites a suggestion to this effect made by Einstein in 1910.

Popper considers a large surface of water initially at rest into which a stone is dropped, thereby producing an outgoing concentric wave of decreasing amplitude.[31] And he argues that the irreversibility of this process is attributable to the physical impossibility of the *uncoordinated* concatenation on all points of a circle of the initial conditions requisite to the occurrence of a *contracting* wave. Now, one might wish to object to this argument by pointing out that the entropy law is *not irrelevant* to the irreversibility of the outgoing wave propagation, pointing out that the diminution in the amplitude of this wave is due to the superposition of *two* independent effects: (1) the requirements of the law of conservation of energy (*first* law of thermodynamics), and (2) an entropy increase as a result of dissipative viscosity in an essentially closed system. While this retort is right to the extent that the entropy increase is a *sufficient* condition for the statistical irreversibility of this process, Popper's case is nonetheless sound in the sense that another, *independent sufficient condition* for irreversibility is provided by the physical impossibility of realizing the initial conditions required for the occurrence of a contracting wave process which was *not* set off by a prior *outgoing* wave motion or by any other influence first emanating from a *central source*. He admits that in a closed system of finite size, there is an entropy increase here as a result of viscous losses. But he goes on to strengthen his argument decisively by explaining that if a thin gas, for example, expands from a center in a system having no bounding walls (presumably in an infinite universe), then the expansion constitutes an irreversible process *without* entropy increase.

The existence of processes whose irreversibility is *not* entropic (cf. Note 30) but derives solely from the fact that their temporal inverses would require a *deus ex machina* has been affirmed as a generalization of Popper's examples in the form of a general principle of nature by reference to a *spatially infinite* universe.[32] This claim concerning non-entropic irreversibility can be made more precise by considering illustratively a light wave emitted at a center and going out into *infinite* space. Its temporal inverse would be a light wave which has been *contracting* for *all* infinite past time. Now, in claiming that the latter would require a *deus ex machina* for its occurrence, the principle of non-entropic irreversibility is not laying down the self-contradictory condition (akin to Kant's fallacious procedure in his First Antinomy) that a process which has been going on for all infinite past time must have a finite beginning (production by past *initial* conditions) after all. What this principle does assert, however, is the *de facto* physical *non*-occurrence of contracting wave processes which have been in progress through an infinite space for all past eternity, and therefore the assumption of their existence would involve a *deus ex machina* in that sense.

The temporal anisotropy defined by this non-entropic irreversibility is more pervasive than either the anisotropy depending on the formation of branch systems or any purely cosmological anisotropy of time such as is defined by the speculative expansion of a spherical space whose radius increases monotonically beginning with a singular state in the finite past having no temporal predecessor. For the irreversibility assured by the *deus ex machina* principle assures uniform temporal anisotropy both for local intervals in the time continuum and in the large.[33]

IV. The Physical Status of "Becoming"

The distinction between earlier and later dealt with so far in the context of the anisotropy of time makes no reference to a transient present. But the time of human conscious awareness exhibits not only the earlier-later distinction but also a flow, passage or becoming in the sense of a transiency of the "now." And this apparent flux of events gives rise to the constantly *shifting* division of the time continuum into the *past* and the *future*, a division which involves more than the "*static*" one into earlier and later. It must be strongly emphasized that what is a factual property of psychological time is the existence and transiency of the "now" in the sense of the diversity and order of the Now-contents. But the shifting of the "now" *in the future direction cannot* be deemed a factual property, since this *directional* affirmation of shifting *is a mere tautology.*

Having found a physical basis for the anisotropy of time, our final concern in the consideration of the time problem is the physical status, *if any,* of "becoming." Our earlier characterization of the difference between the two directions of time does not, as such, affirm the existence of a *transient,* threefold division of events into those that have already "spent their existence," as it were, those which actually exist, and those which are yet to "come *into* being." And the relativistic picture of the world makes no allowance for such a division.[34] It conceives of events not as "coming into existence" but as simply being and thus allowing us to "come across" them and produce "the formality of their taking place" by our "entering" into their absolute future. This view, which some writers *mistakenly* believe to depend on determinism, as we shall see, has been expressed by H. Weyl in the following partly metaphorical way: "the objective world simply *is,* it does not *happen.* Only to the gaze of my consciousness crawling upward along the life- [world-] line of my body does a section of this world come to life as a fleeting image."[35] . . .

. . . If we consider *any one* of the temporally successive regions of space-time, we can assert the following: the events belonging to its particular absolute past could be (more or less) uniquely specified in records which are a part of that region, whereas its particular absolute future is thence quantum-mechanically unpredictable. Accordingly, *every* "now," be it the "now" of Plato's birth or that of Carnap's, *always* constitutes a divide in Reichenbach's sense between its own recordable past and its unpredictable future, thereby satisfying his definition of the "present." But this fact is fatal to his aim of providing a physical basis for a *unique,* transient "now" and thus for "becoming." Reichenbach's recent characterization of the determinacy of the past as recordability as opposed to the quantum-mechanical indeterminacy of the future can therefore not serve to vindicate his conception of becoming any more than did his paper of 1925, which was penetratingly criticized by Hugo Bergmann as follows:

Thus, according to Reichenbach, a cross-section in the state of the world is distinguished from all others; the now has an objective significance. Even when no man is alive any longer, there is a now. "The present state of the planetary system" would even then be just as precise a descriptive phrase as "the state of the planetary system in the year 1000."

Concerning this definition one must ask: Which now is intended, if one says: the present state of the planetary system? That of the year 1800 or 2000 or which other one? Reichenbach's reply is: The now is the threshold of the transition from the state of indeterminacy to that of determinacy. But (if Reichenbach's indeterminism holds) this transition has *always* occurred and will always occur. And if the rejoinder would be: The indeterminacy of the year 1800 has already been transformed into a determinacy, then one must ask: For whom? Evidently for us, for the present, for our now. Accordingly, this definition by Reichenbach seems to refer after all to a now which it must first define. What is the objective difference between the now of the year 1800 and the now of the present instant? The answer must be: Now is the instant of the transition from indeterminacy to determinacy, that is, one explains the present now . . . by reference to itself.

. . . Reichenbach writes: The problem can be formulated as the question concerning the difference between the past and the future. For determinism, there is no such difference. . . . But the reproach which Reichenbach directs at determinism here should be aimed not at it but at the world view of physics, which does not take cognizance of any psychological categories, for which there is no "I," . . . a concept which is inextricably intertwined with the concept "now." Even those who regard the supplanting of determinism by indeterminism as admissible, as we do, will not be willing to admit that the concept of "now" can be assigned a legitimate place within indeterministic physics. Even if one assumes—as we wish to do along with Reichenbach—that the future is not uniquely determined by a temporal cross-section, one can say only that this indeterminacy prevails just as much for Plato as for myself and that I cannot decide by physical means who is living "now." For the difference is a psychological one.

. . . "Now" is the temporal mode of the experiencing ego.[36]

Bergmann's demonstration here that an indeterminist universe fails to define an objective (non-psychological) transient now can be extended in the following sense to justify his contention that the concept "now" involves features peculiar to consciousness: the "flux of time" or transiency of the "now" has a meaning only in the context of the egocentric perspectives of *sentient* organisms and does *not also* have relevance to the relations between purely inanimate individual recording instruments and the environmental physical events they register, as Reichenbach claims. For what can be said of every state of the universe can also be said, *mutatis mutandis*, of every state of a given inanimate recorder. Moreover, the dependence of the meaning of now on the presence of properties peculiar to consciousness emerges from William James' and Hans Driesch's correct observations that a simple isomorphism between a succession of *brain traces* and a succession of states of awareness does not explain the temporal features of such psychological phenomena as melody awareness. For the hypothesis of isomorphism renders only the succession of states of awareness but not the *instantaneous awareness of succession*.[37] But the *latter* awareness is an essential ingredient of the meaning of "now": the flux of time consists in the *instantaneous awarenesses* of *both* the temporal order *and* the *diversity* of the membership of the set of remembered (recorded) or forgotten events, awarenesses in each of which the instant of its own occurrence constitutes a *distinguished element*.

I cannot see, therefore, that the accretion of time-tagged marks or traces on an inanimate recording tape so as to form an expanding spatial series can also be held to define a flux of time. Thus, Bergmann's exclusively psychologistic conception of this flux or becoming must be upheld against Reichenbach.

A brief comment needs to be added concerning an *unjustified* criticism of the concept of the flux of time. We see from our characterization that this idea is a *qualitative* concept without any metrical ingredients. It therefore will not do to offer a *metrical* refutation via a *reductio ad absurdum* in an endeavor to show that metaphorical discourse involving reference to "the flow of time" or to "the direction of time" commits a breach of logical grammar. Such a *reductio* was recently offered by J. J. C. Smart, who writes: "The concept of the flow of time or of the advance of consciousness is, however, an illusion. How fast does time flow or consciousness advance? In what units is the rate of flow or advance to be measured? Seconds per—?"[38]

NOTES

[1] H. Reichenbach, *The Direction of Time* (Berkeley, Calif., 1956), p. 32.

[2] For details on the "new cosmology," see H. Bondi, *Cosmology* (Cambridge, Eng., 1961); a brief digest is given in A. Grünbaum, "Some Highlights of Modern Cosmology and Cosmogony," *Review of Metaphysics, V* (1952), 493-498.

[3] A. S. Eddington, *The Nature of the Physical World* (New York, 1928), pp. 69 ff.

[4] Examples are the definition of the metric of time on the basis of the empirical law of inertia, and the definition of congruence for spatially separated bodies on the basis of the fact that two bodies which are congruent at a given place will be so everywhere, independently of the respective paths along which they are transported individually. Cf. M. Schlick, "Are Natural Laws Conventions?" in H. Feigl and M. Brodbeck, eds., *Readings in the Philosophy of Science* (New York, 1953), p. 184; H. Reichenbach, "Ziele und Wege der physikalischen Erkenntnis," *Handbuch der Physik*, vol. IV, 1929, 52-53 and *The Philosophy of Space and Time* (New York, 1958), pp. 16, 17.

[5] Cf. K. G. Denbigh, "Thermodynamics and the Subjective Sense of Time," *British Journal for the Phil. of Science, IV* (1953), 183-186, and E. Zilsel, "Über die Asymmetrie der Kausalität und die Einsinnigkeit der Zeit," *Naturwissenschaften, XV* (1927), 282.

[6] See A. Grünbaum, *Philosophical Problems of Space and Time* (New York: Knopf, 1963), pp. 219-220, n. 8, for his reply to Mehlberg's objection that Carathéodory's rigorous axiomatization of the phenomenological second law of thermodynamics has undercut its claim of irreversibility and temporal anisotropy.

[7] *The Philosophy of Space and Time*, p. 118.

[8] Bridgman, *Reflections of a Physicist* (New York, 1950), p. 165. Lest Bridgman's point here be misunderstood, it must be emphasized that he is explicitly *not* resting his case on the giant entropy decreases affirmed by statistical mechanics.

[9] Eddington, *op. cit.*, pp. 67, 87-110.

¹⁰ H. Poincaré, *The Foundations of Science*, tr. Halsted (Lancaster, Pa., 1913), pp. 399-400; O. Costa de Beauregard, "L'Irréversibilité Quantique, Phénomène Macroscopique," in A. George, ed., *Louis de Broglie, Physicien et Penseur* (Paris, 1953), and *Théorie Synthétique de la Relativité Restreinte et des Quanta* (Paris, 1957), Ch. XIII, esp. pp. 167-171.

¹¹ M. Schlick, *Grundzüge der Naturphilosophie* (Vienna, 1948), pp. 106-107. J. J. C. Smart in "The Temporal Asymmetry of the World" (*Analysis, XIV* [1954], 80) also discusses the significance of traces but reaches the following unwarrantedly agnostic conclusion: "So the asymmetry of the concept of trace has something to do with the idea of formlessness or chaos. But it is not easy to see what." See also his paper in *Australasian Journal of Philosophy, XXXIII* (1955), 124.

¹² For a refutation of purported counterexamples to the temporal asymmetry of the recordability of interactions as set forth here, cf. A. Grünbaum, "Temporally-Asymmetric Principles, Parity Between Explanation and Prediction, and Mechanism versus Teleology," *Philosophy of Science, XXIX* (April 1962), p. 155.

¹³ H. Reichenbach, "Die Kausalstruktur der Welt und der Unterschied von Vergangenheit und Zukunft," *Ber. d. Bayer, Akad. München, Math.-Naturwiss. Abt.* (1925), 157, and "Les Fondements Logiques de la Mécanique des Quanta," *Annales de l'Institut Poincaré, XIII* (1953), p. 146. Cf. also C. F. von Weizsäcker, "Der Zweite Hauptsatz und der Unterschied von Vergangenheit und Zukunft," *Annalen d. Physik, XXXVI* (1939), 279. By noting in his later publications (especially in *The Direction of Time*, pp. 157-167) that the temporal asymmetry involved here has an entropic basis, Reichenbach abandoned his earlier view that it provides an *independent* criterion for the anisotropy of time. Thus, he has essentially admitted the validity of H. Bergmann's telling criticisms (*Der Kampf um das Kausalgesetz in der jüngsten Physik*, Braunschweig, 1929, pp. 19-24.)

¹⁴ See Smart, "The Temporal Asymmetry . . . ," *op. cit.*, p. 81.

¹⁵ The entropy S is related to the thermodynamic probability W representing the corresponding number of microscopic complexions by the equation $S = k \log W$. And the quantity H of Boltzmann's theorem is connected with S by the relation $S = -kH$. Thus, an increase in the entropy is equivalent to a decrease in H.

¹⁶ Cf. R. C. Tolman, *The Principles of Statistical Mechanics* (Oxford, 1938), pp. 102-104.

¹⁷ J. Loschmidt, "Über das Wärmegleichgewicht eines Systems von Körpern mit Rücksicht auf die Schwere," *Sitzungsber. Akad. Wiss. Wien, LXXIII,* (1876), 139 and *LXXV* (1877), 67.

¹⁸ H. Poincaré, "Sur le problème des trois corps et les équations de la dynamique," *Acta Mathem., XIII* (1890), 67.

[19] E. Zermelo, "Über einen Satz der Dynamik und der mechanischen Wärmetheorie," *Wied. Ann.* (*Ann. d. Phys. u. Chem.*), *LVII* (1896), 485.

[20] For additional details on these objections and references to Boltzmann's replies, see P. Epstein, "Critical Appreciation of Gibbs' Statistical Mechanics," in A. Haas, ed., *A Commentary on the Scientific Writings of J. Willard Gibbs,* II (New Haven, 1936), pp. 515-519.

[21] P. and T. Ehrenfest, "Begriffliche Grundlagen der statistischen Auffassung in der Mechanik," *Encykl. d. math. Wiss., IV,* 2, II, 41-51. See also Tolman, *op. cit.,* pp. 152-158, esp. p. 156; R. Fürth, "Prinzipien der Statistik," in H. Geiger and K. Scheel, eds., *Handbuch der Physik, IV* (Berlin, 1929), 270-272, and H. Reichenbach, "Ziele und Wege der physikalischen Erkenntnis," *op. cit.,* pp. 62-63.

[22] Cf. Fürth, *op. cit.,* p. 272.

[23] *The Direction of Time,* pp. 116-117.

[24] See the discussion of open kinds of time given by Grünbaum in *Philosophical Problems of Space and Time, op. cit.,* chap. 7, sec. B.—Ed.

[25] E. Schrödinger, "Irreversibility," *Proc. Royal Irish Acad., LIII,* Sect. A (1950), 189, and "The Spirit of Science" in *Spirit and Nature* (New York, 1954), pp. 337-341.

[26] Cf. *The Direction of Time, op. cit.,* pp. 118-143.

[27] Cf. Tolman, *op. cit.,* p. 149.

[28] Cf. Fürth, *op. cit.,* pp. 270 and 192-193. The next-to-last sentence on p. 270 is to be discounted, however, since it is self-contradictory as it stands and incompatible with the remainder of the page.

[29] Although the decisive asymmetry just noted was admitted by H. Mehlberg in his "Physical Laws and Time's Arrow" (*Current Issues in the Philosophy of Science,* H. Feigl and G. Maxwell, eds. [New York, 1961], p. 129), he dismisses it as expressing "merely the factual difference between the two relevant values of probability." But an asymmetry is no less an asymmetry for depending on *de facto,* nomologically-contingent boundary conditions rather than being assured by a *law* alone. Since our verification of laws generally has the same partial and indirect character as that of our confirmation of the existence of certain complicated *de facto* boundary conditions, the assertion of an asymmetry depending on *de facto* conditions is generally no less reliable than one wholly grounded on a law. Hence when Mehlberg (*op. cit.,* p. 117, n. 30) urges against Schrödinger's claim of asymmetry that for every pair of branch systems which change their entropy in one direction, "there is nothing to prevent" another pair of closed subsystems from changing their entropy in the opposite direction, the reply is: Mehlberg's criticism can be upheld only by gratuitously

neglecting the statistical asymmetry admitted but then dismissed by him as "merely" factual.

30 This conclusion departs significantly from Reichenbach's "hypothesis of the branch structure" (*The Direction of Time*, p. 136): by (1) *not* assuming that the entropy is defined for the entire universe such that the universe as a whole can be presumed to exhibit the entropic evolution of the statistical entropy curve for a permanently closed, *finite* system, an assumption which leads Reichenbach to affirm the parallelism of the direction of entropy increase of the universe and of the branch systems; and by therefore (2) *not* concluding that cosmically the statistical anisotropy of time "fluctuates" in the following sense: the alternations of epochs of entropy increase and decrease of the universe go hand-in-hand with the alternations of the direction of entropy increase of the ensembles of branch systems associated with these respective epochs.

In view of the reservations which Reichenbach himself expressed (*The Direction of Time*, pp. 132-133) concerning the reliability of assumptions regarding the universe as a whole in the present state of cosmology, one wonders why he invoked the *entropy* of the universe at all instead of confining himself, as we have done, to the much weaker assumption of the existence of states of disequilibrium in the universe. More fundamentally, it is unclear how Reichenbach thought he could reconcile the assumption that the branch systems satisfy initial conditions of randomness during whatever cosmic epoch they may form—an assumption which, as we saw, makes for the *same* statistical anisotropy on the part of most disequilibrium epochs of the universe—with the following claim of alternation: "When we come to the downgrade [of the entropy curve of the entire universe], always proceeding in the same direction [along the time-axis], the branches begin at states of high entropy . . . and they end at points of low entropy" (*The Direction of Time*, p. 126). For we saw in our statement of the consequences of the postulate of randomness under Group 2 above that, in the vast majority of cases, branch systems beginning in a state of equilibrium (high entropy) will *remain* in equilibrium for the duration of their finite careers instead of decreasing their entropies!

An inherent limitation on the applicability of the Maxwell-Boltzmann entropy concept to the entire universe lies in the fact that it has no applicability at all to a *spatially infinite* universe for the following reasons. If the infinite universe contains a denumerable *infinity* of atoms, molecules or stars, the number of complexions W becomes infinite, so that the entropy is not defined, and *a fortiori* no increase or decrease thereof (cf. K. P. Stanyukovic, "On the Increase of Entropy in an Infinite Universe,"

Doklady, Akad. Nauk. SSSR, N.S. [1949], *LXIX* 793, in Russian, as summarized by L. Tisza in *Math. Reviews, XII* [1951], 787). And if the number of particles in the infinite universe is only finite, then (a) the equilibrium state of maximum entropy cannot be realized by a *finite* number of particles in a phase-space of *infinitely* many cells, since these particles would have to be *uniformly* distributed among these cells, and (b) the quasi-ergodic hypothesis, which provides the essential basis for the probability metric ingredient in the Maxwell-Boltzmann entropy concept, is presumably false for an infinite phase space. For additional doubts concerning the cosmological relevance of the entropy concept, cf. E. A. Milne, *Sir James Jeans* (Cambridge, Eng., 1952), pp. 164-165, and *Modern Cosmology and the Christian Idea of God* (Oxford, 1952), pp. 146-150; also L. Landau and E. Lifshitz, *Statistical Physics* (2d ed.; New York, 1958), pp. 22-27.

[31] K. R. Popper, "The Arrow of Time," *Nature, CLXXVII* (1956), 538 and *CLXXVIII* (1956), 382.

[32] E. L. Hill and A. Grünbaum, "Irreversible Processes in Physical Theory," *Nature, CLXXIX* (1957), 1296-1297. For a rebuttal of Popper's purported counterexample to this general principle (Popper, *Nature, CLXXIX* [1957], 1297-1299), cf. A. Grünbaum, "Popper on Irreversibility," in M. Bunge, ed., *The Critical Approach: Essays in Honor of Karl Popper* (New York: The Free Press of Glencoe, 1964).

[33] For an interesting proposed cosmological explanation on the basis of the "steady state" theory of why it is "that the universe is a non-reflecting sink for radiation," cf. T. Gold's "The Arrow of Time" in R. Stoops, ed., *La Structure et l'Évolution de l'Univers* (Brussels, 1958), pp. 81-91.

[34] Cf. E. Cassirer, *Zur Einsteinschen Relativitätstheorie* (Berlin, 1921), pp. 120-121, and A. Grünbaum, *Philosophical Problems of Space and Time, op. cit.,* pp. 318-319 and 326.

[35] H. Weyl, *Philosophy of Mathematics and Natural Science,* p. 116. This metaphor involving "crawling" must *not,* of course, be taken to suggest the "metaphysical error" charged against it by J. J. C. Smart ("Spatializing Time," *Mind, CXIV* [1955], 240) that psychologically time itself "flows" spatially at a certain rate measured in some non-existent hypertime. We shall see presently that the concept of "becoming" does *not* involve *this* logical blunder.

On the other hand, defenders of the objectivity of becoming such as M. Capek have been guilty of other misunderstandings of Weyl's metaphor on which they have then sought to erect a *reductio ad absurdum* of Weyl's thesis. Thus, Capek (*The Philosophical Impact of Contemporary Physics* [New York, 1961], p. 165) writes that: "although the world scheme of Minkowski eliminates succession in the physical world, it recognizes at least the *movement of our consciousness* to the future. Thus arises an absurd dualism of the timeless physical world and temporal consciousness, that is, a dualism of two altogether disparate realms whose correlation becomes completely unintelligible . . . in such a view . . . we are already dead without realizing it now; but our consciousness creeping along the world line of its own body will certainly reach any pre-existing and nominally future event which in its completeness *waits* to be finally reached by our awareness. . . . To such strange consequences do both spatialization of time and strict determinism lead." But it is a careless and question-begging falsehood to declare that on Weyl's view the physical world is "timeless." For what Weyl is contending is only that the physical world is devoid of becoming, while fully granting that the states of physical systems are ordered by an "earlier than" relation which is isomorphic, in important respects, with its counterpart in consciousness. Capek's claim of the unintelligibility of the correlation between physical and psychological time within Weyl's framework is therefore untenable, especially in the absence of an articulation of the kind (degree) of correlation which Capek requires and also of a justification of that requirement. More unfortunate still is the grievous mishandling of the meaning of Weyl's metaphor in Capek's attempt at a *reductio ad absurdum* of Weyl's view, when Capek speaks of our "already" being dead without realizing it now and of our completed future death *waiting* to be finally "reached" by our awareness. This gross distortion of Weyl's metaphorical rendition of the thesis that coming *into* being is only coming into present awareness rests on an abuse of the temporal and/or kinematic components of the meanings of the words "already," "completed," "wait," "reach," etc.

36 H. Bergmann, *op. cit.*, pp. 27-28. Wilfrid Sellars has independently developed the basis for similar criticisms as part of his penetrating study of a complex of related issues; cf. his "Time and the World Order," in H. Feigl and G. Maxwell, eds., *Scientific Explanation, Space, and Time*, "Minnesota Studies in the Philosophy of Science," III (Minneapolis, 1962).

37 Cf. W. James, *The Principles of Psychology* (New York, 1890 and 1950), pp. 628-629, and H. Driesch, *Philosophische Gegenwartsfragen* (Leipzig, 1933), pp. 96-103.

38 J. J. C. Smart, *op. cit.*, p. 81. A similar reductio is presented by Max Black in "The 'Direction' of Time," *Analysis, XIX* (1959), 54. On the other hand, Smart has offered valid criticisms of Reichenbach's contention that we can "change the future" but not the past in his critical notice of Reichenbach's *The Direction of Time*, in *Phil. Quarterly, VIII* (1958), esp. 76. And H. Mehlberg (*op. cit.*, pp. 109-111) has given a telling refutation of Reichenbach's attempt to characterize "the" direction of time or becoming on the basis of his concept of an "undirectional" relation.

BIBLIOGRAPHICAL
NOTES

In order to avoid repetition, this bibliography does not in general contain references to the works from which selections in this volume have been taken. For these works see the acknowledgements for the various selections.

General Works on Space and Time

Though it is now slightly dated, the book *Space and Time* by the famous French mathematician E. Borel (New York: Dover, 1960) can be recommended as a first-rate popular introduction. H. Reichenbach's *The Philosophy of Space and Time*, from which our selection in Part II is taken, was first published in German in 1928 and has been partly superseded by his *The Direction of Time* (University of California Press, 1956). A comprehensive book on the philosophy of time is *The Natural Philosophy of Time* by the well-known cosmologist G. J. Whitrow (London and Edinburgh: Nelson, 1961). Some other important works are *Philosophical Problems of Space and Time* by A. Grünbaum (New York: Knopf, 1963) and two books by O. Costa de Beauregard, *La Notion de Temps* (Actualités Scientifiques et Industrielles 1300), (Paris: Hermann, 1963), and *Le Second Principe de la Science du Temps* (Paris: Editions du Seuil, 1963).

An essay by A. Grünbaum on "The Nature of Time" appears in *Frontiers of Science and Philosophy*, edited by R. G. Colodny (University of Pittsburgh Press, 1962). An important essay, "Time and the World Order," by Wilfrid Sellars, appears in H. Feigl and G. Maxwell, eds., *Scientific Explanation, Space, and Time*, "Minnesota Studies in the Philosophy of Science," Vol. 3 (University of Minnesota Press, 1962). An older essay which should not be forgotten is the article "Time" by C. D. Broad in the *Encyclopedia of Religion and Ethics*, edited by James Hastings and others, Vol. 12 (Edinburgh and New York: Scribner's, 1922). Some of the essays, especially those by W. R. Dennes and P. Marhenke, in *The Problem of Time*, "University of California

Publications in Philosophy," Vol. 18 (1935), are valuable.

Much of interest will be found in *From Euclid to Eddington*, by the eminent applied mathematician E. T. Whittaker (New York: Dover, 1958). See also Part I, Chapter 3 and Part II, Chapter 1 of *Philosophy of Mathematics and Natural Science* by the great mathematical physicist Hermann Weyl, (Princeton University Press, 1949). An important source book on time is being prepared by Robert S. Cohen and Max Jammer, to appear in the "Classics of Modern Science" series. Max Jammer is also preparing a historico-philosophical monograph on *Concepts of Time* (to be published by Harvard University Press), which will need no recommendation to those who know his *Concepts of Space*, from which our first selection is taken, and his *Concepts of Force* and *Concepts of Mass* (Harvard University Press, 1957 and 1961). An account of the history of philosophical thought about time may be found in M. F. Cleugh, *Time and Its Importance for Modern Thought* (London: Methuen, 1937).

Works of Interest Mainly in Connection with Part I

On pre-Socratic ideas about space and time, consult J. Burnet, *Early Greek Philosophy* (3d ed.; London: Black, 1930), and K. Freeman, *Companion to the Pre-Socratic Philosophers* (3d ed.; Oxford: Blackwell, 1953). A good modern discussion of Zeno's paradoxes will be found in A. Grünbaum, "Modern Science and the Refutation of the Paradoxes of Zeno," *Scientific Monthly, 81*, 1955. Since 1951 numerous articles on the paradox of Achilles and the tortoise have appeared in *Analysis*. See also V. C. Chappell, "Time and Zeno's Arrow," *Journal of Philosophy, 59*, 1962. Chapter 3 of Gilbert Ryle's *Dilemmas* (Cambridge University Press, 1954) is concerned with this paradox. An older article about Zeno's paradoxes is "The Flying Arrow," by P. E. B. Jourdain, *Mind, 25*, 1916. In this article Jourdain refers to the chapter by Russell from which our selection is taken. Plato's account of space is mainly in the *Timaeus*, which is translated with a running commentary by F. M. Cornford in *Plato's Cosmology* (New York: Harcourt, Brace, 1937). For Aristotle see especially his *Metaphysica*, *Physica*, and *De Caelo* in *The Works of Aristotle, Translated into English*, W. D. Ross, ed. (Oxford: Clarendon Press, Vol. 8, 1905, Vol. 2, 1930).

There has been a voluminous literature on Aristotle's problem about the sea fight. Much of it stems from Donald Williams' exciting article "The Sea Fight Tomorrow," in *Structure, Method and Meaning*, P. Henle and others, eds. (New York: Liberal Arts Press, 1951). There is a discussion note on Williams' article by L. Linsky and a rejoinder by Williams in *Philosophical Review*, *63*, 1954. Williams differs from Miss Anscombe (see our second selection) in supposing that Aristotle held that some propositions about the future are neither true nor false. For this reason three-valued logics have come into the discussion. See A. N. Prior "Three-valued Logics and Future Contingents," *Philosophical Quarterly*, *3*, 1953, and R. J. Butler, "Aristotle's Sea-Fight and Three-valued Logic," *Philosophical Review*, *64*, 1955. A relevant article which is earlier than Williams' one is C. A. Baylis, "Are Some Propositions Neither True Nor False?" *Philosophy of Science*, *3*, 1936. Thomas Hobbes also wrote on the sea fight. See Hobbes' *English Works*, W. Molesworth, ed., 1839–40, Vol. IV, p. 277. Hobbes' views are ably discussed by A. G. N. Flew, "Hobbes and the Sea Fight," *Graduate Review of Philosophy*, *2*, 1959. Recent discussions of Aristotle's problem include "The Problem of Future Contingents," by Richard Taylor, *Philosophical Review*, *66*, 1957, with a reply by Rogers Albritton, *ibid.*, and articles by C. K. Grant (*Mind*, *66*, 1957), J. T. Saunders (*Philosophical Review*, *67*, 1958), J. King-Farlow (*Analysis*, *19*, 1958–59), and R. D. Bradley (*Mind*, *68*, 1959). On closely related topics see Chapter 2 of Ryle's *Dilemmas*, already mentioned, and the article "Fatalism," by Richard Taylor, with the discussion by B. Aune, *Philosophical Review*, *71*, 1962. The article by Colin Strang, "Aristotle and the Sea Battle" (*Mind*, *69*, 1960), includes comments on Miss Anscombe's article.

For evidence of Wittgenstein's interest in St. Augustine's questions about time see L. Wittgenstein, *The Blue and Brown Books* (Oxford: Blackwell, 1958), p. 26. W. H. Watson, a physicist who had attended lectures on philosophy by Wittgenstein, uses a passage from St. Augustine on time as a paradigm case of the statement of a philosophical problem in Chapter 1 of his *On Understanding Physics* (Cambridge University Press, 1938). A discussion of St. Augustine by reference to Wittgenstein's ideas will be found in an article by R. Suter, "Augustine on Time, with Some Criticism from Wittgenstein" (*Revue Internationale de Philosophie*, *16*, 1962).

Descartes' philosophy is best understood by reading his

Discourse on Method, Meditations, Principles of Philosophy, and *Replies to Objections*, but most of Leibniz's published writings were of a popular character, and his true views may be got best from Bertrand Russell's very interesting book *The Philosophy of Leibniz* (London: Allen and Unwin, 1900). Russell supports his interpretation by reference to many fragments from Leibniz's writings which are printed in translation in his book. On the topics of our volume the introduction by H. G. Alexander to his edition of *The Leibniz-Clarke Correspondence* (see our Leibniz selection) is well worth reading. On Newton's theories of absolute space and time see the article by Stephen Toulmin in *The Philosophical Review*, 68, 1959, in which Toulmin disentangles the physical grain from the metaphysical chaff. An interesting eighteenth-century defense of Newton's views on space and time is to be found in an article which Euler contributed to the *Histoire de l'Académie des Sciences et Belles Lettres* in 1748. The clearest account of the Newtonian conception of an inertial system is still perhaps that to be found in W. H. Macaulay, "Newton's Theory of Kinetics," *Bulletin of the American Mathematical Society*, 3, 1896–97. Newton's absolute theory of space and time is criticized by `Berkeley in his *Principles of Philosophy*, sections 110–116, and in his *De Motu*. K. R. Popper has drawn attention to interesting anticipations of Mach in various of Berkeley's writings in an article "A Note on Berkeley as Precursor of Mach," *British Journal for the Philosophy of Science*, 4, 1953–54.

In connection with our selections from Kant, see two articles by C. D. Broad: "Kant's Theory of Mathematical and Philosophical Reasoning," *Proceedings of the Aristotelian Society*, 42, 1941–42, and "Kant's Mathematical Antinomies," *ibid.*, 55, 1954–55. Kant's first antinomy is discussed by Milton K. Munitz in "Kantian Dialectic and Cosmology," *Journal of Philosophy*, 48, 1951. In relation to Kant's problem of the left and right hands, see D. F. Pears, "The Incongruity of Counterparts," *Mind*, 61, 1952. In my introduction I related this problem to the modern discovery of the nonconservation of parity. On this matter there is a beautiful popular exposition by the famous Cambridge physicist O. R. Frisch, "Parity Not Conserved, a New Twist to Physics?" *Universities Quarterly* (England), 11, 1957. See also P. Morrison, "The Overthrow of Parity," *Scientific American*, 196,

April 1957; L. D. Rodberg and V. F. Weisskopf, "Fall of Parity," *Science*, *125*, 1957; and the unsigned note "Parity not conserved" on p. 185 of the same volume.

Our selection from F. H. Bradley is destructively criticized by G. E. Moore in the essay "The Conception of Reality," reprinted in his *Philosophical Studies* (London: Routledge and Kegan Paul, 1948).

For Bergson, in addition to the work from which our selection is taken, see especially *Time and Free Will* (New York: Macmillan, 1910). Concise criticisms of Bergson are Bertrand Russell, "The Philosophy of Bergson," *Monist*, *22*, 1912, and E. H. Strange, "Bergson's Theory of Intuition," *Monist*, *25*, 1915. Russell's article is reprinted with very slight changes as Chapter 28 of his *History of Western Philosophy* (New York: Simon and Schuster, 1945). A philosopher who—unlike Bergson—was a very good mathematician, but who resembled Bergson in his rather romantic style of metaphysical writing, was A. N. Whitehead. His views on space and time may be found in the following works: *An Enquiry concerning the Principles of Natural Knowledge* (Cambridge University Press, 1919), especially Part III; *The Concept of Nature* (Cambridge University Press, 1920), Chapters 3–6; *Science and the Modern World* (Cambridge University Press, 1926); *Process and Reality* (Cambridge University Press, 1929); and his unorthodox book on relativity, *The Principle of Relativity* (Cambridge University Press, 1926). See also W. Palter, *Whitehead's Philosophy of Science* (University of Chicago Press, 1960), and A. Grünbaum's critical notice of that book (*Philosophical Review*, *71*, 1962), in which Whitehead's theory of space is trenchantly criticized. Whitehead's notion of "becoming" is criticized by V. C. Chappell in his article, "Whitehead's Theory of Becoming," *Journal of Philosophy*, *58*, 1961.

Works of Interest Mainly in Connection with Part II

An excellent account of the relationship between geometry and physics is given by A. Grünbaum in his "Geometry, Chronometry and Empiricism," published in H. Feigl and G. Maxwell, eds., *Scientific Explanation, Space and Time*, "Minnesota Studies in the Philosophy of Science," Vol. 3 (University of Minnesota Press, 1962). Part I of C. D. Broad,

Scientific Thought (London: Kegan Paul, Trench, Trubner, 1923), is still valuable for its theories of space and of physical geometry. D. Londey's article, "The Concept of Space," *Philosophical Review*, *64*, 1955 (especially the first half), is written with almost pre-Socratic freshness and verve. On the modern astronomical approach to chronometry see G. M. Clemence, "Time and Its Measurement," *American Scientist*, *40*, 1952.

A simple account of the nature of geometry is to be found in C. G. Hempel, "Geometry and Empirical Science," in H. Feigl and W. Sellars, eds., *Readings in Philosophical Analysis* (New York: Appleton-Century-Crofts, 1949). For a history of the development of non-Euclidean geometry see R. Bonola, *Non-Euclidean Geometry* (La Salle, Ill.: Open Court, 1912). A good nontechnical account of Cayley's and Klein's approach to metric geometry from the side of projective geometry may be found in Chapter 9 of Ernest Nagel's *Structure of Science*, another chapter of which is reprinted in this volume. A more technical account will be found in Part III of Felix Klein, *Elementary Mathematics from an Advanced Standpoint—Geometry* (New York: Dover, 1939). Some of Poincaré's views on the relation of geometry to physics can be found in Part II, Sections 4 and 5 of his *Science and Hypothesis* (New York: Dover, 1952). According to Grünbaum, Poincaré's conventionalism has been exaggerated; he has pointed out that the above sections of Poincaré's book have been lifted from the wider context of his article "Des Fondements de la Géometrie, à propos d'un Livre de M. Russell," *Revue de Métaphysique et de Morale*, 7, 1899. The book by Bertrand Russell in question is his *Essay on the Foundations of Geometry* (Cambridge University Press, 1897). It was written from a semi-Kantian standpoint quite uncharacteristic of Russell's later position, and Russell later came to admit that the book has little value. See also Russell's "Sur les Axiomes da la Géometrie," in *Revue de Métaphysique et de Morale*, 7, 1899, and Poincaré's "Sur les Principes de la Géometrie, Réponse à M. Russell," *ibid.*, 8, 1900. Other relevant writings by Poincaré are Chapters 3–4 of *The Value of Science* (New York: Dover, 1958), Part II, Chapter 1 of *Science and Method* (New York: Dover, n.d.), and in his *Les Dernières Pensées* (Paris: 1913). Discussions of Poincaré's views will be found in Grünbaum's book mentioned above.

Works of Interest Mainly in Connection with Part III

Excellent popular introductions to the theory of relativity are Eddington's *Space, Time and Gravitation*, from which our selection is taken, and *Einstein's Theory of Relativity* by Max Born, revised edition prepared with the collaboration of Günther Leibfried and Walter Biem (New York: Dover Inc., 1962). So is A. Einstein's *Relativity, the Special and General Theory* (London: Methuen, 1960). Einstein's *Meaning of Relativity* (Princeton University Press, 1945) is a difficult book requiring a fair amount of mathematical sophistication. An excellent introductory textbook on the special theory of relativity is W. Rindler, *Special Relativity* (New York: Interscience, 1960).

A. Grünbaum cogently attacks anthropocentric and operationalist misinterpretations of relativity in his "Logical and Philosophical Foundations of the Special Theory of Relativity," in A. Danto and S. Morgenbesser, eds., *Philosophy of Science* (New York: Meridian, 1960). A discussion of part of Einstein's "Autobiographical Notes" (see selection in this volume) occurs in Grünbaum's "The Genesis of the Special Theory of Relativity," in *Current Issues in the Philosophy of Science* (see above), and in his "The Special Theory of Relativity as a Case Study of the Importance of the Philosophy of Science for the History of Science," *Annali di Matematica* (Rome) *56*, 1962. *The Principle of Relativity*, by Einstein and others, from which the paper by Minkowski in this volume is taken, contains additional notes by A. Sommerfeld on the various papers. These are chiefly of mathematical interest.

Works of Interest Mainly in Connection with Part IV

Broad's views about time have changed during his life. His varying views are well discussed by C. W. K. Mundle in "Broad's Views About Time," in P. A. Schilpp, ed., *The Philosophy of C. D. Broad* (La Salle, Ill.: Open Court, 1959), together with Broad's comments on this in the same volume. Broad's views in his *Scientific Thought* (see above) are criticized by R. M. Blake, "On Mr. Broad's Theory of Time," *Mind*, *34*, 1925. This article is indebted to Bertrand Russell, "On the Experience of Time," *Monist*, *25*, 1915. In relation to our selection from Nelson Goodman, see the account of token-reflexiveness and of tenses in sections 50–51 of

H. Reichenbach, *Elements of Symbolic Logic* (New York: The Macmillan Company, 1947); see also Y. Bar-Hillel, "Indexical Expressions," *Mind*, *63*, 1954; Jonathan Cohen, "Tense Usage and Propositions," *Analysis*, *11*, 1950–51; Richard M. Gale, "Tensed Statements," *Philosophical Quarterly*, *12*, 1962; and the ensuing discussion in volumes 12 and 13 of this journal. Zeno Vendler in "Verbs and Times," *Philosophical Review*, *66*, 1957, shows that tenses have more functions than one might at first suspect.

In the footnote on p. 355, above, J. N. Findlay had suggested that the logic of tenses should be developed as part of modal logic. This has since been done by A. N. Prior in his *Time and Modality* (Oxford University Press, 1957). R. M. Martin in his review of this book in *Mind*, *68*, 1959, questions whether this is legitimately part of logic. See also Jonathan Cohen's critical notice of the same book in *Philosophical Quarterly*, *8*, 1958. J. N. Findlay has given arguments in favor of tensed language in his essay "An Examination of Tenses," in H. D. Lewis, ed., *Contemporary British Philosophy, Third Series* (New York: The Macmillan Company, 1956).

Goodman, in our selection, rightly objects to the idea that time flows or passes. A criticism of the same notion will be found in Donald Williams' brilliant article "The Myth of Passage," *Journal of Philosophy*, *48*, 1951, and in J. J. C. Smart, "The River of Time," reprinted in A. G. N. Flew, ed., *Essays in Conceptual Analysis* (London: Macmillan and Co., 1956), and "Spatialising Time," *Mind*, *64*, 1955.

See also: R. G. Collingwood, "Some Perplexities About Time," *Proceedings of the Aristotelian Society*, *26*, 1925–26; the symposium "Time and Change," *ibid*., supplementary Vol. 8, 1928; and D. F. Pears, "Time, Truth and Inference," reprinted in *Essays in Conceptual Analysis*. Other articles are A. N. Prior, "Time After Time," *Mind*, *67*, 1958, and "Thank Goodness That's Over," *Philosophy*, *34*, 1959, with Jonathan Cohen's reply, *ibid*. Another essay by Prior is *Changes in Events and Changes in Things*, the Lindley Lecture (University of Kansas, 1962). See also C. W. Webb, "Could Time Flow, If So How Fast?" *Journal of Philosophy*, *57*, 1960; B. Mayo, "Objects, Events and Complementarity," *Philosophical Review*, *70*, 1961; and Anthony Quinton, "Spaces and Times," *Philosophy*, *37*, 1962. The logical possibility of time travel is defended by Hilary Putnam in an article "It Ain't

Necessarily So," *Journal of Philosophy*, *59*, 1962. This paper also contains interesting remarks on physical geometry. J. J. C. Smart, "Is Time Travel Possible?" (*ibid.*, *60*, 1963), comments on Putnam's paper.

If not explicitly, at least by implication, several of the above articles refute McTaggart's well-known argument for the unreality of time. This argument will be found in J. M. E. McTaggart, *Philosophical Studies* (London: Arnold, 1934), and in his *The Nature of Existence*, Vol. II (Cambridge University Press, 1927), Chapter 33. This is criticized in C. D. Broad, *Examination of McTaggart's Philosophy* (Cambridge University Press, 1938), Vol. II, Chapter 35. The selection from Broad on pages 321-338 of this volume is taken from that work. Other criticisms of McTaggart include P. Marhenke's article in the book *The Problem of Time*, "University of California Publications in Philosophy," Vol. 18, mentioned earlier, and D. W. Gotshalk, "McTaggart on Time," *Mind*, *39*, 1930. However, see also M. Dummett, "A Defense of McTaggart's Proof of the Unreality of Time," *Philosophical Review*, *69*, 1960, and L. O. Mink, "Time, McTaggart and Pickwickian Language," *Philosophical Quarterly*, *10*, 1960.

A suggestion made by Richard Taylor on p. 391 above is amplified in his "Moving About in Time," *Philosophical Quarterly*, *9*, 1959, and criticized by W. J. Huggett, *ibid.*, *10*, 1960. On the topic of our Selection 29 see also N. L. Wilson, "Space, Time and Individuals," *Journal of Philosophy*, *52*, 1955.

On the status of the past, see A. J. Ayer, "Statements about the Past," in his *Philosophical Essays* (London: Macmillan and Co., 1954), and G. E. Hughes, *On Having the Past All Over Again*, Inaugural Lecture (Victoria University College, N. Z., 1951).

On the questions concerned with the so-called "direction of time," besides Reichenbach's book of that name, already mentioned, see L. Boltzmann, "On Certain Questions of the Theory of Gases," *Nature*, *51*, 1895; E. Schrödinger, "Irreversibility," *Proceedings of the Royal Irish Academy*, *52*, 1950; and N. Wiener, "Newtonian Time and Bergsonian Time," which is Chapter 1 of his *Cybernetics* (2d ed.; New York: Wiley, 1961). In spite of its title this latter chapter has little to do with Newton and less with Bergson. The problem is well posed, though not solved, by C. Ehrenfels in his *Cosmogony*

(New York: Philosophical Library, 1948). See especially Chapter 1 and the review of the book by J. N. Findlay in *Philosophy*, *25*, 1950. A particularly important contribution is the paper by A. Grünbaum from which our final selection is extracted. A different solution to the problem is to be found in notes by K. R. Popper in *Nature*, *177* and *178*, 1956; *179*, 1957; and *181*, 1958. In connection with these see the note by E. L. Hill and A. Grünbaum, *ibid.*, *179*, 1957. Popper's view is discussed by Grünbaum in "Popper on Irreversibility," in *The Critical Approach, Essays in Honor of Karl Popper* (New York: The Free Press of Glencoe, 1964). There is a beautifully written article on "The Arrow of Time" by the cosmologist T. Gold in *La Structure et L'Évolution de L'Univers, Proceedings of the 11th Solvay Conference* (Brussels: R. Stoops, 1958). Another article by Gold on the same theme and with the same title is to be found in the *American Journal of Physics, 30*, 1962. See also H. Bondi, "Physics and Cosmology," *The Observatory, 82*, 1962. Other relevant articles are the following: J. J. C. Smart, "The Temporal Asymmetry of the World," *Analysis, 14*, 1952–53; B. Mayo, "Professor Smart and Temporal Asymmetry," *Australasian Journal of Philosophy, 33*, 1955; J. J. C. Smart, "Mr. Mayo on Temporal Asymmetry," *ibid.*; the answers by J. E. McGechie, J. R. Searle, and R. Taylor to Analysis Problem No. 9, *Analysis, 16*, 1955–56; and Max Black, "The Direction of Time," in his *Models and Metaphors* (Cornell University Press, 1962). Also relevant is part of the final Appendix to M. Schlick's *The Philosophy of Nature*, the book from which our Selection 21 was taken. Henryk Mehlberg's important "Essai sur la théorie causale du temps" appeared in *Studia Philosophica, 1*, 1935.

This volume has been concerned with space and time themselves, and so psychological questions of our *experience* of space and time have been as far as possible avoided. Nevertheless two articles on the subject are worth mentioning: J. D. Mabbott, "Our Direct Experience of Time," *Mind, 60*, 1951, and C. W. K. Mundle, "How Specious Is the Specious Present?" *Mind, 63*, 1954. Broad, in the works already mentioned, has written a good deal on this subject. The discussion goes back at least to William James, *Principles of Psychology* (New York: Holt, 1890), Vol. I, Chapter 15; and even beyond—for evidence of this see G. J. Whitrow's *Natural Philosophy of Time* (mentioned above), p. 78.